AF766357

Methods in Molecular Biology

Series Editor
John M. Walker
School of Life and Medical Sciences
University of Hertfordshire
Hatfield, Hertfordshire, AL10 9AB, UK

For further volumes:
http://www.springer.com/series/7651

Microbial Steroids

Methods and Protocols

Editors

José-Luis Barredo

Department of Biotechnology, Crystal Pharma,
A Division of Albany Molecular Research Inc. (AMRI),
Parque Tecnológico de León, León, Spain

Ignacio Herráiz

Research and Development Department, Gadea Pharmaceutical Group,
A Division of Albany Molecular Research Inc. (AMRI), Parque Tecnológico de Boecillo,
Boecillo, Valladolid, Spain

Humana Press

Editors
José-Luis Barredo
Department of Biotechnology, Crystal Pharma
A Division of Albany Molecular
Research Inc. (AMRI)
Parque Tecnológico de León
León, Spain

Ignacio Herráiz
Research and Development Department
Gadea Pharmaceutical Group
A Division of Albany Molecular
Research Inc. (AMRI)
Parque Tecnológico de Boecillo
Boecillo, Valladolid, Spain

ISSN 1064-3745 ISSN 1940-6029 (electronic)
Methods in Molecular Biology
ISBN 978-1-4939-7182-4 ISBN 978-1-4939-7183-1 (eBook)
DOI 10.1007/978-1-4939-7183-1

Library of Congress Control Number: 2017944137

© Springer Science+Business Media LLC 2017
This work is subject to copyright. All rights are reserved by the Publisher, whether the whole or part of the material is concerned, specifically the rights of translation, reprinting, reuse of illustrations, recitation, broadcasting, reproduction on microfilms or in any other physical way, and transmission or information storage and retrieval, electronic adaptation, computer software, or by similar or dissimilar methodology now known or hereafter developed.
The use of general descriptive names, registered names, trademarks, service marks, etc. in this publication does not imply, even in the absence of a specific statement, that such names are exempt from the relevant protective laws and regulations and therefore free for general use.
The publisher, the authors and the editors are safe to assume that the advice and information in this book are believed to be true and accurate at the date of publication. Neither the publisher nor the authors or the editors give a warranty, express or implied, with respect to the material contained herein or for any errors or omissions that may have been made. The publisher remains neutral with regard to jurisdictional claims in published maps and institutional affiliations.

Printed on acid-free paper

This Humana Press imprint is published by Springer Nature
The registered company is Springer Science+Business Media LLC
The registered company address is: 233 Spring Street, New York, NY 10013, U.S.A.

Preface

Steroids are terpenoid lipids with specific structure that occur widely in living systems. Over 250 sterols and related compounds have been reported in plants, insects, vertebrates, and lower eukaryotes such as yeasts. Steroid-based drugs have a wide range of therapeutic purposes, such as anti-inflammatory, immunosuppressive, progestational, diuretic, anabolic, and contraceptive, as well as other applications. As a result, about 300 approved steroid drugs exist to date, and the global market for steroid-containing products is in excess of $10 billion and more than one million tons annually.

Scientific research on steroid chemistry began in the early twentieth century and was encouraged in the 1950s, with the discovery of the pharmacological effects of cortisol and progesterone. Their production is being done by chemical synthesis pathways, but the replacement of some of the chemical steps by bioconversions is allowing, in many cases, fewer stages, higher yields, and new modifications leading to more competitive and robust processes.

One of the major raw materials for steroid industry is the natural steroid sapogenin called diosgenin. The conversion of diosgenin to valuable steroids is done by a well--established chemical route, despite presenting several shortcomings such as higher costs, multistep syntheses, waste of land resources, and exhaustion of wild plant resources. Alternative starting materials for the steroid industry are the natural sterols, e.g., phytosterols, a mixture of sterols industrially obtained as a by-product of the oil refining process or from the cellulose production process.

Microbial bioconversion of phytosterols into steroid intermediates remains a focus of research in the field of steroids. Growing numbers of microbial biotransformations of steroids have been reported, with an emphasis mainly on steroid hydroxylation, Δ1-dehydrogenation, and sterol side-chain cleavage. Many of these biotransformation reactions, in combination with chemical synthesis, enabled the production of large quantities of steroid compounds. Both phytosterol bioconversions and new steroid modifications are two areas of special interest for the industry.

This book is intended to provide practical experimental laboratory procedures for a wide range of steroid bioconversions. Although not an exhaustive treatise, it provides a detailed "step-by-step" description of the most recent developments in these biotechnological processes. The detailed protocols are cross-referenced in the Notes section, providing special details, minor problems, troubleshooting, and safety comments that may not normally appear in journal articles and can be particularly useful for those not familiar with specific techniques.

The two lead chapters of this volume are overviews on microbial bioconversions and chemical synthesis pathways, respectively. The following chapters show comprehensive experimental methods on strain characterization (genomics, transcriptomics, and proteomics) and improvement, bioconversion methods from sterols to androstenedione (AD) and androstadienedione (ADD), steroid hydroxylations, biocatalysis, downstream processes to purify steroid intermediates, and analysis.

This book has been written by outstanding experts in their field and provides a reference source for laboratory and industrial professionals, as well as for graduate students in a number of biological disciplines (biotechnology, microbiology, genetics, molecular biology, etc.).

We are indebted to the authors who, in spite of their professional activities, agreed to participate in this book, to Dr. J. Walker, Series Editor, for his encouragement and advice in reviewing the manuscripts, and to the staff of Springer for their assistance in assembling this volume and their efforts in keeping this project on schedule. Last but not least, we warmly acknowledge our families for their patience and support.

León, Spain *José-Luis Barredo*
Boecillo, Valladolid, Spain *Ignacio Herráiz*

Contents

Contributors

ESTHER BAHÍLLO • *Department of Biotechnology, Crystal Pharma, A Division of Albany Molecular Research Inc. (AMRI), Parque Tecnológico de León, León, Spain*

ABHAY BAJAJ • *Molecular Biology Laboratory, Department of Zoology, University of Delhi, Delhi, India*

JOSÉ-LUIS BARREDO • *Department of Biotechnology, Crystal Pharma, A Division of Albany Molecular Research Inc. (AMRI), Parque Tecnológico de León, León, Spain*

CARLOS BARREIRO • *Instituto de Biotecnología de León (INBIOTEC), Parque Científico de León, León, Spain; Área de Microbiología, Departamento de Biología Molecular, Campus de Ponferrada, Universidad de León, Ponferrada, Spain*

JOSÉ BECERRA • *Natural Products Chemistry Laboratory, Faculty of Natural and Oceanographic Sciences, University of Concepción, Concepción, Chile*

JOSÉ BERENGUER • *Centro de Biología Molecular Severo Ochoa (UAM-CSIC), Facultad de Ciencias, Universidad Autónoma de Madrid, Madrid, Spain*

SOLEDAD CHAMORRO • *Engineering and Environmental Biotechnology Group, Environmental Science Faculty & Center EULA-Chile, University of Concepción, Concepción, Chile*

RABIH DARWICHE • *Division of Biochemistry, Department of Biology, University of Fribourg, Fribourg, Switzerland*

CARLA C.C.R. DE CARVALHO • *IBB-Institute for Bioengineering and Biosciences, Department of Bioengineering, Instituto Superior Técnico, Universidade de Lisboa, Lisbon, Portugal*

MARINA V. DONOVA • *G.K. Skryabin Institute of Biochemistry & Physiology of Microorganisms, Russian Academy of Sciences, Pushchino, Russia; Pharmins LTD, Pushchino, Russia*

DMITRY DOVBNYA • *G.K. Skryabin Institute of Biochemistry & Physiology of Microorganisms, Russian Academy of Sciences, Pushchino, Russia; Pharmins LTD, Pushchino, Russia*

CARMEN FELPETO-SANTERO • *Centro de Investigaciones Biológicas CSIC, Madrid, Spain*

PEDRO FERNANDES • *IBB – Institute for Bioengineering and Biosciences, Department of Bioengineering, Instituto Superior Técnico, Universidade de Lisboa, Lisboa, Portugal; Faculty of Engineering, Universidade Lusófona de Humanidades e Tecnologias, Lisboa, Portugal*

LORENA FERNÁNDEZ-CABEZÓN • *Centro de Investigaciones Biológicas CSIC, Madrid, Spain*

JOSÉ M. FERNÁNDEZ-CAÑÓN • *Instituto de Biología Molecular, Genómica y Proteómica (INBIOMIC), Dpto. Biología Molecular, Universidad de León, León, Spain*

BEATRIZ GALÁN • *Centro de Investigaciones Biológicas CSIC, Madrid, Spain*

JOSÉ-LUIS GARCÍA • *Centro de Investigaciones Biológicas CSIC, Madrid, Spain*

JULIA GARCÍA-FERNÁNDEZ • *Centro de Investigaciones Biológicas CSIC, Madrid, Spain*

GOVINDA GUEVARA • *Department of Biochemistry and Molecular Biology I, Universidad Complutense de Madrid, Madrid, Spain*

VÍCTOR HERNÁNDEZ • *Natural Products Chemistry Laboratory, Faculty of Natural and Oceanographic Sciences, University of Concepción, Concepción, Chile*

IGNACIO HERRÁIZ • *Research and Development Department, Gadea Pharmaceutical Group, A Division of Albany Molecular Research Inc. (AMRI), Parque Tecnológico de Boecillo, Boecillo, Valladolid, Spain*

AURELIO HIDALGO • *Centro de Biología Molecular Severo Ochoa (UAM-CSIC), Facultad de Ciencias, Universidad Autónoma de Madrid, Madrid, Spain*
SABRINA HOEBENREICH • *Fachbereich Chemie, Philipps-Universität Marburg, Marburg, Germany*
KJELL D. JOSEFSEN • *SINTEF Materials and Chemistry, Department of Biotechnology and Nanomedicine, Trondheim, Norway*
SERGEY KHOMUTOV • *Institute of Biochemistry & Physiology of Microorganisms, Russian Academy of Sciences, Moscow, Russia*
VYACHESLAV KOLLEROV • *Institute of Biochemistry & Physiology of Microorganisms, Russian Academy of Sciences, Moscow, Russia*
HUI LI • *School of Pharmaceutical Science, Jiangnan University, Wuxi, China*
JESSICA K. LORAINE • *Department of Biology, University of York, Wentworth Way, York, UK*
JOSÉ M. LUENGO • *Departmento de Biología Molecular (Área de Bioquímica y Biología Molecular), Facultad de Veterinaria, Universidad de León, León, Spain*
RODRIGO A. MANCILLA • *Escuela de Ingeniería Bioquímica, Pontificia Universidad Católica de Valparaíso, Valparaíso, Chile*
MARCO P.C. MARQUES • *Department of Biochemical Engineering, University College London, London, UK*
IGOR MARTÍNEZ • *Centro de Investigaciones Biológicas CSIC, Madrid, Spain*
SONIA MARTÍNEZ-CÁMARA • *Department of Biotechnology, Crystal Pharma, A Division of Albany Molecular Research Inc. (AMRI), Parque Tecnológico de León, León, Spain*
JULIANE MERZ • *Department of Biochemical and Chemical Engineering, Laboratory of Plant and Process Design, TU Dortmund University, Dortmund, Germany*
MARÍA-ANGÉLICA MONDACA • *Microbiology Department, University of Concepción, Concepción, Chile*
ALEJANDRO MORALES • *INBIOTEC (Instituto de Biotecnologia de León), Parque Cientifico de León, León, Spain*
ALEJANDRO AMOROSO • *Facultad de Ciencias, Universidad San Sebastián, Santiago, Chile*
JUANA-MARÍA NAVARRO-LLORENS • *Department of Biochemistry and Molecular Biology I, Universidad Complutense de Madrid, Madrid, Spain*
NATHALIE NETT • *Fachbereich Chemie, Philipps-Universität Marburg, Marburg, Germany*
ANNA NORDBORG • *SINTEF Materials and Chemistry, Department of Biotechnology and Nanomedicine, Trondheim, Norway*
ELÍAS R. OLIVERA • *Departmento de Biología Molecular (Área de Bioquímica y Biología Molecular), Facultad de Veterinaria, Universidad de León, León, Spain*
LIDIA ORTEGA-DE LOS RÍOS • *Instituto de Biología Molecular, Genómica y Proteómica (INBIOMIC), Dpto. Biología Molecular, Universidad de León, León, Spain*
ROBERTO PAVEZ-DÍAZ • *Departamento de Ingeniería Química y Ambiental, Universidad Técnica Federico Santa María, Valparaíso, Chile*
JULIÁN PERERA • *Department of Biochemistry and Molecular Biology I, Universidad Complutense de Madrid, Madrid, Spain*
ROSARIO PÉREZ-REDONDO • *Instituto de Biotecnología de León (INBIOTEC), Parque Científico de León, León, Spain*
SURYA PRAKASH • *Hunan Yuxin Pharmaceutical Co. Ltd, Shaoyang, Hunan, China*
ANA-LUISA RIBEIRO • *Centro de Biología Molecular Severo Ochoa (UAM-CSIC), Facultad de Ciencias, Universidad Autónoma de Madrid, Madrid, Spain*

ANTONIO RODRÍGUEZ-GARCÍA • *Instituto de Biotecnología de León (INBIOTEC), Parque Científico de León, León, Spain; Área de Microbiología, Departamento de Biología Molecular, Campus de Ponferrada, Universidad de León, León, Spain*
MARTA RODRÍGUEZ-SÁIZ • *Department of Biotechnology, Crystal Pharma, A Division of Albany Molecular Research Inc. (AMRI), León, Spain*
MERCEDES SÁNCHEZ • *Centro de Biología Molecular Severo Ochoa (UAM-CSIC), Facultad de Ciencias, Universidad Autónoma de Madrid, Madrid, Spain*
MIKHAIL I. SCHELKUNOV • *Institute for Information Transmission Problems, Russian Academy of Sciences, Moscow, Russia*
ROGER SCHNEITER • *Division of Biochemistry, Department of Biology, University of Fribourg, Fribourg, Switzerland*
VICTORIA Y. SHTRATNIKOVA • *A.N. Belozersky Research Institute of Physico-Chemical Biology, M.V. Lomonosov Moscow State University, Moscow, Russia*
HÅVARD SLETTA • *SINTEF Materials and Chemistry, Department of Biotechnology and Nanomedicine, Trondheim, Norway*
MARGARET C.M. SMITH • *Department of Biology, University of York, Wentworth Way, York, UK*
ALBERTO SOLA-LANDA • *Instituto de Biotecnología de León (INBIOTEC), Parque Científico de León, León, Spain*
MARTIN SPINCK • *Fachbereich Chemie, Philipps-Universität Marburg, Marburg, Germany*
JIN SUN • *School of Pharmaceutical Science, Jiangnan University, Wuxi, China*
FABIAN B. THYGS • *Department of Biochemical and Chemical Engineering, Laboratory of Plant and Process Design, TU Dortmund University, Dortmund, Germany*
INÉS VÁZQUEZ-IGLESIAS • *Instituto de Biotecnología de León (INBIOTEC), Parque Científico de León, León, Spain*
GLADYS VIDAL • *Engineering and Environmental Biotechnology Group, Environmental Science Faculty & Center EULA-Chile, University of Concepción, Concepción, Chile*
MARICEL VIDAL • *Microbiology Department, University of Concepción, Concepción, Chile*
CLAUDIA XAVIER • *Engineering and Environmental Biotechnology Group, Environmental Science Faculty & Center EULA-Chile, University of Concepción, Concepción, Chile; Federal University of Technology – Paraná – UTFPR, Curitiba, PR, Brazil*
ZHENGHONG XU • *School of Pharmaceutical Science, Jiangnan University, Wuxi, China*

Chapter 1

Steroid Bioconversions

Marina V. Donova

Abstract

Steroid modifications by selected wild-type and engineered strains of microorganisms became an effective tool for the production of high-valued steroidal drugs and their precursors for the pharmaceutical industry. Some microorganisms are effective at the performance of sterol side-chain degradation, oxyfunctionalization of steroid core, and redox reactions at different positions of the steroid molecule. A number of bioprocesses using steroid-transforming microbial strains are well established on an industrial level. Although a range of biocatalytic methods has been developed, selection of suitable microorganisms, as well as creation of new engineered strains, is of great importance for generation of improved bioprocesses and production schemes for obtaining known and new metabolites with potent biological activity. The achievements in genetic and metabolic engineering of steroid-transforming strains in combination with novel approaches in the enzymatic and whole-cell biocatalysis provide a platform for highly effective and selective biotransformations.

Here, we briefly review the current state and prospects in the field of microbial bioconversions with special attention to the application of molecular microbiology methods for the generation of new whole cell biocatalysts.

Key words Steroid, Microbial transformation, Bioconversion, Phytosterol, Side-chain degradation, Sterol catabolism, Hydroxylation, Dehydrogenation, Whole-cell biocatalysis

1 Introduction

Steroid compounds represent a specific class of terpenoid lipids that contain in their structure gonane core of four cycloalkane rings (A-D) (Fig. 1). Steroids fulfill essential vital functions in the higher organisms: control various aspects of cell proliferation and differentiation, regulate signal transduction pathways by the binding to the respective intracellular receptors, some of them serve also as signaling molecules in cell-cell interactions, etc. [1, 2].

Androgens, estrogens, and progestagens play a role as sex hormones in the reproduction of vertebrates. Mineralocorticoids and glucocorticoids are related to adrenal steroids and provide electrolyte and glucose homeostasis in animals. Bile acids are steroid compounds with diverse important biological functions in the

José-Luis Barredo, Ignacio Herráiz (eds.), *Microbial Steroids: Methods and Protocols*, Methods in Molecular Biology vol. 1645, DOI 10.1007/978-1-4939-7183-1_1, © Springer Science+Business Media LLC 2017

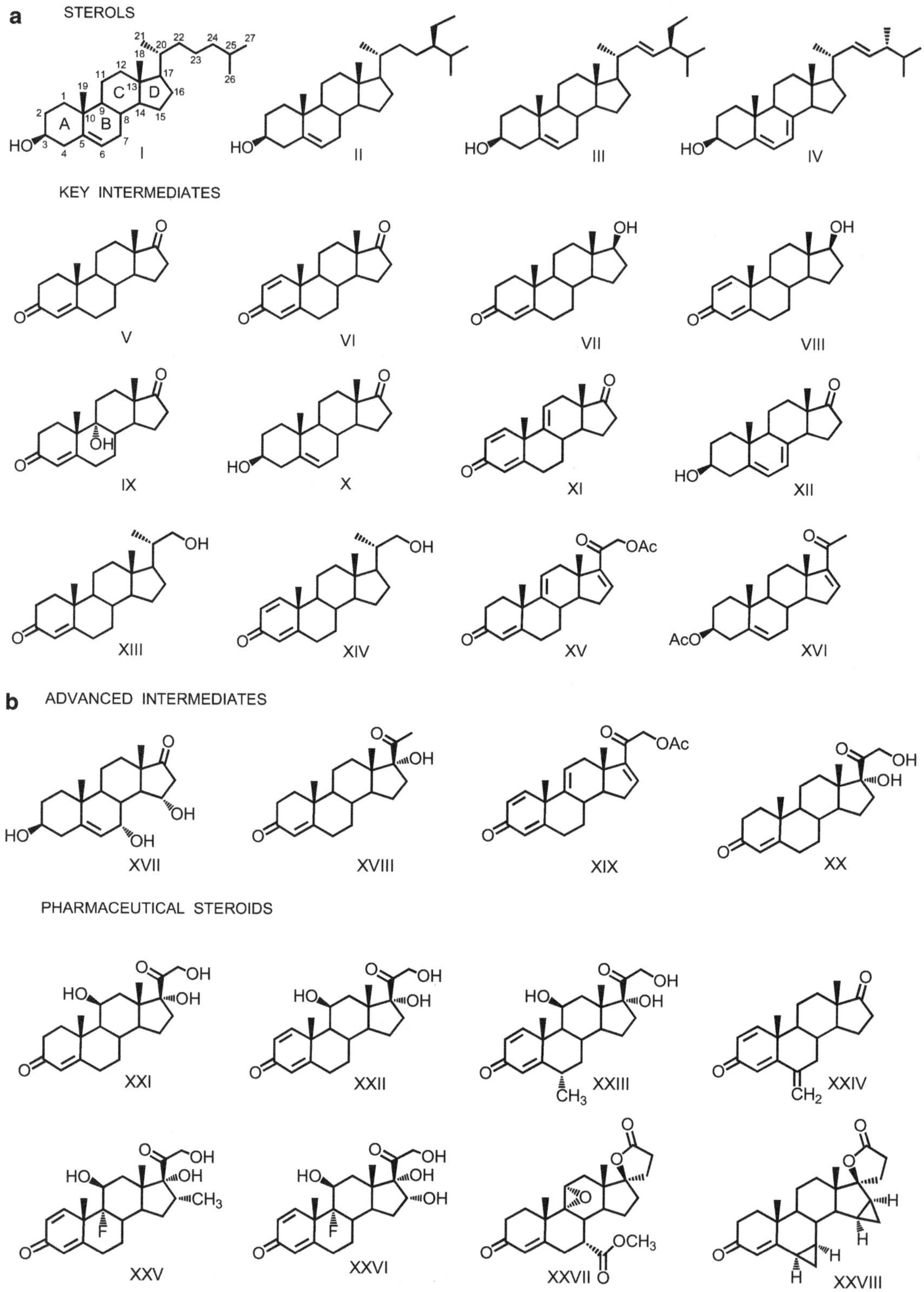

Fig. 1 Structures of some sterols, key intermediates (**a**), advanced intermediates and most marketed APIs (**b**) (adopted from [48]): **Sterols (I–IV): I**, cholesterol; **II**, β-sitosterol; **III**, stigmasterol; **IV**, ergosterol; **Androstanes, C_{19} intermediates (V–XII): V**, androst-4-ene-3,17-dione (AD); **VI**, androsta-1,4-diene-3,17-dione (ADD); **VII**, 17β-hydroxy-androst-4-ene-3-one (testosterone); **VIII**, 17β-hydroxy-androst-1,4-diene-3,17-dione (Δ^1-dehydrotestosterone); **IX**, 9α-hydroxy-4-androstene-3,17-dione (9-OH-AD); **X**, 3β-hydroxy-androst-5-ene-17-one

vertebrate digestion. They play a role in the solubilization and adsorption of fats, cholesterol, and lipid-soluble vitamins, and participate in the regulation of systemic endocrine functions [3]. The so-called neurosteroids, such as dehydroepiandrosterone, dehydroepiandrosterone sulfate, or allopregnanolone, function as allosteric modulators of neurotransmitter receptors to modulate neuronal excitability through direct interaction with the cell surface [4, 5]. Neuroprotective efficacy has been also shown for progesterone at the therapy of some brain disorders and traumatic brain injuries [6], and demonstrated for pregnenolone and pregnenolone-like molecules [7].

In eukaryotes, the hormones, bile acids, and other essential steroids are produced from cholesterol (Fig. 1) which is also an important component of the cell membranes playing a role in membrane fluidity, and cell differentiation and proliferation. Sitosterol, stigmasterol, campesterol, and brassicasterol are major plant sterols, and ergosterol is an important component of cell wall of low eukaryotes (yeasts, fungi) [8, 9].

Pharmaceutical formulations based on steroids are widely used in medicine as anti-inflammatory, antineoplastic, anti-allergic agents, sex hormones, anabolic, etc., in the prevention and treatment of many serious diseases, such as cancer, obesity, diabetes, rheumatoid arthritis, asthma, hormone metabolic syndrome, neurodegenerative diseases, and many others [10–12]. Along with antibiotics, pharmaceutical steroids represent one of the most best-selling categories being a significant sector of the global pharmaceutical market [13]. Forecasts of the global pharmaceutical industry indicate that many drugs on the basis of steroid compounds retain market leadership in future perspectives. The examples of pharmaceutical steroids and their key precursors are given in Fig. 1.

The production processes of steroid hormones may be classified as: (1) extraction from plants, or animal tissues; (2) full organic synthesis; or (3) combination of chemical and enzymatic syntheses. Evidently, with several exceptions, the first one can hardly be applied for large-scale production. Full chemical, or chemical enzymatic syntheses are often multistage, time-, labor- and energy consuming, environmentally risky, and in some cases are even

Fig. 1 (continued) (dehydroepiandrosterone, DHEA); **XI**, androst-1,4,9(11)-triene-3,17-dione; **XII**, 3β-hydroxy-5,7-androsta-diene-17-one; **Pregnanes, C_{21} and C_{22}-intermediates (XIII–XX)**: **XIII**, 20-hydroxymethyl-pregna-4-ene-3-one (HMP); **XIV**, 20-hydroxymethyl-pregna-1,4-diene-3-one (HMPD); **XV**, 21-acetoxypregna-4,9(11),16-triene-3,20-dione; **XVI**, 16-dehydropregnenolone acetate (16-DPA); **XVII**, 7α,15α-dihydroxydehydroepiandrosterone; **XVIII**, 17α-hydroxyprogesterone; **XIX**, 21-acetoxypregna-1,4,9(11),16-tetraene-3,20-dione; **XX**, 17α,21-dihydroxy-4-pregnene-3,20-dione (cortexolone, Reichstein's substance S, 11-deoxycortisol); **APIs (XXI–XXVIII)**: **XXI**, hydrocortisone (cortisol); **XXII**, prednisolone; **XXIII**, 6α-methylprednisolone; **XXIV**, exemestane; **XXV**, dexamethasone; **XXVI**, triamcinolone; **XXVII**, eplerenone; **XXVIII**, drospirenone

impossible. More and more bioconversions became the methods of choice for industrial production, and growing demand for steroid pharmaceuticals stimulates the development of new cost-effective and environmentally friendly biotechnologies.

In general, the term bioconversion, or biotransformation means the chemical modification of a compound by an organism, i.e., microbial steroid bioconversion is a structural modification of steroids by microorganisms, especially by whole cells of microorganisms. It can be a single reaction, or cascade reactions being part of a catabolic pathway intrinsic to the microorganism. Microbial conversion provides steroid production by multi-step reactions with cofactor regeneration and under mild conditions.

Microbial transformations expand the toolbox of organic synthesis, thus enabling obtaining of both well-established and new steroid derivatives of potential biological and pharmacological activity inaccessible otherwise, e.g., [14]. The similarity of some microbial enzymes to mammalian ones may be used as the so-called microbial models of mammalian metabolism in order to predict the transformation routes of the tested compounds (mainly, new bioactive molecules) in the preclinical investigations. Besides, the knowledge on microbial degradation of natural and synthetic steroid hormones is of importance for ecotoxicology and medical research [15, 16].

Microorganisms from different habitats have been described to transform, or fully degrade steroid compounds. Steroid substrate spectrum for microorganisms is rather broad: along with plenty natural steroids that are synthesized in plants, insects, vertebrates, and lower eukaryotes, it includes industrially relevant steroids discharged into the environment. Diverse bacteria are capable of catabolizing steroids as carbon and energy sources, while fungi mainly detoxify steroids as fungitoxic molecules by their structural modification. However, non-sufficient selectivity of steroid transformation reactions by wild-type strains often results in a mixture of products, and along with steroid core degradation by wild-type strains may restrict their industrial application.

Innovative solutions in the field of steroid biotechnology are based on the application of genetically engineered microorganisms. An outstanding achievement is the development of a yeast strain enabling production of hydrocortisone from glucose by reconstructing the fully self-sufficient biosynthetic pathway involving 13 engineered genes encoding several mammalian steroid hydroxylases [17, 18]. More than 10 years required for the development of the production bioprocess by Sanofi-Aventis Co., but it is a unique and impressive experience showing a revolutionary era in the steroid field [19].

2 Steroid Catabolism in Bacteria

The distribution of the steroid catabolism pathways has been recently analyzed among 8000 microorganisms whose genomes are available in the databases (NCBI Rf_Seq) [20]. The bioinformatical research allowed identification of 265 putative steroid degraders within only *Actinobacteria* and *Proteobacteria* from different habitats (eukaryote hosts, soil, and aquatic environments).

Actinobacteria are catabolically versatile aerobic Gram-positive GC-rich bacteria capable of utilizing a wide range of organic substrates. These bacteria were shown to provide different reactions of steroid modification: sidechain degradation, hydroxylations (e.g., at positions 9α, 16α, 7α, and 11β), insertion of C-C-double bonds (e.g., Δ^1 and Δ^4), hydrogenation and isomerization of double bonds (e.g., $\Delta^5 \rightarrow \Delta^4$), carbonyl group oxidation/reduction (e.g., at positions 17 and 20), hydrolysis of steroid esters (e.g., 21-deacetylation) [21]. As a rule, these reactions are the parts of steroid catabolic pathway.

Steroid catabolism pathways (cholesterol degradation) have been shown to be conserved in certain different *Actinobacteria* taxa [20], and characterized best for the representatives of suborder *Corynebacterineae*, such as *Mycobacterium*, *Rhodococcus*, and *Gordonia* species. It was assumed that mycolic acid rich cell wall of these *Actinobacteria* may contribute to the effective transportation of lipophilic substances such as steroids. Cholate degradation pathway was assumed to be conserved within *Rhodococcus* species, while patchy distributed within *Proteobacteria* along with testosterone degradation pathway. For several proteobacterial genera, a horizontal transfer of the pathways has been evidenced [20].

The genetics of steroid catabolism has been mainly studied in several actinobacterial strains, including for example *Mycobacterium tuberculosis* H37Rv, *Mycobacterium neoaurum* NRRL 3805B, *Mycobacterium neoaurum* VKM Ac-1815D, 1817D, *Mycobacterium smegmatis* mc2 155, *Rhodococcus jostii* RHA1, *Gordonia neofelifaecis* NRRL B-59395, and *Nocardioides simplex* VKM Ac-2033D [22–30]. More studies are necessary to confirm its diversity among other taxa.

Presumably, the steroid catabolic pathways are organized according to the parts of the steroid molecule and include the enzyme systems related to the aliphatic side chain degradation, rings A/B and rings C/D degradation, as well as steroid uptake system (ABC-type transporter, mce4) [31] (Fig. 2). In *Mycobacterium* strains sterol-degrading genes were shown to be distributed across the chromosome, while in *Rhodococcus* species they were identified also in plasmids. The genes related to steroid catabolism form different clusters, e.g., the genome of *R. jostii* RHA1 harbors four clusters of paralogous steroid catabolic genes

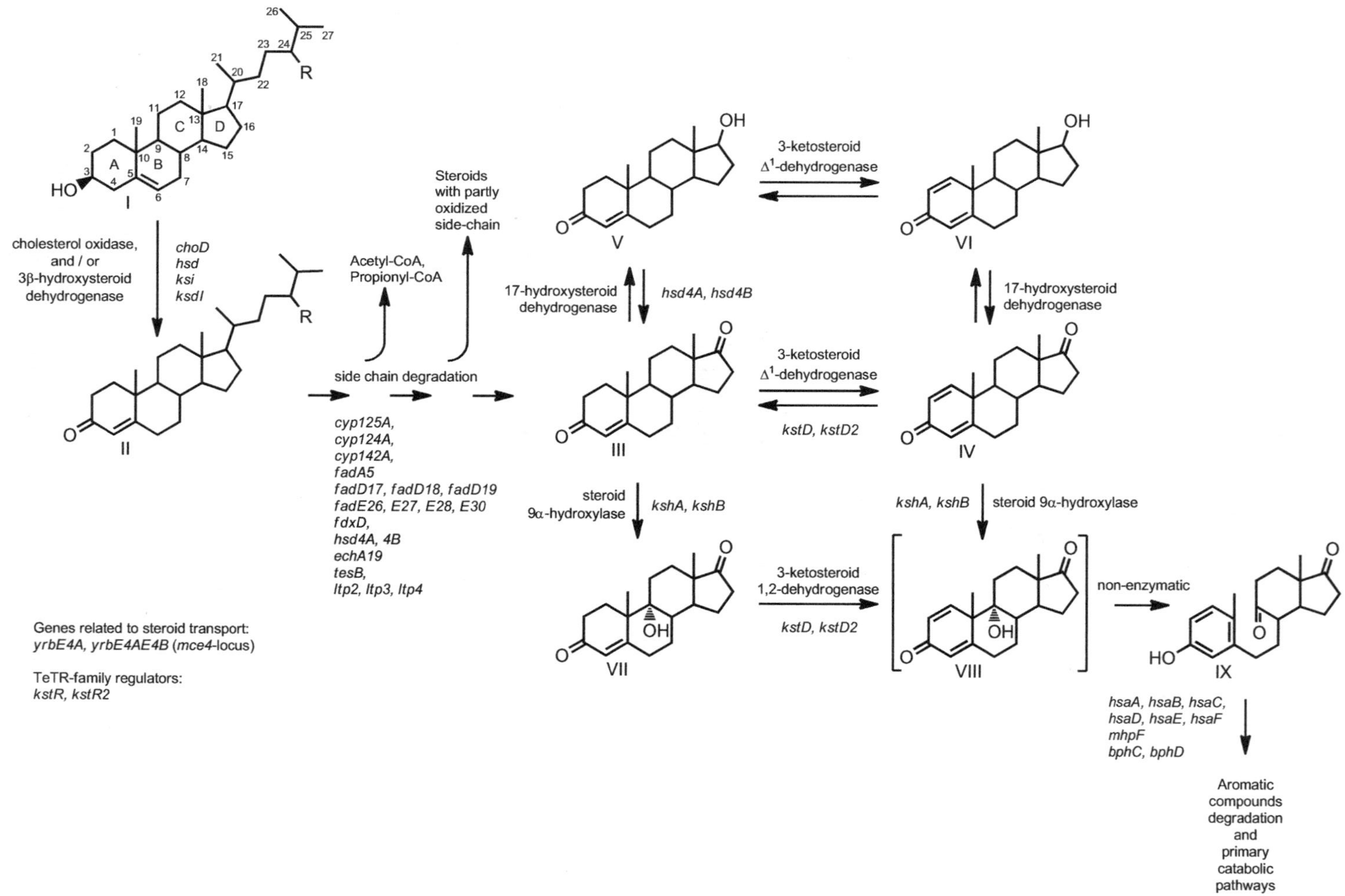

Fig. 2 Biotechnologically relevant steps of sterol catabolism by mycobacteria, enzymes and genes involved (adopted from [48]): **I**—R=H–Cholesterol, R=C_2H_5–β-Sitosterol; **II**—R=H–Cholest-4-ene-3-one, R=C_2H_5–Sitost-4-ene-3-one; **III**—Androst-4-ene-3,17-dione; **IV**—Androsta-1,4-diene-3,17-dione; **V**—Testosterone; **VI**—Δ^1-Dehydrotestosterone; **VII**—9α-Hydroxyandrost-4-ene-3,17-dione; **VIII**—Unstable intermediate; **IX**—9,10-Secoandrost-1,3,5(10)-triene-9,17-dione.

with one of them on a plasmid [32], while in *M. smegmatis* and *M. neoaurum* the genes related to sterol catabolism form three major clusters. In *N. simplex*, one of the predicted clusters related to steroid catabolism has no analogs in other *Actinobacteria* studied so far [23, 29, 33]. Expression of the genes related to steroid catabolism in *Actinobacteria* is controlled by transcription factors KstR and KstR2 [33, 34]. No other regulatory genes of steroid catabolic pathways in *Actinobacteria* have been so far reported.

Bacterial degradation of steroidal A/B rings proceeds via the so-called 9,10-seco pathway (Fig. 2) [35, 36]. Initial steps of the core oxidation result in the 3-keto-4-ene (or 1,4-diene) structure of the A-ring and are catalyzed by the 3-hydroxysteroid dehydrogenase (3-HsD) and cholesterol oxidases, but the role of cholesterol oxidases in the pathway has not been confirmed for some *Actinobacteria*, e.g., [37]. Simultaneous action of 3-ketosteroid-1-dehydrogenase (KstD) and 9α-hydroxylase (KshAB) further leads to 9(10)-secosteroids with an aromatized A-ring. The perhydroindane derivatives are then catabolized with the oxygenases HsaAB, HsaC, and a C–C bond hydrolase to yield an indanone that includes the intact rings C and D. The catabolism of C/D rings remained largely unknown for many *Actinobacteria*, but considerable progress has been made in the last years [38].

Along with the core degradation, the concurrent degradation of the side chain of cholesterol and other sterols may occur by the mechanism that is generally similar to the β-oxidation of fatty acids and proceeds via CoA-thioester intermediates (Fig. 2). It is initiated by cytochromes P450s, and initial steps differ for cholesterol and sterols with branched side chain [39].

The Mce4 steroid transporter system seems to be involved in steroid uptake in diverse *Actinobacteria* [22, 29, 40]. As shown for *R. jostii* RHA1, it is encoded by an 11-gene operon, upregulated with cholesterol and also essential for growth on sitosterol, 5α-cholestanol, and 5α-cholestanone [40].

Significant progress in the understanding of steroid catabolism pathways promoted generation of effective biocatalysts capable of producing valuable steroids. Different tools are applied for the deletion of specific genes, or their augmentation, thus enabling enhanced production of valuable steroids derived from cholesterol catabolic pathways (Fig. 1). Most attention was paid to the generation of effective producers of key androstane steroids, including androstenedione (AD), androstadienedione (ADD), and 9α-hydroxyandrostenedione (9-OH-AD) from phytosterol, which is a well-recognized cheap and available raw material for steroid production. For example, inactivation of three KstDs and augmentation of reductase component (KshB) of 9α-hydroxylase in *M. neoaurum* ATCC 25795 provided effective production of 9-OH-AD from phytosterol [41]. The enhancement of ADD production from phytosterol has been achieved by the augmentation of *kstDs* from *M. neoaurum* NwIB-01 and *Arthrobacter simplex* in the *M. neoaurum* NwIB-01 host [42]. Recombinant strains of *Bacillus subtilis* enabling effective ADD production from AD have been constructed by the expression of *ksdd* encoding KstD from *M. neoaurum* JC-12 [43, 44]. Methods for generation and fermentation of strains capable of effective conversion of phytosterols into AD and ADD are described in detail in the following chapters.

The mutations in the sterol catabolic pathways allow also obtaining other useful intermediates for the synthesis of steroid therapeutic drugs such as those derived from the partial oxidation of the sitosterol side chain, e.g., [45]. Expression of *choM* gene encoding cholesterol oxidase from *M. neoaurum* JC-12 in *Bacillus subtilis* resulted in cholest-4-en-3-one production from cholesterol [46].

There are just few examples illustrating the construction of effective biocatalysts for production of valuable steroids on the base of non-pathogenic species. Generation of engineered strains capable of single-step selective bioconversions of phytosterol and other available and low-cost starting materials to valuable steroid compounds remains one of the most important tasks in the field of steroid biotechnology.

3 Oxyfunctionalization of Steroids

Oxyfunctionalization, and especially, regio- and stereoselective hydroxylation of nonactivated carbon centers is one of the most troublesome challenges in classical organic chemistry, and it is a field where microbial conversions are especially effective. Fungi, which are one of the most diverse kingdoms of eukaryotes, have been reported to carry out hydroxylations at almost all stereogenic centers of the steroid molecule and are widely applied for the production of steroids with broad range of biological activity [47–49].

Oxyfunctionalization of steroids is catalyzed by cytochromes P450 (CYPs) which are the external monooxygenases catalyzing insertion of a single oxygen atom into a non-activated C–H bond of the substrate with a concomitant reduction of other atom to water [19]. Bioconversion of steroids by wild-type strains often results in a mixture of products. Synthetic biology is considered now as the most promising approach for the improvement of the selectivity of steroid hydroxylations.

There is a growing body of research on the recombinant coexpression of mammalian CYPs and their natural redox partners in yeasts (such as *Pichia pastoris*, *Saccharomyces cerevisiae*, and *Schizosaccharomyces pombe*), or bacterial hosts (*Escherichia coli*, *Bacillus subtilis*, and *Bacillus megaterium*). Unlike fungal hydroxylases, mammalian CYPs may provide regio- and stereospecific steroid hydroxylations. For example, current industrial production of cortisol (hydrocortisone) is based on its semichemical synthesis with a final step, microbial transformation by fungal strains of *Curvularia lunata (Cochliobolus lunatus)* in a scale of about 100 tons per year [50]. However, the bioconversion of 11-deoxycortisol with these fungi is accompanied with undesired 14α-hydroxylation [51]. The expression in the yeasts of human mitochondrial CYP11B1 and its natural redox partners, adrenodoxin and adrenodoxin reductase, allows selective hydrocortisone production from 11-deoxycortisol [52, 53]. The engineered strains producing pregnenolone from cholesterol [54, 55], aldosterone from 11-deoxycorticosterone [56], and others [19, 57] have been constructed based on mammalian CYPs. However, commercial application of the whole-cell biocatalysts based on mammalian CYPs is still limited due to several reasons, such as low productivity and, sometimes, insufficient stability [57]. In general, fungal and bacterial P450 systems exhibit higher catalytic rates. Besides, the solubility of bacterial P450s allows higher expression level in heterologous hosts [58, 59]. The design of novel industrial strains is expected by using bacterial hydroxylases with high catalytic activity such as CYP102A1 from *B. megaterium* and CYP101A1 from *Pseudomonas putida* [60].

Along with yeasts, bacterial strains are more and more used as recombinant expression hosts for the construction of steroid transforming strains. Among them, gram-positive bacilli such as *B. megaterium* and *B. subtilis* are mainly used due to effective transport of hydrophobic compounds into the cells. For example, expression of recombinant human CYP11A1 and its redox partners in *B. megaterium* allowed effective bioconversion of cholesterol to pregnenolone [55]. The coexpression of the human mitochondrial cytochrome P450 CYP27A1 with its redox partners provided efficient bioconversion of cholesterol, 7-dehydrocholesterol, and vitamin D3 to their corresponding hydroxylated metabolites [61].

4 Conclusions

Construction and fermentation of engineered microorganisms open a new era in applied microbiology and industrial biotechnology of steroids. However, relatively low productivity, and sometimes, instability of the genetically modified strains, as well as scaling-up problems could be the reason why some bright laboratory bench developments, which are highly important for further practical use, still do not find wide commercial application. Finding solutions to the fundamental problems in the field of synthetic biology of steroid transforming strains would allow creating the scientific basis for a new generation of biocatalysts and innovative bioprocesses for the pharmaceutical industry.

Acknowledgment

Russian Science Foundation is gratefully acknowledged for a support (Grant No. 14-24-00169).

References

1. Baker ME (2011) Origin and diversification of steroids: co-evolution of enzymes and nuclear receptors. Mol Cell Endocrinol 334:14–20. doi:10.1016/j.mce.2010.07.013
2. Waters CM, Bonnie L (2005) Bassler quorum sensing: cell-to-cell communication in bacteria. Annu Rev Cell Dev Biol 21:319–346. doi:10.1146/annurev.cellbio.21.012704.131001
3. Wollam J, Antebi A (2011) Sterol regulation of metabolism, homeostasis, and development. Annu Rev Biochem 80:885–916. doi:10.1146/annurev-biochem-081308-165917
4. Charalampopoulos I, Remboutsika E, Andrew N, Margioris AN, Achille Gravanis A (2008) Neurosteroids as modulators of neurogenesis and neuronal survival. Trends Endocrinol Metab 19(8):300–307. doi:10.1016/j.tem.2008.07.004
5. Melcangi RC, Panzica G, Garcia-Segura LM (2011) Neuroactive steroids: focus on human brain. Neuroscience 15(191):1–5. doi:10.1016/j.neuroscience.2011.06.024
6. Schumacher M, Deniera C, Oudinet J-P, Adams D, Rachida Guennouna R (2016) Progesterone neuroprotection: the background of clinical trial failure. J Steroid Biochem Mol Biol 160:53–66. doi:10.1016/j.jsbmb.2015.11.010
7. Vallée M (2016) Neurosteroids and potential therapeutics: focus on pregnenolone. J Steroid Biochem Mol Biol 160:78–87. doi:10.1016/j.jsbmb.2015.09.030
8. Piironen V, Lindsay DG, Miettinen TA, Toivo J, Lampi AM (2000) Plant sterols: biosynthesis, biological function and their importance to human nutrition. J Sci Food Agric 80:939–966. doi:10.1002/(SICI)1097-0010(20000515)80:7<939::AID-JSFA644>3.0.CO;2-C
9. Fernandes P, Cabral JMS (2007) Phytosterols: applications and recovery methods. Bioresour Technol 98:2335–2350. doi:10.1016/j.biortech.2006.10.006
10. Craigie E, Mullins JJ, Bailey MA (2009) Glucocorticoids and mineralocorticoids. In: Bader M (ed) Cardiovascular hormone systems: from molecular mechanisms to novel therapeutics. Wiley-VCH Verlag GmbH & Co., Weinheim, pp 1–64. doi:10.1002/9783527626236.ch1
11. Finocchi C, Ferrari M (2011) Female reproductive steroids and neuronal excitability. Neurol Sci 32(Suppl 1):S31–S35. doi:10.1007/s10072-011-0532-5
12. Rugutt JK, Rugutt KJ (2012) Antimycobacterial activity of steroids, long-chain alcohols and lytic peptides. Nat Prod Res 26(11):1004–1011. doi:10.1080/14786419.2010.539977
13. Tong W-Y, Dong X (2009) Microbial biotransformation: recent developments on steroid drugs. Recent Pat Biotechnol 3:141–153. doi:10.2174/187220809788700157
14. Swizdor A, Panek A, Milecka-Tronina N (2014) Microbial Baeyer–Villiger oxidation of 5α-steroids using *Beauveria bassiana*. A stereochemical requirement for the 11α-hydroxylation and the lactonization pathway. Steroids 82:44–52. doi:10.1016/j.steroids.2014.01.006
15. Silva CP, Otero M, Esteves V (2012) Processes for the elimination of estrogenic steroid hormones from water: a review. Environ Pollut 165:38–58. doi:10.1016/j.envpol.2012.02.002
16. Ridlon JM, Kang D, Hylemon P (2006) Bile salt biotransformations by human intestinal bacteria. J Lipid Res 47:241–259. doi:10.1194/jlr.R500013-JLR200
17. Kelly D, Kelly S (2003) Rewiring yeast for drug synthesis. Nat Biotechnol 27:133–134. doi:10.1038/nbt0203-133
18. Szczebara FM, Chandelier C, Villeret C, Masurel A, Bourot S, Duport C, Blanchard S, Groisillier A, Testet E, Costaglioli P, Cauet G, Degryse E, Balbuena D, Winter J, Achstetter T, Spagnoli R, Pompon D, Dumas B (2003) Total biosynthesis of hydrocortisone from a simple carbon source in yeast. Nat Biotechnol 21:143–149. doi:10.1038/nbt775
19. Bernhardt R (2006) Cytochromes P450 as versatile biocatalysts. J Biotechnol 124:128–145. doi:10.1016/j.jbiotec.2006.01.026
20. Bergstrand LH, Cardenas E, Holert J, van Hamme JD, Mohn WW (2016) Delineation of steroid-degrading microorganisms through comparative genomic analysis. MBio 7(2):e00166–e00116. doi:10.1128/mBio.00166-16

21. Donova M (2007) Transformation of steroids by actinobacteria: a review. Appl Biochem Microbiol 43(1):1–14. doi:10.1134/S0003683807010012
22. García JL, Uhía I, Galán B (2012) Catabolism and biotechnological applications of cholesterol degrading bacteria. J Microbial Biotechnol 5:679–699. doi:10.1111/j.1751-7915.2012.00331.x
23. Uhía I, Galán B, Kendall SL, Stoker NG, Garcia JL (2012) Cholesterol metabolism in *Mycobacterium smegmatis*: cholesterol pathway. Environ Microbiol Rep 4:168–182. doi:10.1111/j.1758-2229.2011.00314.x
24. Shtratnikova VY, Bragin EY, Dovbnya DV, Pekov YA, Schelkunov MI, Strizhov NI, Ivashina TV, Ashapkin VV, Donova MV (2014) Complete genome sequence of sterol transforming *Mycobacterium neoaurum* strain VKM 1815D. Genome Announc 2(1):e01177–e01113. doi:10.1128/genomeA.01177-13
25. Shtratnikova VY, Schelkunov MI, Dovbnya DV, Pekov YA, Bragin EY, Ashapkin VV, Donova MV (2015) Complete genome sequence of *Mycobacterium* sp. strain VKM 1817D, which is capable of producing 9α-hydroxyandrost-4-ene-3,17-dione from phytosterol. Genome Announc 3(1):e01447–e01414. doi:10.1128/genomeA.01447-14
26. Shtratnikova VY, Schelkunov MI, Pekov YA et al (2015) Complete genome sequence of steroid-transforming *Nocardioides simplex* VKM Ac-2033D. Genome Announc 3(1):e01406–e01414. doi:10.1128/genomeA.01406-14
27. Bragin EY, Shtratnikova VY, Dovbnya DV, Schelkunov MI, Pekov YA, Malakho SG, Egorova OV, Ivashina TV, Sokolov SL, Ashapkin VV, Donova MV (2013) Comparative analysis of genes encoding key steroid core oxidation enzymes in fast-growing *Mycobacterium* spp. strains. J Steroid Biochem Mol Biol 138:41–53. doi:10.1016/j.jsbmb.2013.02.016
28. Rodríguez-García A, Fernández-Alegre E, Morales A, Sola-Landa A, Lorraine J, Macdonald S, Dovbnya D, Smith MCM, Donova M, Barreiro C (2016) Complete genome sequence of *Mycobacterium neoaurum* NRRL B-3805, an androstenedione (AD) producer for industrial biotransformation of sterols. J Biotechnol 224:64–65. doi:10.1016/j.jbiotec.2016.03.021
29. Shtratnikova VY, Schelkunov MI, Fokina VV, Pekov YA, Donova MV (2016) Genome-wide bioinformatics analysis of steroid metabolism-associated genes in *Nocardioides simplex* VKM Ac-2033D. Curr Genet 62:1–14. doi:10.1007/s00294-016-0568-4
30. Ge F, Li W, Chen G, Liu Y, Zhang G, Yong B, Wang Q, Wang N, Huang Z, Li W, Wang J, Wu C, Xie Q, Liu Q (2011) Draft genome sequence of *Gordonia neofelifaecis* NRRL B-59395, a cholesterol-degrading actinomycete. J Biotechnol 193(18):5045–5046. doi:10.1128/JB.05531-11
31. Haußmann U, Wolters DA, Franzel B, Eltis LD, Poetsch A (2013) Physiological adaptation of the *Rhodococcus jostii* RHA1 membrane proteome to steroids as growth substrates. J Proteome Res 12:1188–1198. doi:10.1021/pr300816n
32. McLeod MP, Warren RL, Hsiao WW, Araki N, Myhre M, Fernandes C, Miyazawa D, Wong W, Lillquist AL, Wang D, Dosanjh M, Hara H, Petrescu A, Morin RD, Yang G, Stott JM, Schein JE, Shin H, Smailus D, Siddiqui AS, Marra MA, Jones SJ, Holt R, Brinkman FS, Miyauchi K, Fukuda M, Davies JE, Mohn WW, Eltis LD (2006) The complete genome of *Rhodococcus* sp. RHA1 provides insights into a catabolic powerhouse. Proc Natl Acad Sci USA 103(42):15582–15587. doi:10.1073/pnas.0607048103
33. Kendall S, Withers M, Soffair CN, Moreland NJ, Gurcha S, Sidders B, Frita R, ten Bokum A, Besra GS, Lott JS, Stoker NG (2007) A highly conserved transcriptional repressor controls a large regulon involved in lipid degradation in *Mycobacterium smegmatis* and *Mycobacterium tuberculosis*. Mol Microbiol 65(3):684–699. doi:10.1111/j.1365-2958.2007.05827.x
34. Kendall S, Burgess P, Balhana R, Withers M, ten Bokum A, Lott JS, Gao C, Uhia-Castro I, Stoker NG (2010) Cholesterol utilization in mycobacteria is controlled by two TetR-type transcriptional regulators: kstR and kstR2. Microbiology 156:1362–1371. doi:10.1099/mic.0.034538-0
35. Szentirmai A (1990) Microbial physiology of sidechain degradation of sterols. J Ind Microbiol Biotechnol 6:101–116. doi:10.1007/BF0157642938
36. Petrusma M, Hessels G, Dijkhuizen L, van der Geize R (2011) Multiplicity of 3-ketosteroid-9α-hydroxylase enzymes in *Rhodococcus rhodochrous* DSM43269 for specific degradation of different classes of steroids. J Bacteriol 193:3931–3940. doi:10.1128/JB.00274-11
37. Ivashina TV, Nikolayeva VM, Dovbnya DV, Donova MV (2012) Cholesterol oxidase ChoD is not a critical enzyme accounting for oxidation of sterols to 3-keto-4-ene steroids in fast-growing *Mycobacterium* sp. VKM Ac-1815D. J Steroid Biochem Mol Biol 129:47–53. doi:10.1016/j.jsbmb.2011.09.008
38. Casabon I, Crowe AM, Liu J, Eltis LD (2013) FadD3 is an acyl-CoA synthetase that initiates

catabolism of cholesterol rings C and D in actinobacteria: role of FadD3 in cholesterol catabolism. Mol Microbiol 87:269–283. doi:10.1111/mmi.12095

39. Capyk JK, Kalscheuer R, Stewart GR, Liu J, Kwon H, Zhao R, Okamoto S, Jacobs WR Jr, Eltis LD, Mohn WW (2009) Mycobacterial cytochrome p450 125 (cyp125) catalyzes the terminal hydroxylation of c27 steroids. J Biol Chem 284:35534–35542. doi:10.1074/jbc.M109.072132

40. Mohn WW, van der Geize R, Stewart GR, Okamoto S, Liu J, Dijkhuizen L, Eltis LD (2008) The actinobacterial mce4 locus encodes a steroid transporter. J Biol Chem 283:35368–35374. doi:10.1074/jbc.M805496200

41. Yao K, Xu LQ, Wang FQ, Wei DZ (2014) Characterization and engineering of 3-ketosteroid-△1-dehydrogenase and 3-ketosteroid-9α-hydroxylase in *Mycobacterium neoaurum* ATCC 25795 to produce 9α-hydroxy-4-androstene-3,17-dione through the catabolism of sterols. Metab Eng 24:181–191. doi:10.1016/j.ymben.2014.05.005

42. Wei W, Fan SY, Wang FQ, Wei DZ (2014) Accumulation of androstadienedione by overexpression of heterologous 3-ketosteroid D1-dehydrogenase in *Mycobacterium neoaurum* NwIB-01. World J Microbiol Biotechnol 30(7):1947–1954. doi:10.1007/s11274-014-1614-3

43. Li Y, Lu F, Sun T (2007) Expression of ksdD gene encoding 3-ketosteroid-Δ1-dehydrogenase from *Arthrobacter simplex* in *Bacillus subtilis*. Lett Appl Microbiol 44:563–568. doi:10.1111/j.1472-765X.2007.02134.x

44. Zhang WQ, Shao ML, Rao ZM, Xu MJ, Zhang X, Yang TW, Li H, Xu ZH (2013) Bioconversion of 4-androstene-3,17-dione to androst-1,4-diene-3,17-dione by recombinant *Bacillus subtilis* expressing ksdd gene encoding 3-ketosteroid-Δ^1-dehydrogenase from *Mycobacterium neoaurum* JC-12. J Steroid Biochem Mol Biol 135:36–42. doi:10.1016/j.jsbmb.2012.12.016

45. Andor A, Jekkel A, Hopwood DA, Jeanplong F, Ilkoy E, Konya A, Kurucz I, Ambrus G (2006) Generation of useful insertionally blocked sterol degradation pathway mutants of fast-growing mycobacteria and cloning, characterization, and expression of the terminal oxygenase of the 3-ketosteroid 9α-hydroxylase in *Mycobacterium smegmatis* mc^{2}155. Appl Environ Microbiol 72(10):6554–6559. doi:10.1128/AEM.00941-06

46. Shao ML, Rao ZM, Zhang X, Xu MJ, Yang TW, Li H, Xu ZH, Yang ST (2015) Bioconversion of cholesterol to 4-cholesten-3-one by recombinant *Bacillus subtilis* expressing choM gene encoding cholesterol oxidase from *Mycobacterium neoaurum* JC-12. J Chem Technol Biotechnol 90(10):1811–1820. doi:10.1002/jctb.4491

47. Mahato SB, Garai S (1997) Advances in microbial steroid biotransformations. Steroids 62:332–345. doi:10.1016/S0039-128X(96)00251-6

48. Donova MV, Egorova OV (2012) Microbial steroid transformations: current state and prospects. Appl Microbiol Biotechnol 94:1423–1447. doi:10.1007/s00253-012-4078-0

49. Kristan K, Lanisnik Rizner T (2012) Steroid-transforming enzymes in fungi. J Steroid Biochem Mol Biol 129:79–91. doi:10.1016/j.jsbmb.2011.08.012

50. Van Beilen JB, Duetz WA, Schmid A, Withold B (2003) Practical issues in the application of oxygenases. Trends Biotechnol 21:170–177. doi:10.1016/S0167-7799(03)00032-5

51. Kollerov VV, Shutov AA, Fokina VV, Sukhodol'skaya GV, Gulevskaya SA, Donova MV (2010) Bioconversion of C19- and C21-steroids with parent and mutant strains of *Curvularia lunata*. Appl Biochem Microbiol 46:198–205. doi:10.1134/S0003683810020122

52. Hakki T, Zearo S, Drăgan C-A, Bureik M, Bernhard R (2008) Coexpression of redox partners increases the hydrocortisone (cortisol) production efficiency in CYP11B1 expressing fission yeast *Schizosaccharomyces pombe*. J Biotechnol 133(3):351–359. doi:10.1016/j.jbiotec.2007.06.022

53. Schiffer L, Anderko S, Hobler A, Hannemann F, Kagawa N, Bernhardt R (2015) A recombinant CYP11B1 dependent *Escherichia coli* biocatalyst for selective cortisol production and optimization towards a preparative scale. Microb Cell Factories 14:25. doi:10.1186/s12934-015-0209-5

54. Makeeva D, Dovbnya D, Donova M, Novikova L (2013) Functional reconstruction of bovine P450scc steroidogenic system in *Escherichia coli*. American J Mol Biol 3:173–182. doi:10.4236/ajmb.2013.34023

55. Gerber A, Kleser M, Biedendieck R, Bernhardt R, Hannemann F (2015) Functionalized PHB granules provide the basis for the efficient side-chain cleavage of cholesterol and analogs in recombinant *Bacillus megaterium*. Microb Cell Fact 14:107. doi:10.1186/s12934-015-0300-y

56. Hobler A, Kagawa N, Hutter MC, Hartmann MF, Wudy SA, Hannemann F, Bernhardt R (2012) Human aldosterone synthase: recombi-

nant expression in *E. coli* and purification enables a detailed biochemical analysis of the protein on the molecular level. J Steroid Biochem Mol Biol 132:57–65. doi:10.1016/j.jsbmb.2012.03.002

57. Bernhardt R, Urlacher VB (2014) Cytochromes P450 as promising catalysts for biotechnological application: chances and limitations. Appl Microbiol Biotechnol 98:6185–6203. doi:10.1007/s00253-014-5767-7
58. Petric S, Hakki T, Bernhardt R, Zigon D, Cresnar B (2010) Discovery of a steroid 11α-hydroxylase from *Rhizopus oryzae* and its biotechnological application. J Biotechnol 150:428–437. doi:10.1016/j.jbiotec.2010.09.928
59. Bracco P, Janssen DB, Schallmey A (2013) Selective steroid oxyfunctionalisation by CYP154C5, a bacterial cytochrome P450. Microb Cell Factories 12:95. doi:10.1186/1475-2859-12-95
60. Schmitz D, Zapp J, Bernhardt R (2014) Steroid conversion with CYP106A2 – production of pharmaceutically interesting DHEA metabolites. Microb Cell Fact 13:81. doi:10.1186/1475-2859-13-81
61. Ehrhardt M, Gerber A, Hannemann F, Bernhardt R (2016) Expression of human CYP27A1 in *B. megaterium* for the efficient hydroxylation of cholesterol, vitamin D3 and 7-dehydrocholesterol. J Biotechnol 218:34–40. doi:10.1016/j.jbiotec.2015.11.021

Chapter 2

Chemical Pathways of Corticosteroids, Industrial Synthesis from Sapogenins

Ignacio Herráiz

Abstract

Corticosteroids are products of high industrial and commercial importance. There are dozens of different synthesis published for all of them. Some are coming from academia and some from industry. Here, industrial processes for the synthesis of prednisone, prednisolone, hydrocortisone, dexamethasone, betamethasone, and methylprednisolone are described. The starting material is diosgenin and the desired molecules are reached due to a good combination of chemistry and biotechnology that was developed along the second part of the twentieth century.

Key words Corticosteroid, Microbial transformation, Industrial, Hydroxylation, Dehydrogenation, Diosgenin, Prednisone, Hydrocortisone, Betamethasone, Dexamethasone, Methylprednisolone

1 Introduction

Steroids are a group of molecules with a high structural complexity. This complexity has encouraged organic chemists to find total and partial syntheses of these challenging and attractive molecules. Research on this field started back to the nineteenth century but it was in the mid-twentieth century when it suffered a dramatic acceleration. This boost was provoked by several factors. The first one was the discovery by Hench [1] at the Mayo clinic in 1949 that cortisone is quite useful in treating rheumatoid arthritis. The second one was the achievement of Merck Development Department on getting corticosteroids from ox-bile cholic acids. The third one was the discovery by Marker [2, 3] of the degradation of the side chain of sapogenins that made available a cheap source of precursors for steroid synthesis. The above discoveries, among others, should be linked to the huge economical potential that was obvious for pharmaceuticals companies at the time. This "magic" mix opened the golden era of steroids that lasted almost for the rest of the twentieth century.

José-Luis Barredo, Ignacio Herráiz (eds.), *Microbial Steroids: Methods and Protocols*, Methods in Molecular Biology vol. 1645, DOI 10.1007/978-1-4939-7183-1_2, © Springer Science+Business Media LLC 2017

Having ox-bile cholic acids available and routes to the final desired steroids, this could have been a chemistry era, but it was not. Multidisciplinary approaches were on the table since the very beginning. The combination of chemistry and biology provided sapogenins approach advantage. Although multidisciplinary approaches and enzymatic reactions sound quite modern and updated, it should be noted that in the field of steroids the "wedding" started in the late years of 1940 decade, just after World War II.

This article will not be focused on academic Chemistry/Biology but industrial processes. It will be reviewed how biotransformations and enzymatic processes joined chemistry to produce corticosteroids in the pharmaceutical industry starting from sapogenins, more specifically diosgenin.

2 The Sapogenin Approach

Three main sapogenins have been the selected starting materials for steroid synthesis: hecogenin, tigogenin, and diosgenin (*see* Fig. 1).

While hecogenin has an "activated" C ring, the other two have pure aliphatic C ring that becomes a challenge in terms of chemistry to be "modified." Among these three compounds, the industrial winner was diosgenin, and the other, although used for some time, were latterly abandoned in favor of diosgenin.

2.1 The Synthesis of 16-Epoxy-11α-hydroxy Progesterone

Diosgenin is extracted from roots of different yams. The extraction business started in Mexico [4] and led the industry till the introduction of Chinese companies in the steroids field.

The story starts in a root that contains disogenin as a glucoderivative. After an initial fermentation, diosgenin is obtained as a pure white solid. It is subjected to the Marker degradation (*see* Fig. 2) yielding in three steps a key intermediate 16-dehydropregnenolone (**IV**). Nowadays, the degradation has been slightly modified in its last step yielding the corresponding acetate (**V**), usually called 16-DPA (16-dehydro pregnenolone acetate).

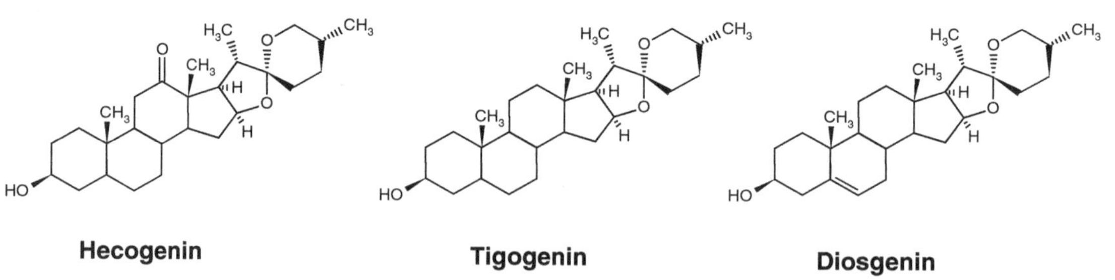

Fig. 1 Most common sapogenins used in industrial production of corticosteroids

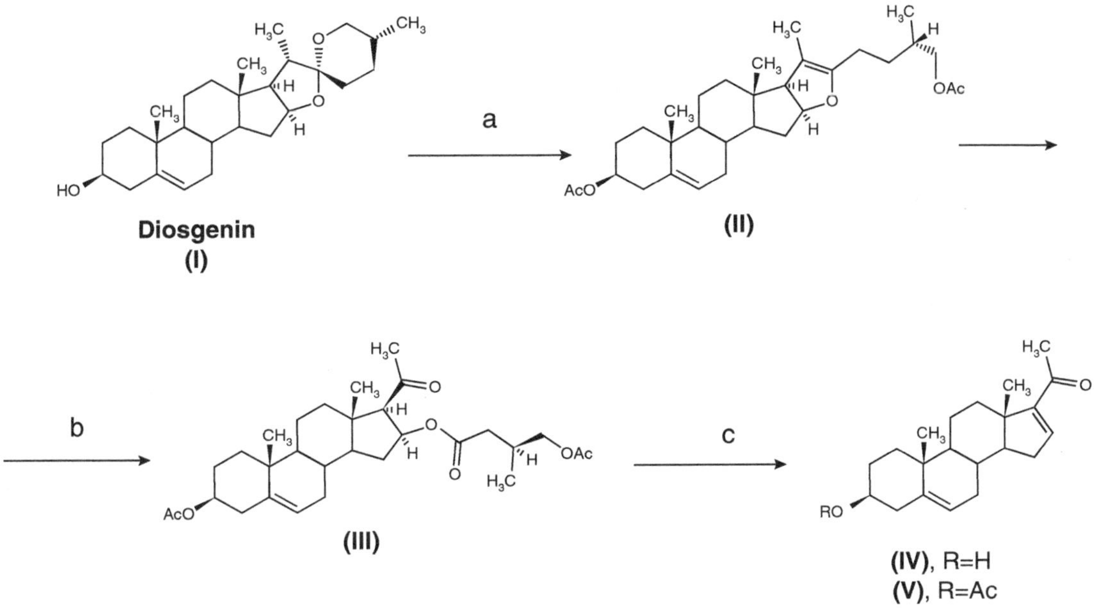

Fig. 2 Marker's degradation. (**a**) Ac_2O, 200 °C. (**b**) CrO_3, AcOH. (**c**) NaOH, EtOH

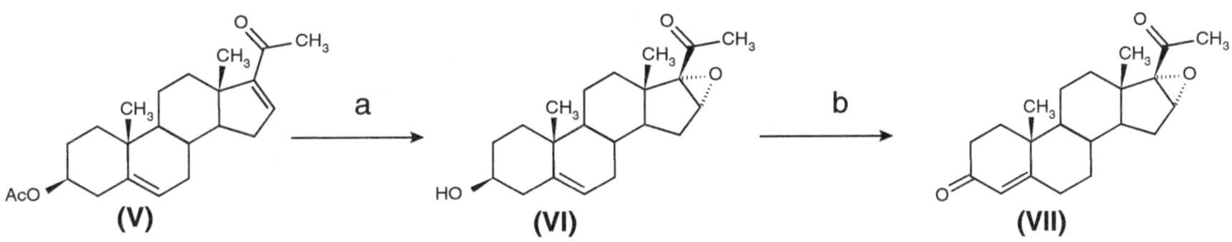

Fig. 3 (**a**) H_2O_2, NaOH, MeOH. (**b**) $Al(^iPrO)_3$, cyclohexanone, toluene

Compound (**V**) is further reacted with H_2O_2/NaOH/MeOH to get the 16,17-epoxy group with a concomitant saponification of the acetoxy group (**VI**). The hydroxy group is oxidized with the Oppenauer protocol [5] that leads to the 3-keto unsaturated intermediate (**VII**) (*see* Fig. 3). This intermediate is crucial in the synthesis of corticosteroids from diosgenin.

Now it is time to introduce the discovery of 11α-hydroxylation. We need to go back to 1950 when Murray and Petersen [6–8] were able to introduce a hydroxyl group in position 11 by fermentation of progesterone with a mold of the genus *Rhizopus*. This key discovery opened the door to an easy synthesis of corticosteroids from progesterone.

In the case of 16-epoxyprogesterone (**VII**) the fermentation with *Rhizopus nigricans* has been extensively used at industrial scale. The fermentation itself does not work so well, as the conversion is only around 50%. This low conversion is not a problem due to the downstream process. Once the fermentation is finished, the broth is filtered and the water phase discarded. The solid phase is extracted with a mixture of toluene and chloroform getting a mixture of the epoxyprogesterone **(VII)** and the 11α-hydroxy compound **(VIII)**. Surprisingly, this mixture is easily crystallized by distillation yielding **(VIII)** and keeping **(VII)** fully in the mother liquor. The starting material is recovered from this mother liquor and recycled to the next fermentation batch. In such a way, the combined yield for the fermentation process is higher than 88% giving the process a full economic sense.

From this intermediate several routes to corticosteroids start.

2.2 Prednisone and Prednisolone

Prednisone and prednisolone are traditionally synthesized following a very efficient route (*see* Fig. 4) which starts with the oxidation of hydroxyl group at carbon-11 **(IX)**. Then the epoxide is opened with hydrobromic acid **(X)** and subsequent Ni-Ra reduction of bromine yields **(XI)**. Next, the 21-methyl group is transformed following the Ringold-Stork protocol [9, 10]. The product obtained is cortisone-21-acetate **(XII)**. At this point, microbiology comes again to solve a problem.

The introduction of a double bond at 1,2 position of the steroid skeleton could be performed by different chemical reagents. Among the options used, DDQ (2,3-Dichloro-5,6-dicyano-1,4-benzoquinone) [11] and SeO_2 [12] are the most common choices. Although the yields are not the best, workup of DDQ reactions is troublesome and toxicity of selenium is high; these options were used in the industry for certain time. These methods became obsolete due to an enzymatic reaction discovered by chance at Schering [13, 14]. While looking for a bacteria capable of promoting the hydrolysis of 11-acetoxy group, it was found that the treatment of 11,21-hydrocortisone diacetate with *Corynebacterium simplex* (ATCC 6946) yielded a new product. This product was identified as 11,21-prednisolone diacetate. It was proven later that the dehydrogenation is performed by an enzymatic system. Cells could be dead and the broth could be used directly or even also concentrated partially. Over the years this enzymatic system has become the only choice for 1,2-dehydrogenation in steroids, and *Corynebacterium simplex* (ATCC 6946) a universal source for it. In Prednisone synthesis, the enzymatic reaction works well with the only requirement of having a pretty low level of the starting material at the end of transformation. Prednisone acetate **(XIII)** is later transformed into Prednisone **(XIV)** with a basic treatment. On the other hand, prednisone acetate **(XIII)** could be transformed into prednisolone **(XVII)** following the protocol by Oliveto [15]. Protection of prednisone acetate **(XIII)** with semicarbazide yields the 3,20-bissemicarbazide prednisone acetate **(XV)**. A further reduction with KBH_4, followed by $NaNO_2$ mediated deprotection afforded prednisolone **(XVII)** in very good yield. Most of prednisone and prednisolone manufactured in the world are being made with the above combination of chemistry and biology.

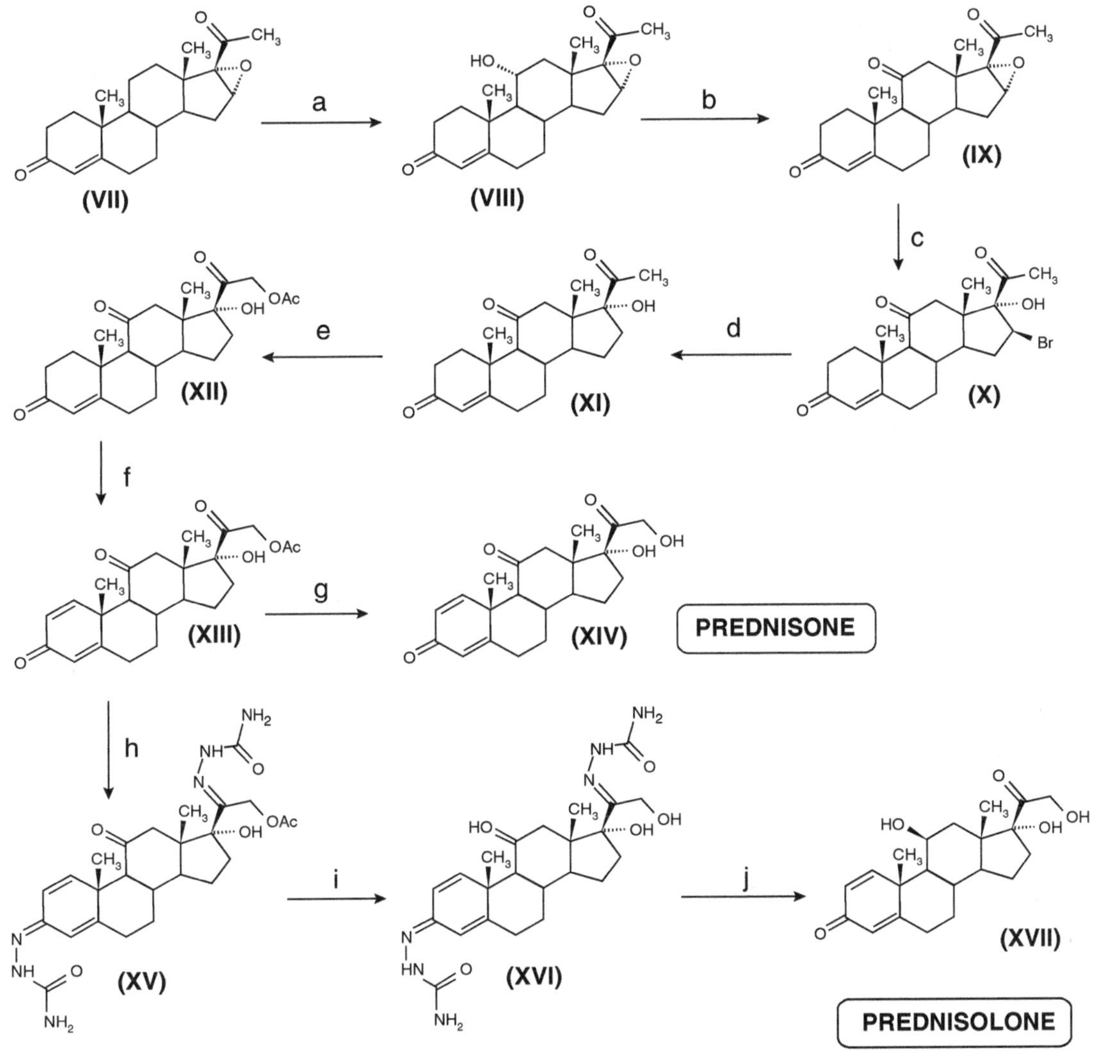

Fig. 4 Prednisone and prednisolone synthesis. (**a**) *Rhizopus nigricans* fermentation. (**b**) CrO_3, AcOH. (**c**) HBr. (**d**) H_2, Ni-Ra. (**e**) i. I_2, $CaCl_2$, CaO, MeOH. ii. AcOK, DMF (**f**) *Corynebacterium simplex* (ATCC 6946). (**g**) KOH, MeOH. (**h**) $NH_2CONHNH_2 \cdot HCl$, $NaHCO_3$, MeOH. (**i**) KBH_4, MeOH. (**j**) $NaNO_2$, HCl

2.3 Dexamethasone

11α-Hydroxy-16-epoxy progesterone (**VIII**) could be also treated with the enzymatic extract of *Corynebacterium simplex* to yield 1,2-dehydro-11α-hydroxy-16-epoxy progesterone (**XVIII**) which is a key intermediate in the synthesis of Dexamethasone, especially in China (*see* Fig. 5).

The dehydro derivative (**XVIII**) is treated with chromium to regenerate the 16,17 double bond (**XIX**) and then the subsequent elimination of the 11α-hydroxyl group through the mesylate yields (**XX**) with four double bond in its structure (this product is usually called 5ST or tetraene-21-methyl). (**XX**) is transformed

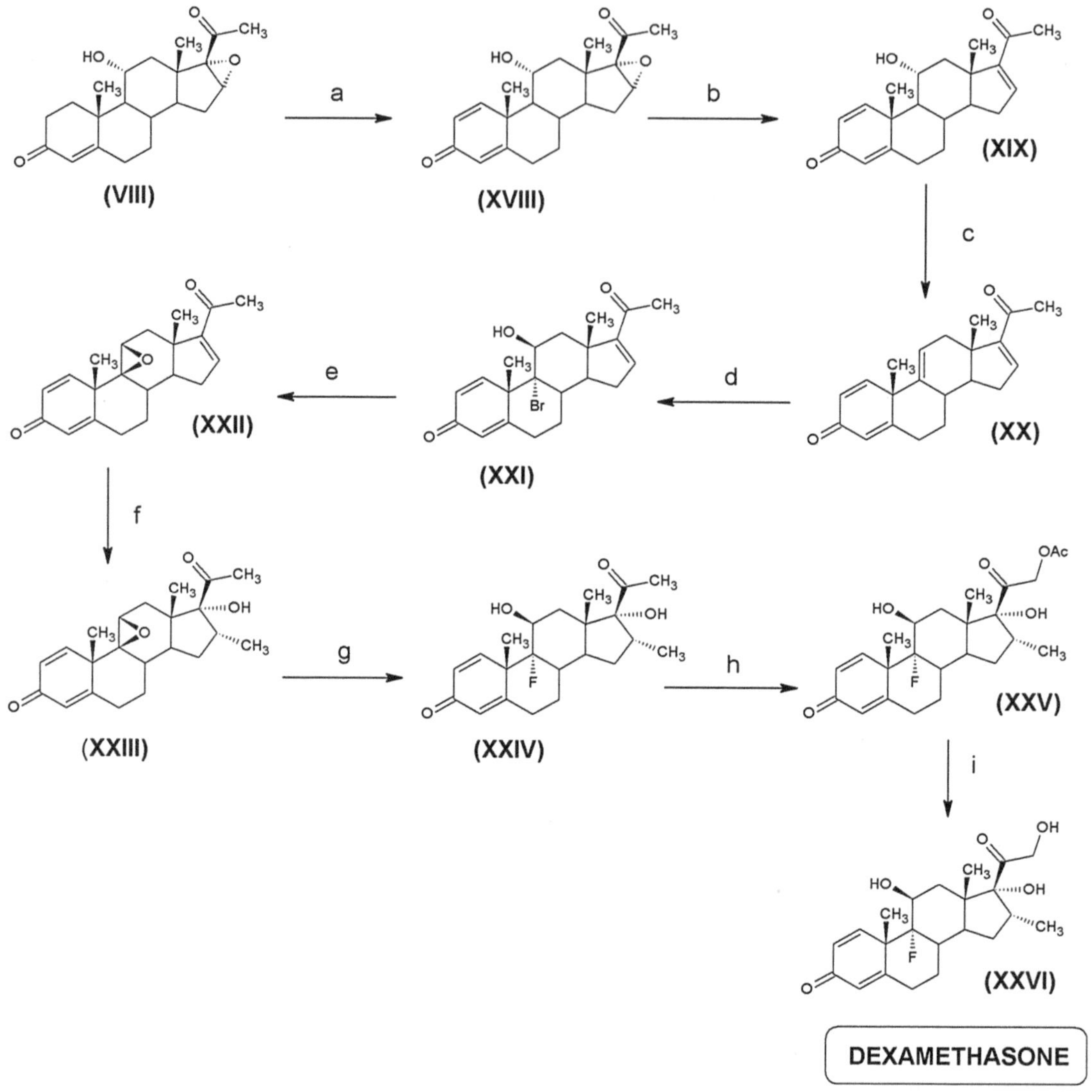

Fig. 5 Dexamethasone synthesis. (**a**) *Corynebacterium simplex* (ATCC 6946). (**b**) Chromium, HCl. (**c**) MsCl, Et_3N. AcONa, AcOH. (**d**) NBS, $HClO_4$, Acetone. (**e**) K_2CO_3, MeOH. (**f**) CH_3MgCl, CuCl, THF. O_2, $(MeO)_3P$, DMF. (**g**) HF, DMF. (**h**) I_2, $CaCl_2$, CaO, MeOH. AcOK, AcOH, Acetone. (**i**) K_2CO_3, MeOH

into bromohydrine **(XXI)** which collapses into the 9(11)β-epoxide **(XXII)** under basic treatment. Now copper catalyzed addition of a methyl group at 16 position and oxidation of the in situ captured TMS enol derivative leads, after hydrolysis, to the 16α-methyl-17-hydroxy-20-keto intermediate **(XXIII)**. Epoxide opening with HF/DMF yields dexamethasone 21-methyl **(XXIV)** which is subjected to Ringold-Stork protocol [10] to provide dexamethasone-21-acetate **(XXV)**. Final hydrolysis yields dexamethasone **(XXVI)**, which is one of the most used corticosteroids in the pharmaceutical market.

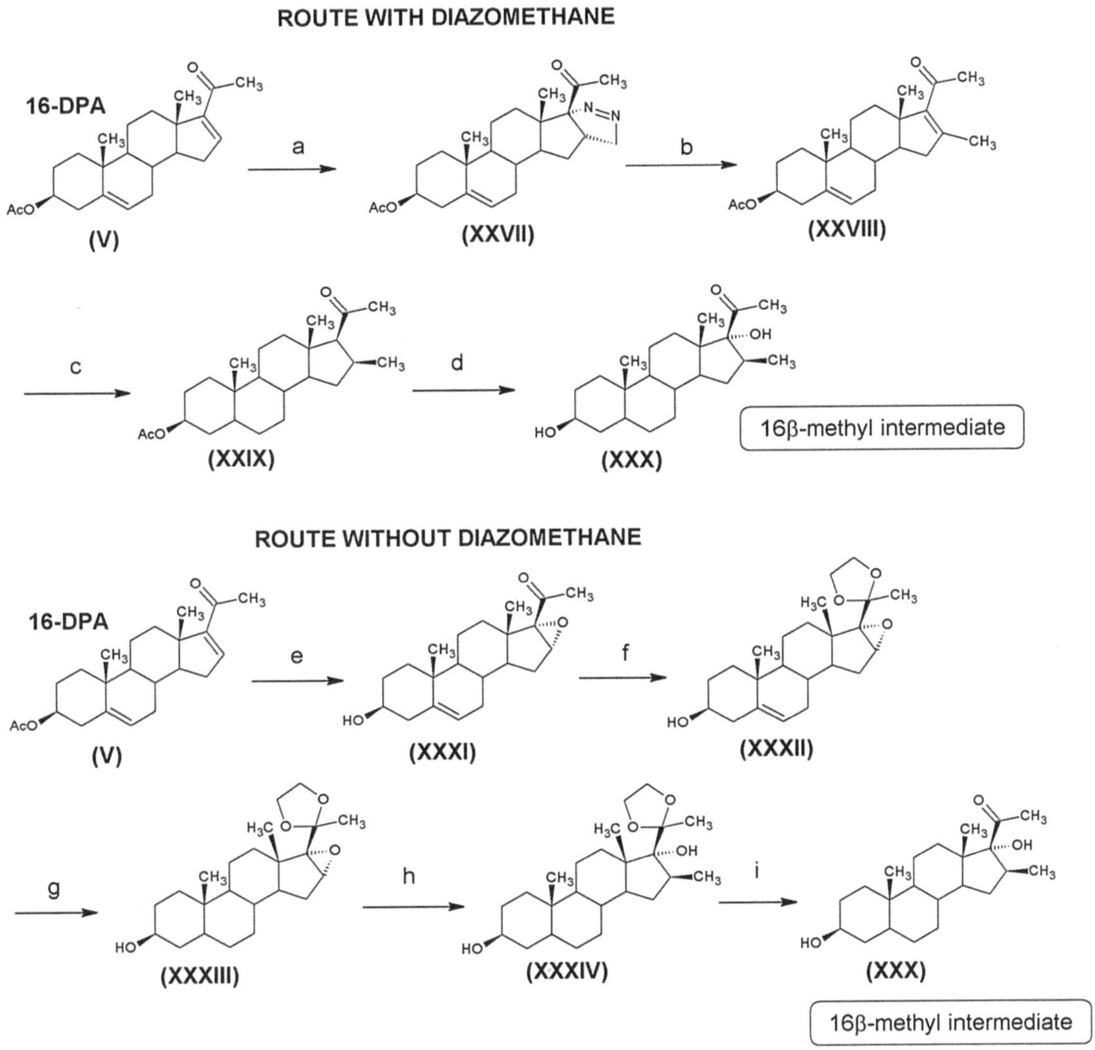

Fig. 6 Synthesis of 16β-methyl intermediates. (**a**) CH_2N_2, DCM, DMF. (**b**) DMF, 135 °C. AcOK. (**c**) H_2, Pd/C, THF. (**d**) O_2, NaH, $(EtO)_3P$, t-BuOH, DMF, −30 °C. (**e**) H_2O_2, NaOH, MeOH. (**f**) Ethylenglycol, TEOF, *p*-TsOH, AcOEt. (**g**) H_2, Pd/$CaCO_3$, THF. (**h**) MeMgCl, THF, toluene, 90 °C. (**i**) HCl

2.4 Betamethasone

Betamethasone is another important corticosteroid made from diosgenin. In this case, the synthesis is also a combination of chemistry and fermentation. At least two different procedures are used to introduce the 16β-methyl group. The older but still use, is the one developed by Schering and Glaxo [16] (*see* Fig. 6).

16-DPA is reacted with diazomethane and the resulting pyrazoline (**XXVII**) subjected to pyrolysis to yield the 16-methyl moiety (**XXVIII**). A hydrogenation step removes both double bonds from the molecule (**XXIX**). Then the oxidation at 17-position following Gardner's protocol [17, 18] yields the desired 16β-methyl-17-hydroxy pregnane structure (**XXX**). This protocol, although quite dangerous due to the use of diazomethane, is still in industrial use.

Another methodology was developed to avoid the use of diazomethane. In this case 16-DPA is first epoxidized with simultaneous cleavage of the acetoxy group at 3 position **(XXXI)**. Then the intermediate is protected at 20-keto group as ethylenediol ketal **(XXXII)**. The protected keto epoxide is hydrogenated yielding the saturated compound **(XXXIII)**, which upon reaction with MeMgCl or MeMgI under harsh conditions [19–21] builds up the key structure in ring D **(XXXIV)**. Final hydrolysis of the ethylene ketal yields the desired intermediate **(XXX)**.

The next step in the synthesis (*see* Fig. 7) will consist of the introduction of 21-acetate through a 21-bromo intermediate **(XXXV)**. This bromine protocol is much cheaper than the Ringold-Stork one but could only be applied when the molecule lacks of double bonds. Reaction of the 21-bromo with sodium acetate yields the 21-acetate compound **(XXXVI)**. It is now the moment for the modification of ring A. The hydroxyl group is oxidized **(XXXVII)** and a very complex bromination process is used to get the 2,4-dibromo compound **(XXXVIII)** which upon elimination leads to the 1,4-dien-3-keto structure **(XXXIX)**. All the previous reactions have allowed us to reach the key step in the synthesis. This intermediate **(XXXIX)** is bioconverted into the 11α-hydroxy derivative **(XL)**. This fermentation procedure works well although some 12-hydroxy derivative is present as impurity.

Concomitant with fermentation the 21-acetoxy group is hydrolyzed. After the fermentation the first step is to protect the 21-hydroxy with a catylate group **(XLI)** and then the hydroxyl group is eliminated [22] with PCl_5 to yield the 9,11 double bond **(XLII)**. This elimination also produces small amounts of the 11,12 double bond as impurity. The double bond is converted into beta epoxide through a sequence of formidohydrobromination/elimination **(XLIV)**, while the 21-hydroxy is once again hydrolyzed. Finally, the epoxide is opened with HF/DMF producing Betamethasone **(XLV)**.

2.5 Hydrocortisone

The synthesis of hydrocortisone is the story of a different fermentation (*see* Fig. 8). The synthesis starts in 16-DPA and goes to 16,17-epoxy progesterone **(VII)** as described above (*see* Fig. 3). The epoxy derivative is opened with HBr **(XLVI)** and then reduced with Ni-Ra to obtain 17-hydroxy progesterone **(XLVII)**. This product is submitted to Ringold-Stork protocol producing the corresponding acetate **(XLIX)** (Reichstein's substance S Acetate or RSA). At this moment fermentation step takes place. This is a particular fermentation. It is performed by *Absidia orchidis* [23–26] and produces a mixture of 11β and 11α-hydroxyl derivatives **(L)**. The mixture of isomers is around 72/17. At the same time, the 11-hydroxylation takes place, the 21-acetate group is

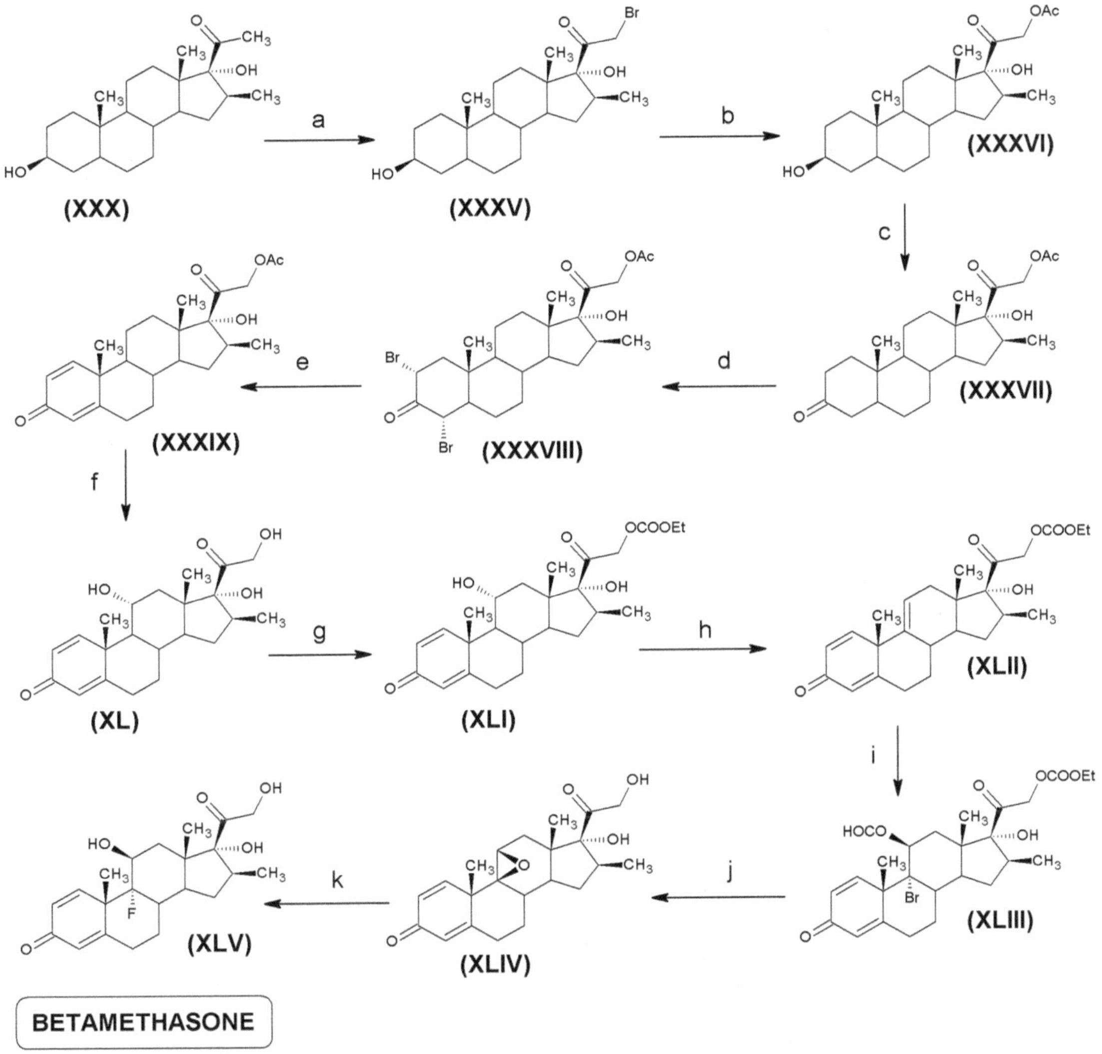

Fig. 7 Betamethasone synthesis. (**a**) Br_2, HCl, MeOH. (**b**) AcONa, HSO_4NBu_4, AcOEt. (**c**) $Ca(ClO)_2$, AcOH. (**d**) Br_2, HBr, AcOH, dioxane. (**e**) $CaCO_3$, LiBr, dimethylacetamide. (**f**) *Rhizopus nigricans* fermentation. (**g**) Ethylchloroformate, Et_3N, DCM. (**h**) PCl_5, THF, −90 °C. (**i**) DDH, $HClO_4$, DMF. (**j**) NaOH, MeOH, DCM, 0 °C. (**k**) HF, DMF

hydrolyzed. The broth is then filtered and the steroids extracted. Then after concentration the isomers are separated by several crystallizations. Finally 11β compound (hydrocortisone, **LI**) is obtained. The mother liquor residue, which contains the 11α isomer plus some 11β isomer, is isolated as a mixture of products. The mixture is latterly oxidized and used in the synthesis of prednisone in an effort to improve the economy of the process through recycling of the residue.

Although the fermentation is not so efficient in terms of stereo selectivity, the direct access to hydrocortisone instead of the traditional approach from 11α-hydroxy, through 11 keto, and then

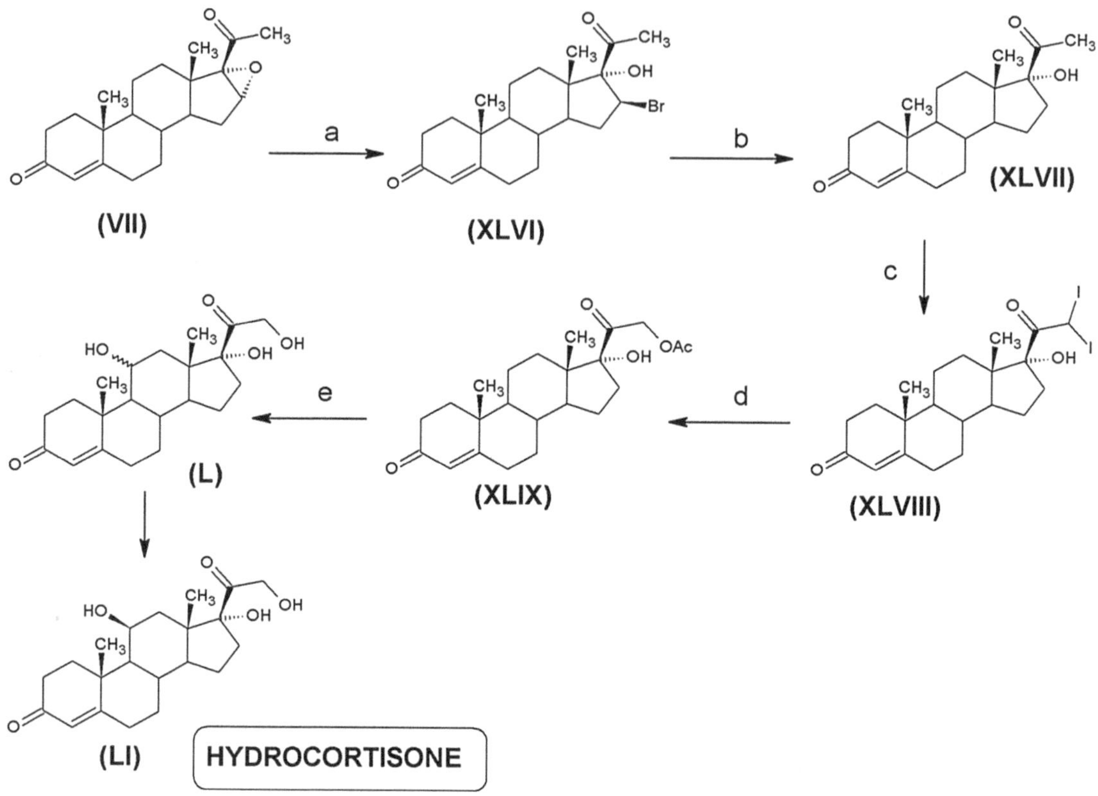

Fig. 8 Hydrocortisone synthesis. (**a**) HBr. (**b**) H_2, Ni-Ra. (**c**) I_2, CaO, $CaCl_2$, $CHCl_3$, MeOH. (**d**) AcOK, DMF. (**e**) *Absidia orchidis* fermentation

11β-hydroxy, has been prevalent in industrial manufacture of this corticosteroid when sapogenins are the starting material.

2.6 Methylprednisolone

The synthesis of methylprednisolone **(LXI)** (*see* Fig. 9) is even longer than dexamethasone or betamethasone ones. It starts from 16-DPA **(V)** and goes through 16,17-epoxy progesterone **(VII)** (*see* Figs. 3 and 4) to 11α-hydroxy-16,17-epoxy progesterone **(VIII)** with *Rhizopus nigricans* fermentation. The 11α-hydroxy group has the wrong stereochemistry as the final product is 11β so the configuration should be inversed in an oxidation/reduction sequence. This is possible because the reduction of 11-keto leads with very high stereoselectivity to the 11β isomer. In the synthesis of methylprednisolone this procedure is done in separate steps. The first step is the oxidation with CrO_3. Then the epoxide at 16–17 is opened and the bromine at 16 position removed to reach 17-hydroxy-11-keto progesterone **(XI)**. In this moment the 3 and 20-keto groups should be protected as ethylene ketal groups [27, 28]. It is possible to do these protections in the presence of the 11-ceto group because the latter is quite unreactive. This third keto group is shielded by the two methyl groups of 18 and 19

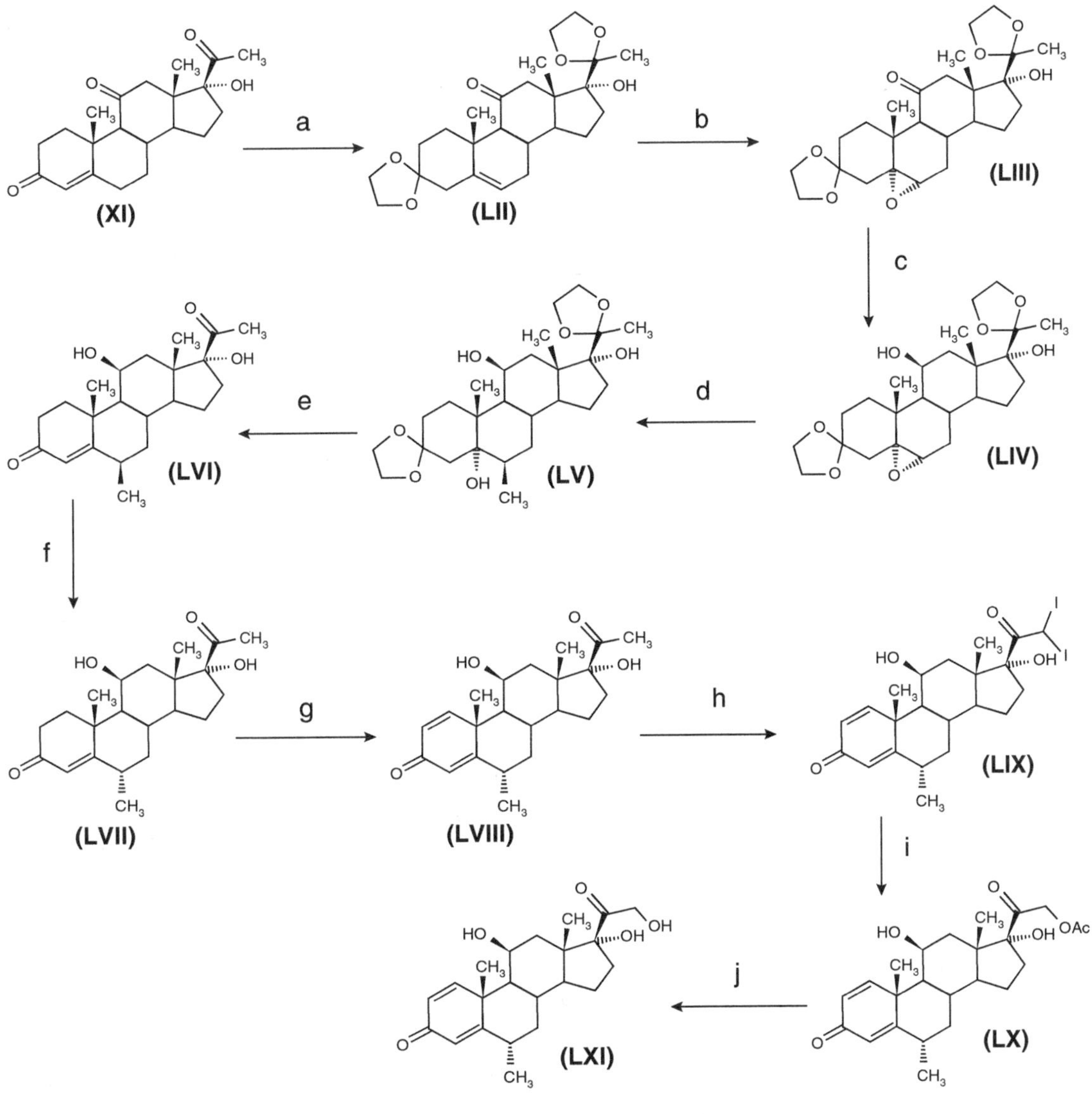

Methylprednisolone

Fig. 9 Methylprednisolone synthesis. (**a**) Methylethyldioxolane, ethylene glycol, *p*-TsOH. (**b**) Phthalic anhydride, H_2O_2, $CHCl_3$. (**c**) $NaBH_4$, THF, H_2O. (**d**) CH_3MgBr, THF, CH_2Cl_2. (**e**) H_2SO_4, MeOH. (**f**) NaOH, 60 °C. (**g**) *Corynebacterium simplex* (ATCC 6946). (**h**) I_2, $CaCl_2$, CaO, MeOH. (**i**) AcOH, Et_3N, acetone. (**j**) K_2CO_3, MeOH

positions, so we could easily achieve a selective protection. The protection of 3-keto group with ethylene ketal group implies also the movement of the double bond into positions 5,6. The synthesis will take fully advantage of this fact. The protected product obtained **(LII)** is ready for 11-keto reduction as well as for methyl group introduction. The double bond is epoxidized giving mainly the alpha-oriented epoxide **(LIII)**. Being both reactive ketones protected it is time to end the inversion of the 11-hydroxy group. The ketone is reduced with $NaBH_4$ **(LIV)**. The next step consists of the opening of the epoxide with methyl magnesium chloride. The opening is also regioespecific leading to a 6β-methyl derivative **(LV)** which after hydrolysis of ketal groups and spontaneously elimination of 5-hydroxy leads to an unsaturated ketone **(LVI)**. The 6β-methyl has the wrong configuration but this is easily corrected as the thermodynamic isomer is the 6α one. Treatment of the 6β-methyl with sodium hydroxyde yields the 6α-methyl one **(LVII)** with a diasteromeric excess or around 94%. This is the moment to once again request biotechnology to provide some help. **(LVII)** is dehydrogenated with *Corynebacterium simplex* yielding the desired double unsaturated ketone **(LVIII)** in very good yields. Ringold-Stork protocol leads to methylprednisolone acetate **(LX)** which is then deacetylated with potassium carbonate into final Methylprednisolone **(LXI)**.

This synthesis is an elegant combination of chemistry and biotechnology.

3 Conclusions and Future Outlook

Sapogenins, as readily available natural products, were called to have a preponderant task in industrial synthesis of steroids. This task was only achieved due to a remarkable combination of chemistry and biotechnology. Chemistry assumed the main tasks, but biotechnology was always ready to solve the most complex problems.

The above-depicted syntheses have been warhorses for many pharmaceutical companies in the steroid field along the second part of the twentieth century and the beginning of the twenty-first century. Nowadays, the spreading of fermentation technologies on phytosterols is moving the syntheses into new ones starting from Androstendione and 9α-hydroxy androstendione. In these new syntheses, once again, we will see a good combination of chemistry and biotechnology to reach the desired goals.

References

1. Hench PS, Kendall EC, Slocumb CH, Polley HF (1949) The effect of a hormone of the adrenal cortex (17-hydroxy-11-dehydrocorticosterone: compound E) and of pituitary adrenocorticotropic hormone on rheumatoid arthritis. Ann Rheum Dis 8:97–104
2. Marker RE, Rohrmann E (1940) Sterols. LXXXVIII. Pregnanediols from sarsasapogenin. J Am Chem Soc 62:518–520
3. Marker RE, Tsukamoto T, Turner DL (1940) Sterols. C. Diosgenin. J Am Chem Soc 62:2525–2532
4. American Chemical Society, Sociedad Química de Mexico (1999) https://www.acs.org/content/dam/acsorg/education/whatischemistry/landmarks/progesteronesynthesis/marker-degradation-creation-of-the-mexican-steroid-industry-by-russell-marker--commemorative-booklet.pdf

5. Oppenauer RV (1937) Eine methode der dehydrierung von sekundaren alkoholen zu ketonen. I, zur herstellung von sterinketonen und sexualhormonen. Recl Trav Chim Pays-Bas 56:137–144
6. Peterson DH, Murray HC (1952) Microbiological oxygenation of steroids at carbon-11. J Am Chem Soc 74:1871–1872
7. Peterson DH, Murray HC (1952) Oxygenation of steroids by *Mucorales* fungi. US Patent US2602769
8. Peterson DH, Murray HC, Eppstein SH, Reineke LM, Weintraub A, Meister PD, Leigh HM (1952) Microbiological transformation of steroids. I. Introduction of oxygen at carbon-11 of progesterone. J Am Chem Soc 75:5933–5936
9. Stork G, Ringold HJ, Sondheimer F, Rosenkranz G (1959) Process for the production of 20-Keto-21-Iodo steroids. US Patent US2874154
10. Joly R, Jolly J (1961) Procédé d'acétoxylation de stéroïdes en 21 par l'intermédiaire de dérivés diiodés. French patent FR1243528
11. Burn D, Kirk DN, Petrow V (1960) A new reagent for the preparation of $\Delta^{1,4}$- and $\Delta^{1,4,6}$-steroidal ketones. Proc Chem Soc 14
12. Meystre C, Frey H, Voser W, Wettstein A (1956) Gewinnung von 1;4-Bisdehydro-3-oxo-steroiden. Helv Chim Acta 34:734–742
13. Nobile A (1958) Process for production of dienes by *Corynebacteria*. US Patent US2837464
14. Herzog HL, Payne CC, Hughes MT, Gentles MJ, Hershberg EB, Nobile A, Charney W, Federbush C, Sutter D, Perlman PL (1962) Microbiological transformations of steroids. X. 1-Dehydro analogs of cortical steroids. Tetrahedron 18:581–589
15. Oliveto EP, Rausser R, Weber L, Shapiro E, Gould D, Hershberg EB (1956) 11-Oxygenated steroids. XVI. The preparation of Hydrocortisone from cortisone acetate. J Am Chem Soc 78:1736–1738
16. Slates HL, Wendler NL (1959) Reaction of diazomethane with Δ^{16}-21-keto steroids. J Am Chem Soc 81:5472–5475
17. Gardner JN, Carlon FI, Gnoj O (1968) A one-step procedure for the preparation of tertiary α-ketols. J Org Chem 33:3294–3297
18. Gardner JN (1967) Method for introducing an oxygen function into a steroid molecule containing a keto group. US Patent US3356696
19. Suvorov NN, Sokolova LV, Makarov NV (1960) Reaction of methyl magnesium iodide with ketoxides of steroids. Izvestiya Academii Nauk SSRR, seriya khimicheskaya 2257–2258
20. Sciaky R (1961) Synthesis of 16β-methyl-17α-acetoxyprogesterone. Gaz Ch Ital 91: 562–570
21. Amiard G, Torelli V, Cerede J (1963) Process of preparation of β-Methasone. US Patent US3104246
22. Fu X, Tann CH, Thiruvengadam TK (2001) Process improvements in the synthesis of corticosteroid 9,11β-epoxides. Org Process Res Dev 5:376–382
23. Hanc O, Capek A, Kakac B (1961) Microbiological transformations of steroids. XV. Transformation of steroid S (Reichstein) by *Absidia orchidis* 310. Folia Microbiol 6:392–397
24. Hanc O, Capek A, Tadra M (1962) Microbiological transformation of steroids. XVI. Preparation of cortisol and cortisone from 17alpha-hydroxycortexone with the use of microbial hydroxylation with the aid of *Absidia orchidis* 310. Cesk Farm 11:181–185
25. Wix G, Weisz E, Albrecht K, Varga M, Kassazan B, Doczi B (1966) Verfahren zur Herstellung von Hydrocortison. Austrian Patent AT 248630
26. Breuker E, Hamann KH, Singer S Wagner B (1968) Process for the manufacture of Δ^4-Pregnene-11β and 11α,17α, 21-triol-3,20-Diones. British Patent GB 1100505
27. Bowers A, Ringold HJ (1958) Synthesis of 6a-Methyl-21-desoxycortisone. A new route to 6a-Methylcortisone. J Am Chem Soc 80:3091–3093
28. (1962) Cyclopentanophenanthrene derivatives. British Patent GB 897732

Chapter 3

Genome Sequencing of Steroid-Producing Bacteria with Illumina Technology

Victoria Y. Shtratnikova, Mikhail I. Schelkunov, and Marina V. Donova

Abstract

Illumina technology is widely used for bacterial whole-genome sequencing due to its simplicity, cheapness, reliability, and abundant software for manipulation with raw data. Illumina technology belongs to a second generation of whole genome sequencing that yields great amount of short reads for genome regions. Genomic DNA is fragmented to short pieces. DNA fragments are amplified for signal increasing, and are read using sequencing-by-synthesis. Millions of short reads up to 100–300 bp in length are assembled in continuous sequences. Mate-pair technology allows resolving a long repeat.

Here, we describe the principles of standard and mate-pair library preparation from DNA samples, library quality control, sequencing with MiSeq instrument and following data bioinformatics treatment. Software for genome assembly and completion are listed that assemble, map, annotate, visualize, edit and allow doing other manipulations with genomic sequences. The whole genomes sequencing of the steroid-producing Actinobacteria using these protocols is exemplified.

Key words Next-generation sequencing, Steroid, Bacteria, Genome sequencing, Mate pair preparation, Genome completion

1 Introduction

Whole genome reading of bacterial genomes with Next-Generation Sequencing (NGS) technologies is a very widely used method now, becoming cheaper and cheaper (it is a joke that in short time bacteria will pay us to sequence their genomes, https://thescience-web.wordpress.com/2015/03/31/bacteria-will-pay-you-to-sequence-them-by-2016-analysis-reveals). Approximately 8,000 complete genomes and 30,000 draft assemblies (*see* **Note 1**) had been submitted in the bacterial genomic data base by January, 2016 (http://www.ncbi.nlm.nih.gov/genomes/MICROBES/microbial_taxtree.html; https://gold.jgi.doe.gov/). Usually, bacteria have genomes of some millions bp, 40–70% GC, and contain copies of genes (or other repeats) up to several thousands bp in length.

José-Luis Barredo, Ignacio Herráiz (eds.), *Microbial Steroids: Methods and Protocols*, Methods in Molecular Biology vol. 1645, DOI 10.1007/978-1-4939-7183-1_3, © Springer Science+Business Media LLC 2017

Nowadays, Illumina technology is considered the best for bacterial genome assembly yielding relatively long reads (*see* **Note 2**), deep coverage (*see* **Note 3**), and minimal quantity of errors [1]. The optimum instrument is MiSeq (Illumina, USA)—a fully automated station that includes all the necessary modules in a desktop device (*see* **Note 4**). During sequencing with Illumina technology [2], a genomic DNA is chopped up into random fragments (a set of such fragments with special adapters on the ends is named a "library"), each several hundreds bp long, which will then be read. Short reads are assembled in genomic sequences.

Mate-pair preparation is applied to genomic DNA of high quantity and quality and allows getting best result with minimal cost. In a mate-pair protocol (*see* Fig. 1), genome DNA is cut into long fragments up to 10,000 bp, these fragments are further circularized on biotin tags, the circles are digested to the fragments of 200–600 bp, the biotin-tagged fragments are selected on streptavidin beads, then the library is prepared as a usual short library and is sequenced.

We assume that the reader is familiar with standard molecular biology operations as PCR, qPCR, gel electrophoresis, etc. and we do not give the detailed protocol of these operations. Examples of whole genomes of steroid producing strains sequenced with Illumina technology and with mate-pair preparation are *Mycobacterium neoaurum* VKM Ac-1815D [3], *Mycobacterium* sp. VKM Ac-1817D [4], and *Nocardioides simplex* VKM Ac-2033D [5].

2 Materials

2.1 Genomic Library Preparation (See Notes 5 and 6)

2.1.1 Major Reagent Kits

1. Nextera mate-pair sample preparation kit (Illumina, USA) for mate pair preparation.
2. TruSeq DNA Sample preparation (Illumina, USA) for mate pair and standard library preparation (*see* **Note 7**).

2.1.2 Other Reagents

We recommend using:

1. Qiagen Min Elute Kit (Qiagen, Netherlands).
2. AMPure XP beads (Beckman Coulter, USA) (*see* **Note 8**).
3. Dynabeads streptavidin beads M-280 (Invitrogen, USA) (only Mate-pair preparation) (*see* **Note 8**).
4. Agilent High Sensitivity DNA Kit.
5. Qubit dsDNA HS Assay Kit.
6. Covaris glass tubes, e.g., microTUBE AFA Fiber Crimp-Cap 6 × 16 mm.
7. Reagents for real-time PCR.
8. Gel electrophoresis reagents.

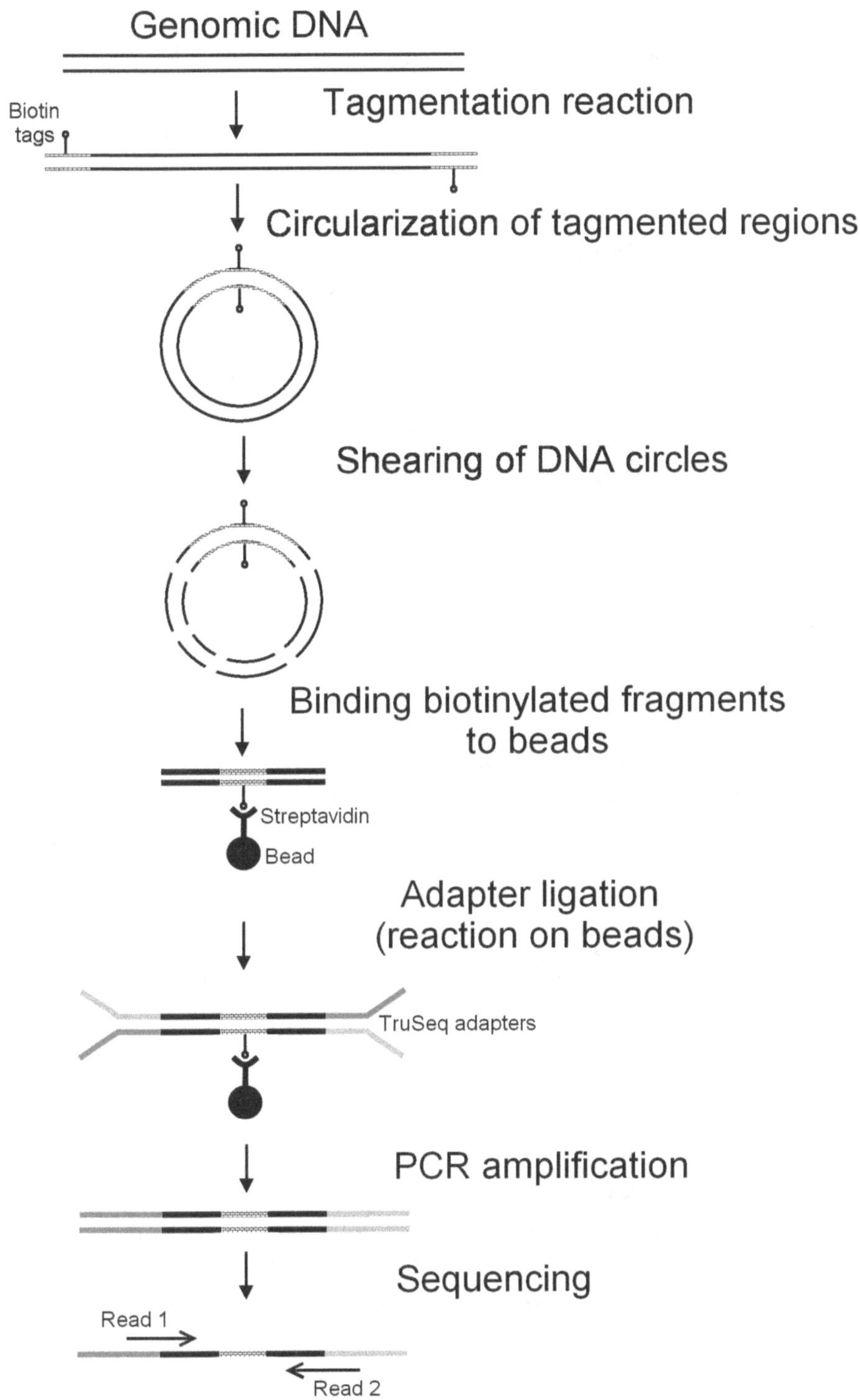

Fig. 1 Mate-pair principle. Some steps are omitted. Details in text

2.1.3 Special Equipment

Several steps of library preparation require the special equipment and corresponding reagents that make the protocol faster and more reliable.

1. Bioanalyzer 2100 (Agilent, USA) (*see* **Note 9**).
2. Qubit fluorometer (Invitrogen, USA) (*see* **Note 10**).
3. Covaris S2 or S220 (Covaris, USA) focused ultrasonicator (*see* **Note 11**).
4. Real-time PCR instrument.

2.1.4 Standard Laboratory Equipment (See Note 12)

1. Thermal cyclers.
2. Heat blocks (20–70°).
3. Centrifuges for 1.7 and 0.5 mL tubes, 14,000 × *g*.
4. Magnetic racks.
5. Gel electrophoresis equipment.
6. Equipment for viewing gel results in UV and cutting DNA bands from gel.
7. Pipettes, tubes, racks, etc.

2.2 Sequencing Equipment and Materials

1. MiSeq (Illumina, USA). A detailed guide to working with this instrument and maintenance procedures can be found at https://support.illumina.com/downloads/miseq_system_user_guide_15027617.html
2. MiSeq Reagent Kit.

2.3 Genome Assembly

2.3.1 Equipment

1. Computer. Requirements on computer resources depend on a tool that performs the assembly and processing. We recommend up to 40 Gb of RAM, a data storage hard drive at least 1 Tb, and the higher number of CPU threads that are possible.
2. PCR instrument.
3. Gel electrophoresis equipment.

2.3.2 Reagents

1. PCR reagents.
2. Gel electrophoresis reagents.
3. Qiagen Min Elute Kit (Qiagen, Netherlands).

3 Methods

3.1 Genomic Library Preparation: Nextera Mate Pair Sample Preparation Procedure

We describe the general procedure of library preparation using Nextera Mate Pair preparation (Illumina, http://www.illumina.com/products/nextera-mate-pair-sample-prep-kit.html), not any specific protocol because the technologies and reagents for Illumina library preparations are constantly optimizing and improving about every half a year. Recommended kits include guides taking into account features of used enzymes and buffers, see manuals for precise amounts of reagents and other technology details.

1. First rule of genomic sequencing: trash in–trash out. Pay careful attention to the high quality and sufficient quantity of the input DNA (*see* **Note 13**).
2. Fragment the genomic DNA sample and tag it with a biotinylated mate pair junction adapter, by incubation with tagmentation enzyme from Nextera mate-pair sample preparation kit at 55 °C for 30 min (*see* **Note 14**).
3. Fill the short single-stranded sequence gaps in the tagmented DNA by incubation at 20 °C for 30 min with the special polymerase from Nextera mate-pair sample preparation kit.
4. Remove the smaller DNA fragments (<1500 bp) by further purification with AMPure beads.
5. Size-select fragments in agarose gel (*see* **Note 15**).
6. Cut as many DNA bands of various lengths as possible, use two bands for further treatment (e.g., 4 kb and 10–11 kb), and leave the rest frozen in gel as a backup.
7. Elute DNA bands from the gel with Qiagen Min Elute Kit.
8. Circularize the size-selected fragments by blunt-ended intramolecular ligation with circularization ligase from Nextera mate-pair sample preparation kit at 30 °C overnight (12–16 h).
9. Remove any linear molecule still remaining in the circularization reaction by DNA exonuclease treatment from Nextera mate-pair sample preparation kit at 37 °C for 30 min.
10. Inactivate the exonuclease at 70 °C for 30 min.
11. Shear the large circularized DNA, e.g., by ultrasonication with Covaris (*see* **Note 11**) to smaller-sized fragments (approximately 300–1000 bp) (*see* instrument manuals for exact conditions).
12. Purify sheared DNA fragments that contain the biotinylated junction adapter (mate pair fragments) on streptavidin magnetic beads, remove non-biotinylated molecules through a series of washes (*see* **Note 16**).
13. Blunt overhangs resulting from the DNA shearing step with end repair enzyme mix from TruSeq DNA Sample preparation kit at 30 °C for 30 min (*see* **Note 17**).
14. Add single "A" nucleotide to the 3′ ends of the blunt fragments with A-tailing mix from TruSeq DNA Sample preparation kit at 37 °C for 30 min (*see* **Note 18**).
15. Ligate indexing adapters (e.g., TruSeq adapters) to the ends of the DNA fragments with ligase from TruSeq DNA Sample preparation kit at 30 °C for 10 min (*see* **Note 19**).

16. Amplify PCR fragments that have DNA adapters on both ends with amplification reagents from TruSeq DNA Sample preparation kit. Reaction conditions: 98 °C, 30″; 15 cycles of PCR: 98 °C, 10″; 60 °C, 30″; 72 °C, 30″; 72 °C, 5′; hold at 4 °C.
17. Clean up the PCR reaction and remove the smallest fragments (<300 bp) from the final library with AMPure XP bead (*see* **Note 20**).
18. Run library quality control (*see* **Note 21**). Concentration of high-quality libraries should be more than 2 nM.

3.2 Sequencing with MiSeq

This protocol is done with MiSeq (*see* **Note 22**) using sequencing-by-synthesis.

1. Combine the samples' libraries (where there are many samples) into one 1.5 mL tube.
2. Dilute multiplexed library to an effective concentration of 8 pM.
3. Load multiplexed and diluted library to the reagent cartridge of MiSeq Reagent Kit.
4. Load the set of all reagents of MiSeq Reagent Kit into the instrument in one box.
5. Fill a sample sheet, a csv-file that contains information required to set up and analyze the sequencing run, including the list of samples and their index (barcode) sequences.
6. At a first stage named cluster generation, run bridge-amplification to generate a cluster of hundreds of copies of the fragment. This is performed by the MiSeq (*see* **Note 23**).
7. Following cluster generation, sequence the attached to flow cell and amplified DNA libraries using Sanger principle. This is performed by the MiSeq (*see* Fig. 3 and **Note 24**).

3.3 Analysis of Raw Data

1. Perform image analysis. This is performed by the MiSeq on-the-fly during a sequencing run by RTA (Real Time Analysis) software. This tool accomplishes a base calling, determining which nucleotides correspond to each cluster in the images (*see* **Note 25**).
2. Run bcl conversion to transform base calling data, stored in the format *.bcl, to FASTQ files (*see* **Note 26**). This process can be performed either locally by Bcl2fastq software or on an Illumina's server BaseSpace. Make demultiplexing during bcl conversion, i.e., if a run was performed for multiple samples, determine to which of the samples each read belongs, using the input sample sheet with barcode sequences.
3. Perform additional quality control to detect a drop in nucleotide quality that often occurs at the 3′ end of reads (*see* **Note 27**) and

fragments of sequencing adapters in reads (*see* **Note 28**). A common tool to perform quality control of reads is FastQC (http://www.bioinformatics.bbsrc.ac.uk/projects/fastqc/).

4. Perform trimming, removing regions with poor quality and adapter fragments, e.g., by Trimmomatic (http://www.usadel-lab.org/cms/index.php?page=trimmomatic) [6].
5. In case of a mate-pair library, remove contamination by paired end reads with short insert sizes, with, e.g., NextClip (https://github.com/richardmleggett/nextclip/) [7].

3.4 Genome Assembly

3.4.1 Draft Assembly

1. Perform the draft genome assembly to contigs (*see* **Note 29**). We recommend using next assemblers: Spades [8], Velvet [9], CLC (www.clcbio.com) [10]. The output of the software is a FASTA file (*see* **Note 30**). Measure the quality of assembly (*see* **Note 31**).
2. Find misassemblies (*see* **Note 32**) with REAPR [11] that breaks contigs in improper contig joints, producing shorter but more precise contigs.

3.4.2 Complete Assembly

Draft contigs are long enough to contain all or almost all of genes. To assemble a genome completely in a contiguous circular sequence (*see* **Note 1**), some additional steps are needed.

1. Perform a manual analysis of read pairs localized at ends of contigs. If one read in a pair maps to the end of one contig and the other read maps to the end of another contig, there is a reason to suppose that these contigs may be close to each other in a genome. Propose order of contigs by analyzing numbers of read pairs that connect various contigs. Read mapping and visualization can be performed, e.g., by a combination of Bowtie2 [12] and Tablet [13] (freeware tools) or by CLC tools (shareware tool).
2. Use a computational approach to join contigs by using "scaffolders" and "gap-closers" that, respectively, propose contig junctions and try to guess what sequence lies in a gap between two adjacent contigs. Use, e.g., SSPACE [14] and GAPFILLER [15].
3. Compare assemblies made by different software (*see* **Note 33**). Manual proof, viewing, and alignment can be done, e.g., with DNAStar (http://www.dnastar.com/), BioEdit (http://www.mbio.ncsu.edu/bioedit/bioedit.html) [16], CLC.
4. Use a reference genome (a close relative's known complete genome) for contigs' or scaffolds' mutual arrangement (*see* **Note 34**).
5. Verify by PCR the proposed order of contigs.
   (a) Pick up primers for unique regions up- and downstream of ambiguous sequence or breaking point.
   (b) Amplify sequences from each pair of primers.

(c) If PCR amplification is successful and the length of the resulting band is approximately equal to the length of the ambiguous sequence plus lengths of regions up to primers, then the join of contigs is uniquely determined.

6. Use Sanger sequencing if an unknown sequence is localized between contigs, if several regions that differ too little are present in the genome (e.g., transposons), in case of many tandem repeats, or in case of homopolymer regions unresolved by assemblers.
   (a) Pick up primers to unique regions up- and downstream of the unknown sequence.
   (b) Amplify the unknown sequence.
   (c) Run an agarose gel electrophoresis with PCR products.
   (d) Extract the desired fragments from gel with Qiagen MinElute Kit.
   (e) Run Sanger sequencing of these fragments (e.g., in commercial companies providing this service).
   (f) If the length of the problem regions is more than the length of a standard Sanger read (some hundreds bp), then several sequencings are required.
      - Pick up primers to internal parts of the investigated region known after the first Sanger sequencing.
      - Repeat **steps 6b–6e** until the entire sequence becomes known.
7. Merge sequences of contigs and Sanger reads into a whole genome sequence manually in editing tools mentioned above. A bacterial chromosome is usually circular, so the ends of your genome contig should be the same. Cut off the repeated end region.
8. Verify the genome that is built and circularized by mapping of the reads to the genome (*see* **Note 35**) with, e.g., Bowtie2 or one of CLC tools. Determine discrepancies (small deletions, insertions, and single nucleotide mistakes) between the mapped reads and the assembled genomes by a CLC tool or by GATK [17] or Samtools [18], if you mapped reads by Bowtie2. Fix the genome sequence according to a list of the discrepancies in positions, which may contain assembly mistakes. Repeat the cycles of mapping and error correction until no discrepancies and replacements are offered by the software.

3.5 Annotation

Annotate assembled genome, whether in contigs or complete (*see* **Note 36**). The annotation includes steps that are organized into pipelines: (1) prediction of open reading frames (ORFs), (2) analysis of various parameters to distinguish random ORFs from those belonging to certain genes, (3) prediction of transfer RNA-coding genes and ribosomal RNA-coding genes, (4) comparison of the predicted protein-coding genes with the genes of related species to propose those genes' functions, etc. Use RAST [19], available as an online service, and PGAP (http://www.ncbi.nlm.nih.gov/genome/annotation_prok/), which is integrated in NCBI GenBank.

4 Notes

1. The main advantage of obtaining only draft genomes is the possibility of complete automation of all processes, starting from DNA extraction. Library preparation, sequencing, and further assembly can be done by a robotic station and sequencing equipment. Information from draft assemblies helps to carry out wide metagenomics investigations, to sequence many genomes simultaneously, to compare certain genes and the numbers of their copies, phylogenetic investigations, etc. But data from draft assemblies are imperfect. Comparing clusters of genes and whole genomes is problematic, many possible deletions, insertions, and translocations remain unknown. Full promoter sequences, number and quality of transcription factor binding sites are not always known. As a practical issue you cannot assign a certain name of locus-tag for a gene because every new assembly assigns its own locus-tag for every gene. So if you want to deeply investigate a genome, a finalized assembly and a complete and annotated genome are required.
2. One characteristic of a sequencing instrument is read length, the length of one output sequence. The read length of Illumina technology is short (50–300 bp) as compared to the length of bacteria genomes. Raw data require further assembly in extended sequences, called contigs. One-end read up to 50 bp in length will give you contigs which are about one gene long (several thousands of bp). Usually, DNA fragments loaded in a sequencer are longer than the read length. If technologies used allow pair-end run then the assembly software gets information about the distance between reads made at two ends of a fragment. This information increases the capacities of assemblers. Pair-end reads up to 50 bp in length will yield contigs up to tens of thousands bp in length. Longer pair-end reads up to 100–300 bp in length will yield contigs up to hundreds of thousands of bp in length. Illumina technology does not allow reading fragments longer than 600–800 bp. But bacteria genomes can contain longer repeats. The mate-pair protocol yields information about distances between the pair-end reads up to several tens of thousands bp. This information allows resolving most of the long repeats and, if one is lucky enough,

to assembly a circularized chromosome of the bacterial genome in one contig.

3. For a genome assembly de novo a 30-fold sequencing depth is required [20]. All modern commercial NGS technologies enable such sequencing depth when bacteria are concerned and this factor is important only when lots of samples are sequenced.
4. The following characteristics should be taken into account when choosing a sequencing method: (1) Long read length and pair-end capabilities. (2) Cost of reagents and equipment. (3) Quantity of output data and sequence depth. The optimum tool meeting these requirements is MiSeq (Illumina, USA). By the January, 2016 MiSeq has the capacity to generate up to 15 Gb of high-quality data from a 250–300 bp run in 65 h; the cost of 1 mln bp sequencing is approximately 0.14$, and MiSeq is capable of sequencing many samples (multiplexing).
5. As a library preparation requires not only reagent kits but special equipment, it should be carried out in specialized centers with all the necessary conditions available.
6. Minimize freeze-thaw cycles of reagents. Divide reagents into aliquots after initial thaw.
7. You may use DNA sample preparation kit (New England Biolabs, USA), NEXTflex DNA Sequencing Kit (Bioo scientific, USA), or other possible kits compatible with Illumina technology, instead.
8. Beads should be kept at above 0 °C! Do not freeze beads!
9. Bioanalyzer 2100 (Agilent, USA) used to define quality, amount and size range of a DNA and of final libraries in nanomolar concentrations and in microliter quantities of a sample.
10. Qubit fluorometer (Invitrogen, USA) used to define amount of DNA in microliter quantities of a sample.
11. Covaris S2 or S220 (Covaris, USA) focused ultrasonicator used to randomly shear a DNA with ultrasound. By the date Covaris ultrasonication is the simplest and fastest technique for such shearing.
12. Avoid PCR product contamination! PCR products can contaminate reagents, instrumentation, and samples, causing inaccurate and unreliable results. The laboratory space and full sets of equipment and supplies to pre-PCR and post-PCR processes should be physically separated. A double set of standard laboratory equipment is therefore essential.
13. The gel-plus version of the mate pair protocol requires 4 μg of DNA for size-selection of fragments (before circularization) with different lengths. The gel-free protocol requires 1 μg of

DNA. But it will yield a fragment distribution ranging from 2 to 15 kb, with a median fragment size between 2.5 and 4 kb. Such ranging complicates the assembler's work and reduces the assembly quality. We recommend gel size-selection as an obligatory step. Degraded genomic DNA will yield fragments with lengths below the desired size range. To assess the quality of the starting material, run it in low-percentage agarose gel. High-quality DNA should run in gel as a band greater than 50 kb and with minimal smearing. When handling DNA samples with fragment sizes above 2 kb, do not vortex samples, as this may further shear the DNA.

14. In later steps the biotin adapter serves to purify mate pair fragments. During post sequencing analysis, the sequence of this junction adapter is used to identify the location of the joined ends of the larger fragments.
15. Size-selection of fragments in agarose gel allows generating libraries with tighter distributions of fragment sizes.
16. A protocol of traditional library preparation for draft assembly begins with random DNA fragmentation into fragments up to 300–600 bp in length by physical means (e.g., Covaris ultrasonication). Further steps are the same for enzymatic treatment. The mate pair protocol uses streptavidin magnetic beads binding the fragments.
17. The 3′ to 5′ exonuclease activity of the end repair enzyme mix removes the 3′ overhangs and the polymerase activity fills in the 5′ overhangs.
18. In the A-tailing step, a single A prevents them from ligating to one another during the adapter ligation reaction; a corresponding single “T” nucleotide on the 3′ end of the adapter provides a complementary overhang for ligating the adapter to the fragment.
19. The adapters include primer sequences for PCR amplification and subsequent hybridization onto a flow cell, as well as the index sequence (barcode) for the multiplexing of samples. Samples are demultiplexed after the run by their own index sequence.
20. Libraries can be stored at −20 °C for years but strict quality control is essential before sequencing.
21. In order to achieve the highest quality of data on Illumina sequencing platforms, it is important to create optimum cluster densities across the flow cell. Libraries require thorough quality and quantity control. Three methods are recommended: (1) Approximate estimation of the amount of DNA with Qubit fluorometer, in microliter quantities. (2) Evaluation of quantity and quality of libraries with Bioanalyzer 2100

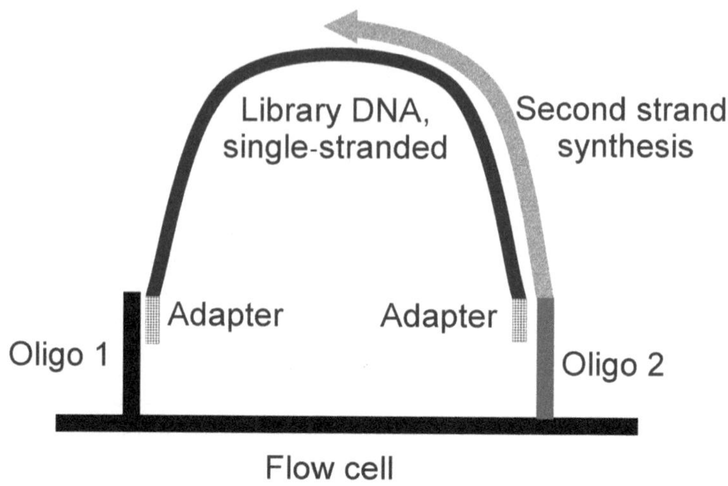

Fig. 2 Bridge amplification. Details in text

(High Sensitivity DNA LabChip). Bands of libraries should be run in a tight range of fragment sizes (e.g., 500 bp) without smearing and without adapter dimers band at 150 nt. Adapter dimers form when sequencing adapters mistakenly join each other without an intervening DNA fragment. 10% of adapters in the total amount of library are permitted; more adapters contaminate a sequence and reduce the useful output of a run. (3) The amount of DNA is not an indicator of effective concentration of a library. For example, fragments without adapters can be present in the final solution. The effective quantity of a library should be evaluated by quantitative PCR with libraries of known effective concentration used as a standard.

22. You can use another Illumina instrument for sequence: HiSeq, NextSeq, etc. They have their advantages, e.g., HiSeq have higher capability of sequencing. But MiSeq provides longer reads (up to 300 bp) than HiSeq (up to 100 bp) because using LEDs instead of lasers that damage clusters.
23. At the bridge amplification (*see* Fig. 2) two regions of adapters are bound to complementary regions of flow cell surface oligos; a bridge is formed. The second strand is extended from a flow cell oligo as a primer. Then the original strand that is not attached to the flow cell is removed. The amplification cycle is repeated with the synthesized strand 40 times to generate a cluster of hundreds of copies of the fragment. Amplification of fragments is essential for signal intensification.
24. At the sequencing-by-synthesis (*see* Fig. 3) dideoxynucleotides with terminated tags are bound to complementary bases of DNA fragments. Each of the four nucleotides has its own fluorescent label. Fluorescent clusters on the surface of flow cell

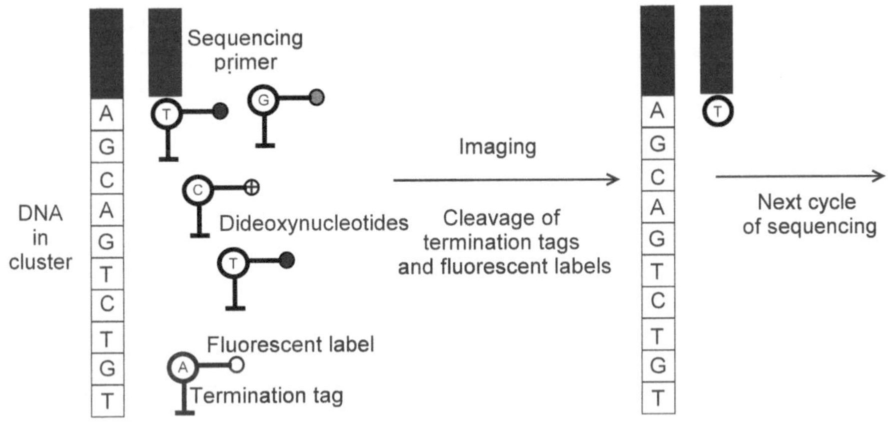

Fig. 3 Sequencing by synthesis. Details in text

are imaged using LED and filter combinations specific to each of the four labels. A flow cell contains several dozens of tiles, units of imaging. After all tiles are imaged, fluorescent labels and termination tags are removed. A new portion of labeled dideoxynucleotides is loaded and the process is repeated.

25. The results of base calling and supporting files usually require a lot of space; each MiSeq run usually produces about 50 Gb of data, so use an external storage device.

26. FASTQ is the most widely used format for storing sequencing reads. Each read in a FASTQ file is represented by the following four lines: (1) The first line is the title of a read. It must start with "@" character. (2) The second line is the nucleotide sequence of the read, i.e., a line of characters A, T, G, C and, in special cases, an unidentified nucleotide N. (3) The third line contains just one character "+" and has no special meaning. (4) The fourth line encodes the quality values of each nucleotide. Each character in the fourth string denotes a quality of the respective nucleotide in the second string. These characters can be English letters, numbers, or other symbols of the ASCII code. They designate the probability that the respective nucleotide was determined incorrectly. For details of Phred encoding system see at https://en.wikipedia.org/wiki/FASTQ_format.

27. A main reason for a drop in nucleotide quality at the 3′ end of reads is gradual accumulation of mistakes in nucleotide incorporation during sequencing-by-synthesis.

28. Sequencing adapters may be present in reads in case of adapter dimers. Also, if DNA of a sample is cut into fragments that are too small, the sequenced DNA fragment can be shorter than the read's length, in this case part of an adapter is found at the 3′ end of a read.

29. Draft assembly is a process in which special software analyzes the reads and tries to reconstruct the genome as completely as possible. Ideally, the result of this process should be a complete genome, but usually the tool is unable to perform such a complete reconstruction and assembles the genome partially. The contiguous reconstructed parts of the genome are called "contigs." Sometimes, it is also possible to reconstruct the order in which some of the contigs are oriented on the chromosome. Such ordered contigs are called "scaffolds." Assemblers perform genome reconstruction by finding overlaps between reads from close regions in a genome, usually using structures called "De Bruijn graphs," see details in [8]. We recommend using next assemblers: (1) Spades [8] usually produces contigs that are longer than those of other programs and contain less assembly mistakes. (2) Velvet [9] had been the most popular tool before the appearance of Spades. (3) Tools by CLC (www.clcbio.com), such as CLC Genomics Workbench, unlike other tools described here, are not freeware, but have a graphical interface that is suitable for non-bioinformaticians. Also, tools by CLC are among the fastest assemblers and require relatively little RAM, although the quality of the resulting assembly is somewhat inferior to Spades [10].
30. FASTA is the text-based format that consists of a description line starting with ">" character followed by the text identifier of a sequence, and the actual sequence in standard one-letter code. Every ">" at the start of a line indicates the start of another sequence.
31. There are several measures of the quality of assembly. One of the most popular is N50. N50 characterizes contig lengths. It is defined as a length of a contig such that all contigs of that length or longer constitute half of the total length of all contigs. As an example, take a genome of 4.5 mbp. Usually, the sum of lengths of all produced contigs is close to the genome size. Say, the assembly is fragmented in contigs, the largest of which is 1300 kbp, the second largest is 700 kbp, the third is 400 kbp, and all others are smaller. Since (400 + 700 + 1300) is 2400 kbp, which is larger than a half of 4.5 mbp, we say that N50 is 400 kbp. So, N50 is a measure of how long the contigs in a produced assembly are. A similar measure is NG50, which compares the contig lengths not with the total length of the assembly, but with the length of a genome. It is hard to calculate NG50, since no one usually knows the exact genome size before assembling it. Anyway, NG50 is almost always identical to N50, since the total length of all contigs is typically close to the genome size, i.e., there are very little genome fragments that are not represented in some of the produced contigs. A

similar measure is NA50, which is calculated by breaking contigs in places in which a scientist supposes that the software that performed the assembly could produce assembly mistakes. Since the resulting contigs are shorter than the initial ones, NA50 is less than N50. A common tool to calculate N50 and other measures of contig statistics is QUAST [21]. To know how good the genome is assembled in biological terms use BUSCO [22] that contains a set of highly conserved genes that are expected to be present in almost all bacteria.

32. Tools that perform genome assembly sometimes mistakenly connect nucleotide sequences that are situated in regions of the genome which are far apart. Such cases are called misassemblies A usual reason for misassemblies is long repeats that represent, for example, mobile elements inserted in bacterial genomes, as it is hard for genome assembly algorithms that operate short Illumina reads to deal with them. The main advantage of mate-pair reads over paired-end reads with short insert sizes is their ability to deal with long repeats. REAPR maps reads to contigs, i.e., determines places in contigs that these reads correspond to, and searches for places with sharp coverage drops and large numbers of improperly paired reads. If such doubtful places are found, REAPR, suspecting improper contig joints, breaks contigs in them, producing shorter but more precise contigs.
33. Comparing of results of different assembling software is a useful method to produce more accurately genome sequence due to different algorithms of assembly.
34. Using of reference genome is not an absolutely reliable method as genomes can have their own order of genes.
35. Only mapping of mate pair reads can identify single nucleotide differences when many copies of a long gene are localized in distant parts of the genome (if the length of the gene is more than the fragment size of the initial library). Thereby, mate pair reads are necessary for correct genome completion. Long Sanger reads or other reads longer than thousand bp can also be used.
36. The genome annotation consists of structural annotation, when the coordinates of genes are determined, and functional annotation, when the functions of these genes are predicted. There is no ideal annotation software by the date. There can be mistakes in gene border definition, gene's start codons may be determined improperly, functions and names may be assigned to genes incorrectly, and there is no universal nomenclature of gene and protein names. Thereby, manual verification of genes which are most important for your investigation is required.

Acknowledgment

This work was supported by Russian Science Foundation grant (14-24-00169).

References

1. Xuan J, Yu Y, Qing T et al (2013) Next-generation sequencing in the clinic: promises and challenges. Cancer Lett 340:284–295. doi:10.1016/j.canlet.2012.11.025
2. Bentley DR, Balasubramanian S, Swerdlow HP et al (2008) Accurate whole human genome sequencing using reversible terminator chemistry. Nature 456:53–59. doi:10.1038/nature07517
3. Shtratnikova VY, Bragin EY, Dovbnya DV et al (2014) Complete genome sequence of sterol-transforming *Mycobacterium neoaurum* strain VKM Ac-1815D. Genome Announc 2:e01177–e01113. doi:10.1128/genomeA.01177-13
4. Shtratnikova VY, Schelkunov MI, Dovbnya DV et al (2015) Complete genome sequence of *Mycobacterium* sp. strain VKM Ac-1817D, capable of producing 9α-hydroxy-androst-4-ene-3,17-dione from phytosterol. Genome Announc doi 3(1):e01447–e01414. doi:10.1128/genomeA.01447-14
5. Shtratnikova VY, Schelkunov MI, Pekov YA et al (2015) Complete genome sequence of steroid-transforming *Nocardioides simplex* VKM Ac-2033D. Genome Announc 3:e01406–e01414. doi:10.1128/genomeA.01406-14
6. Bolger AM, Lohse M, Usadel B (2014) Trimmomatic: a flexible trimmer for Illumina sequence data. Bioinformatics 30:2114–2120. doi:10.1093/bioinformatics/btu170
7. Leggett RM, Clavijo BJ, Clissold L et al (2014) NextClip: an analysis and read preparation tool for Nextera long mate pair libraries. Bioinformatics 30:566–568. doi:10.1093/bioinformatics/btt702
8. Bankevich A, Nurk S, Antipov D et al (2012) SPAdes: a new genome assembly algorithm and its applications to single-cell sequencing. J Comput Biol 19:455–477. doi:10.1089/cmb.2012.0021
9. Zerbino DR, Birney E (2008) Velvet: algorithms for *de novo* short read assembly using de Bruijn graphs. Genome Res 18:821–829. doi:10.1101/gr.074492.107
10. Jünemann S, Prior K, Albersmeier A et al (2014) GABenchToB: a genome assembly benchmark tuned on bacteria and benchtop sequencers. PLoS One 9:e107014. doi:10.1371/journal.pone.0107014
11. Hunt M, Kikuchi T, Sanders M et al (2013) REAPR: a universal tool for genome assembly evaluation. Genome Biol 14:R47. doi:10.1186/gb-2013-14-5-r47
12. Langmead B, Salzberg SL (2012) Fast gapped-read alignment with Bowtie 2. Nat Methods 9:357–359. doi:10.1038/nmeth.1923
13. Milne I, Bayer M, Cardle L et al (2010) Tablet—next generation sequence assembly visualization. Bioinformatics 26:401–402. doi:10.1093/bioinformatics/btp666
14. Boetzer M, Henkel CV, Jansen HJ et al (2011) Scaffolding pre-assembled contigs using SSPACE. Bioinformatics 27:578–579. doi:10.1093/bioinformatics/btq683
15. Boetzer M, Pirovano W (2012) Toward almost closed genomes with GapFiller. Genome Biol 13:R56. doi:10.1186/gb-2012-13-6-r56
16. Hall TA (1999) BioEdit: a user-friendly biological sequence alignment editor and analysis program for Windows 95/98/NT. Nucl Acids Symp Ser 41:95–98
17. McKenna A, Hanna M, Banks E et al (2010) The genome analysis toolkit: a MapReduce framework for analyzing next-generation DNA sequencing data. Genome Res 20:1297–1303. doi:10.1101/gr.107524.110
18. Li H, Handsaker B, Wysoker A et al (2009) The sequence alignment/map format and SAMtools. Bioinforma Oxf Engl 25:2078–2079. doi:10.1093/bioinformatics/btp352
19. Aziz RK, Bartels D, Best AA et al (2008) The RAST server: rapid annotations using subsystems technology. BMC Genomics 9:75. doi:10.1186/1471-2164-9-75
20. Li R, Zhu H, Ruan J et al (2010) *De novo* assembly of human genomes with massively parallel short read sequencing. Genome Res 20:265–272. doi:10.1101/gr.097261.109
21. Gurevich A, Saveliev V, Vyahhi N, Tesler G (2013) QUAST: quality assessment tool for genome assemblies. Bioinformatics 29:1072–1075. doi:10.1093/bioinformatics/btt086
22. Simão FA, Waterhouse RM, Ioannidis P et al (2015) BUSCO: assessing genome assembly and annotation completeness with single-copy orthologs. Bioinformatics 31:3210–3212. doi:10.1093/bioinformatics/btv351

Chapter 4

Genome Sequencing of Steroid Producing Bacteria Using Ion Torrent Technology and a Reference Genome

Alberto Sola-Landa, Antonio Rodríguez-García, Carlos Barreiro, and Rosario Pérez-Redondo

Abstract

The Next-Generation Sequencing technology has enormously eased the bacterial genome sequencing and several tens of thousands of genomes have been sequenced during the last 10 years. Most of the genome projects are published as draft version, however, for certain applications the complete genome sequence is required.

In this chapter, we describe the strategy that allowed the complete genome sequencing of *Mycobacterium neoaurum* NRRL B-3805, an industrial strain exploited for steroid production, using Ion Torrent sequencing reads and the genome of a close strain as the reference. This protocol can be applied to analyze the genetic variations between closely related strains; for example, to elucidate the point mutations between a parental strain and a random mutagenesis-derived mutant.

Key words *Mycobacterium*, B-3805, Steroid production, Next-generation sequencing, Reference genome, Assemble, Reads, Analysis of variants

1 Introduction

Genome sequence technology has exponentially evolved since 1995, when the first complete genome sequence of a free-living organism was published, corresponding to the bacterium *Haemophilus influenzae* with a genome size of 1.83 Mb. For that purpose, 28,643 sequencing reactions were realized, needing an average of 14 ABI 373 DNA Sequencers per day working for 3 months [1]. During the first 10 years, Sanger method was the main technology available and just over 300 bacterial genomes were sequenced by this way, most of them published as a complete sequence [2]. A qualitative jump took place in 2005, with the publication of two different high-throughput technologies, the pyrosequencing [3] (*see* **Note 1**) and the polony (polymerase colony) sequencing [4] (*see* **Note 2**), starting the Next-Generation

José-Luis Barredo, Ignacio Herráiz (eds.), *Microbial Steroids: Methods and Protocols*, Methods in Molecular Biology vol. 1645, DOI 10.1007/978-1-4939-7183-1_4, © Springer Science+Business Media LLC 2017

Sequencing (NGS) era. Currently, there are multiple NGS systems and platforms, being HiSeq 2000 from Illumina, Ion Torrent from Life Technologies, and 454 GS FLX+ from Roche [5, 6] the most commonly used. Second-generation sequencing produces shorter reads than Sanger technology, thus requiring a higher coverage for a good quality genome sequence. However, the presence of repeats in the bacterial genomes makes very difficult, if not impossible, to completely assemble a genome with only short reads, resulting in draft genomes with a higher number of contigs than previously.

Thanks to NGS, the sequencing cost was significantly reduced and more than 30,000 bacterial genomes were sequenced in 2005–2015 period, an increase of two orders of magnitude over the 10 previous years [7]. Due to this reduction the cost ratio between a closed and a draft genome has enormously been increased, representing more than 95% of the total budget the cost of completing the genome, and currently more than 90% of bacterial genomes are not complete [7] (*see* **Note 3**). With the starting third-generation sequencing or single-molecule sequencing, producing long reads of several thousands of bp [6, 8], it will be easier and cheaper to close a genome, and probably the new genomes sequenced by these technologies will be published as complete genomes again.

Here, we describe a protocol to get a complete sequence with short reads obtained through Ion Torrent platform, using another genome as reference, and filling the gaps with Sanger reads obtained from PCR products. For this purpose it is essential to find a suitable reference genome. Initially, the reads, once filtered by quality and stored in FASTQ format, are mapped over the reference genome using the software segemehl [9], one of the best mapping algorithms suitable for Ion Torrent data [10]. The correct mapping is analyzed with the visualizer Gap5 [11], which allows not only the visualization of the alignment but also its edition. The gaps and regions with a low coverage are identified by means of an ad-hoc spreadsheet in order to be sequenced by Sanger technology. The Ion Torrent and Sanger reads are then assembled de novo and analyzed with QUAST [12] in order to confirm the suitability of the reference genome. Finally, the variants from the reference sequence are identified.

2 Materials

2.1 Hardware

For the requirements of a bacterial genome project, the programs used here can be run on current desktop computers. The processes of assembly and mapping are the most demanding of computational resources, especially of RAM (random access memory). A minimum of 32 GB of RAM is recommended. Input FASTQ files and output SAM/BAM files require large disk spaces (>100 GB recommended).

2.2 Software

Bioinformatics programs are most often command line interface programs and require a computer with 64-bit Linux operating system. Alternatively, for Windows and MacOS users, running a Linux distribution as a virtual machine is a convenient solution (*see* **Note 4**).

Several programs are available for each step of the analysis. What follows is the list of the selected programs for the proposed pipeline (*see* **Note 5**):

1. Segemehl mapper [9, 13]. This mapping tool serves to align (map) the reads on the reference sequence (*see* **Note 6**). It is a Linux command line program available at http://www.bioinf.uni-leipzig.de/Software/segemehl/
2. BEDTools utilities [14]. A suite of Linux command line programs used here to compute the coverage (*see* **Note 7**). Available at https://github.com/arq5x/bedtools2
3. SAMtools [15–17]. A suite of Linux command line programs widely used for working with high-throughput sequencing data (*see* **Note 8**). Available at http://www.htslib.org/
4. Gap5 [11]. This alignment and assembly viewer is part of the Staden package. It is also an editing tool for base-level editing as well as for contig rearrangements. There are Windows, Linux, and MacOS versions available at https://sourceforge.net/projects/staden/files/staden/
5. Mauve, Multiple Genome Alignment software [18, 19]. A Java-based program that can be run under Windows, Linux, MacOS operating systems for comparing two or more draft or finished genomes (*see* **Note 9**). Available at http://darlinglab.org/mauve/
6. SPAdes Genome Assembler [20] (*see* **Note 10**). SPAdes requires a 64-bit Linux system or MacOS. Available at http://bioinf.spbau.ru/spades
7. QUAST: Quality Assessment Tool for Genome Assemblies [12]. QUAST evaluates genome assemblies, working both with and without a reference genome. The tool accepts multiple assemblies, thus it is suitable for comparison. Available at http://quast.sourceforge.net/
8. Burrows-Wheeler Alignment tool, BWA [21, 22]. Used for mapping Sanger sequences against the reference genome (*see* **Note 11**). Available at http://bio-bwa.sourceforge.net/
9. Picardtools. A Java-based command line set of programs for manipulating high-throughput sequencing data and formats such as SAM and VCF. Available at http://broadinstitute.github.io/picard/
10. GATK, Genome Analysis Toolkit [23–25]. The main purpose of this set of tools is variant discovery and genotyping. There

are program versions running on Unix, Linux, and MacOS operating systems. Available at https://software.broadinstitute.org/gatk/

2.3 Input Files

1. Ion Torrent sequencing reads in FASTQ format and quality-trimmed (*see* **Note 12**) from the DNA of the strain that is going to be sequenced. The sequence coverage requirement usually ranges from 25× to 100× [26] (*see* **Note 13**).
2. The complete genome of a closely related strain is required as the reference. A file containing the sequence in FASTA format is used for mapping the reads, while a GFF (General Feature Format) file with the genome annotation is optionally used for the QUAST analysis of de novo assemblies (*see* **Note 14**). The choice of the reference genome can be based on a priori knowledge (e.g., if the strain to be sequenced is a random mutant of an already sequenced strain, or from taxonomy analysis) or based on a posteriori analysis (*see* **Note 15**).

3 Methods

Each step of the bioinformatics procedure described here generally consist of: (1) a program; (2) an input file(s); (3) a line of text which is the command to execute the program, including the arguments or specific options for that program, and which is typed or pasted in a terminal emulator window (a terminal emulator is an application that displays a window which allows sending the commands to the operating system); (4) an output file(s) that is(are) generated once the program has completed its task. The output file(s) will be, in turn, the input file(s) of the next step. It is important to keep the files organized in directories and to maintain a simple but meaningful naming system for both files and directories. Depending on the installation process, to run a program the prompt of the terminal may be placed anywhere in the directory tree or must be in the same directory where the program file is. In the following steps, this requirement is indicated by "(*)" at the beginning of the command line. Anyway, to conduct easily the indicated steps it is highly recommended to follow a Linux command line tutorial.

3.1 Mapping of the Ion Torrent Reads on the Reference Genome

1. The FASTA file of the reference genome ("RefGenome.fa") must be indexed (i.e., an index file is generated with the extension "idx") in order to be used by the segemehl mapper program (the same program that will be used to map the reads serves also to create the index file). This step is done entering the following command in a terminal window:

```
(*)./segemehl.x -x RefGenome.idx -d RefGenome.fa
```

2. Mapping of the Ion Torrent trimmed reads (file "trimmed_reads.fastq") on the reference genome. Using the following options of the command (*see* **Note 16**), the program dedicates six cores of the computer to the task and sends the non-mapped reads to a file (please note that the character "\" in the command serves to type a single command as separate lines):

```
(*)./segemehl.x -i RefGenome.idx -d RefGenome.fa -q \
trimmed_reads.fastq -u unmapped_reads.fastq \
--threads 6 >mapped_v1.sam
```

3. A second indexation of the reference fasta file is needed to meet the format requirements of the following programs. The file containing the reads aligned on the reference genome is converted to the binary format (SAM to BAM), which is smaller. Both tasks and the following are conducted using SAMtools:

```
samtools view -bS mapped_v1.sam >mapped_v1.bam
samtools faidx RefGenome.fa
```

4. Reordering and indexation of the alignment (*see* **Note 17**):

```
samtools sort mapped_v1.bam mapped_v1_ord
samtools index mapped_v1_ord.bam
```

3.2 Determination of Regions with Poor Coverage

1. The coverage data (the number of reads that map at each nucleotide) is generated from the BAM alignment with BEDTools. With the following command, the output is in BedGraph format and also the zero coverage regions are reported. The coverage file is a tab-delimited text file containing a line for each genome region with the following values: the start and end coordinates (related to the reference genome), and the coverage of that region (regions are continuous segments of the genome with the same coverage).

```
genomeCoverageBed -ibam mapped_v1_ord.bam -bga
>coverage.txt
```

2. To extract the genome segments with coverage lower than an established threshold (in this case a value of 5 was established to define low coverage), we use the programming language awk that is included in the Linux distribution:

```
awk '$4 <5' coverage.txt>LowCoverageSegments.txt
```

3. The list of the genome regions that must be resequenced by Sanger is not the previous output file, since it contains contiguous or close segments with low but different coverage values. The closely located segments will be merged to define the region that will be the target of the Sanger sequencing. This process is conveniently done using a spreadsheet program,

such as Microsoft Excel, and the following functions. The values contained in the file "LowCoverageSegments.txt" are copy-pasted into a spreadsheet. The columns of the data show the BedGraph structure: "identifier of the reference sequence," "start coordinate," "end coordinate," "coverage." First and last lines should be reviewed and possibly deleted. With the following cell functions the coordinates of fused regions are determined. In order to describe the structure and location of the input data and the output of the functions, we include the Table 1. Its cell values should be copy-pasted into a spreadsheet so the sheet A1 cell contains the upper left cell value of Table 1 (indicated as "A1" CELL). Then, the following functions or values should be typed into the indicated cells: (1) cells D3 and D29, type "1," which means "yes" to the question placed on the column header (a zero means "no"); (2) cell D4, type "=IF(A4>B3+D$1,1,0)," then copy the function to cells D4 to D28; (3) cell E3, type "=IF(D3=1,A3,"_")," then copy the function to cells E4 to E28; (4) cell F3, type "=IF(D3=1,INDEX(B3:B$29,MATCH(1,D4:D$29,0)),"_")," then copy the function to cells F4 to F28 (the bottom ends of the ranges "B3:B$29" and "D4:D$29" are defined by the last row of the table, which in this example is the row 29); (5) cell G3, type "=IF(D3=1,F3-E3+1,"_")," then copy the function to cells G4 to G28. In Table 1, the cells with contents in red are the output of the indicated functions and, as shown in Table 1, the rows with a "1" in the fourth column are the rows that define the merged regions that should be sequenced by Sanger (*see* **Note 18**).

3.3 Validation of the Mapping Strategy Using a De Novo Assembly with All the Ion Torrent Reads and the Sanger Sequences

1. The Sanger sequences, generated to cover the gaps and regions of low coverage, can now be combined with the NGS reads to obtain the final genome sequence. At this point, it is recommended to create a de novo assembly to confirm the similarity between the target and the reference genomes. Thus, if the target genome contains a particular region, absent in the reference genome, this region can be revealed by the unguided assembly of the reads. The recommended software, capable of the hybrid (Ion Torrent and Sanger sequences) assembly, is SPAdes. With the following command line, the files of the assembly are stored in the directory, created by the program, "SPAdes_assembly":

```
(*)./spades.py --careful --sanger Sanger.fastq\
--iontorrent -s\
trimmed_reads.fastq -o SPAdes_assembly
```

2. QUAST is a quality assessment tool for evaluating assemblies and it can be used to compare the SPAdes-created assembly with the reference genome. The input files are the multifasta

Table 1
Example for the determination of genome regions with low coverage that must be resequenced by Sanger

"A1" CELL			400	Threshold (bp) to merge neighbour regions		
Start	**End**	**Coverage**	**Is the first line of a new region?**	**Start, merged region**	**End, merged region**	**Length, merged region**
116578	116580	3	1	116578	116598	21
116580	116584	2	0	–	–	–
116584	116590	0	0	–	–	–
116590	116591	1	0	–	–	–
116591	116598	2	0	–	–	–
909486	909487	1	1	909486	909616	131
909533	909538	3	0	–	–	–
909538	909561	2	0	–	–	–
909561	909604	1	0	–	–	–
909604	909615	0	0	–	–	–
909615	909616	1	0	–	–	–
4136740	4136742	3	1	4136740	4136756	17
4136748	4136750	2	0	–	–	–
4136751	4136756	1	0	–	–	–
4137256	4137258	2	1	4137256	4137277	22
4137263	4137265	1	0	–	–	–
4137266	4137268	2	0	–	–	–
4137269	4137271	1	0	–	–	–
4137272	4137274	2	0	–	–	–
4137275	4137277	3	0	–	–	–
4366657	4366658	3	1	4366657	4366713	57
4366658	4366667	2	0	–	–	–
4366667	4366670	3	0	–	–	–
4366696	4366697	3	0	–	–	–
4366697	4366706	2	0	–	–	–
4366706	4366713	3	0	–	–	–
			1			

file containing the contiguous sequences created by SPAdes ("contigs.fasta"), the reference sequence ("RefGenome.fa"), and, optionally, the annotation of the reference genome ("RefAnnotation.gff"). Several files, containing statistics and graphs, are produced by QUAST and stored in the output directory (indicated in the following command after "-o") (*see* **Note 19**):

```
(*)./quast.py contigs.fasta -R RefGenome.fa \
-G RefAnnotation.gff -o QUAST_results
```

3. Optionally, the Ion Torrent reads not previously mapped with segemehl can also be used to make a de novo assembly and to inspect the contigs for new sequences not present in the reference genome (*see* **Note 20**):

```
(*)./spades.py --careful --iontorrent -s\
unmapped_reads.fastq\
-o SPAdes_unmapped_assembly
```

3.4 Final Mapping of the Ion Torrent Reads and Sanger Sequences

1. Before proceeding to the final mapping of the reads, the suitability of the reference genome should be decided. After the analysis of the genome assembly with QUAST, it should be clear whether between the target and the reference genomes exist only small sequence differences, such as single nucleotide variations (SNVs) or insertions and deletions of a few nucleotides (INDELs), or larger sequence differences, such as regions unique to the target genome, inverted regions or copy number variations of repeats. In the first case, the final mapping can be made with the original reference genome; in the second case, the reference genome should be edited to accommodate the large structural variations. Regions unique to the target genome are inserted from the sequence of assembly contigs (*see* **Note 21**). For clarity, it is supposed in the following steps that a new reference genome sequence is used ("NewRefGenome.fa"). As in the first mapping step, an index file must be created and then the segemehl mapper is used with the Ion Torrent reads:

```
(*)./segemehl.x -x NewRefGenome.idx -d
NewRefGenome.fa
(*)./segemehl.x -i NewRefGenome.idx -d
NewRefGenome.fa\
-q trimmed_reads.fastq> mapped_v2.sam
```

2. A problem found with SAM files created with segemehl is that the file header may contain errors. These errors prevent the functioning of downstream programs. The Picard package can be used to detect and correct such errors. To validate the SAM file:

```
(*)java -jar picard.jar ValidateSamFile\I=mapped_
v2.sam\
O=Picard_Output.txt
```

3. The output of the previous step should be analyzed. The usual detected errors in the SAM header are missing values related to read groups (*see* **Note 22**). To modify the SAM header in order to assign the required values (read group identifier, read group library, read group platform, run barcode, and read group sample name, respectively):

```
(*)java -jar picard.jar AddOrReplaceReadGroups\
I=mapped_v2.sam\
O=mapped_v2_Correct.sam ID=ReadGroupID LB=Library1 \
PL=IONTORRENT PU=1WV2J SM=SAMPLE
```

4. Optionally, the Picard tool can be used to check the corrected SAM file. In addition, the header can be viewed with this command:

```
samtools view -H mapped_v2_Correct.sam
```

5. The longer Sanger sequences are mapped on the reference genome with BWA-MEM algorithm. The following first command line serves to index the reference sequence, and the second command line runs the mapper and assigns a read group header line:

```
bwa index NewRefGenome.fa
bwa mem -R \
'@RG\tID:SangerReads\tLB:Library2\tPL:SANGER\
tPU:sanger\tSM:SAMPLE' \
NewRefGenome.fa Sanger.fastq > bwa_Sanger.sam
```

6. Both Ion Torrent and Sanger alignments are merged with Picard. Then, the merged alignment should be viewed and inspected with Gap5. To load the BAM file in Gap5, an index file ("Merged.bam.bai") needs to be created with SAMtools:

```
(*)java -jar picard.jar MergeSamFiles\I=mapped_v2_
Correct.sam\
I=bwa_Sanger.sam O=Merged.bam
samtools index Merged.bam
```

7. The merged alignment is the sequencing processed data that will be used to detect the sequence variants between the target and the reference genomes in the next section. In order to reduce the number of miscalls of INDELs it is recommended to apply the GATK Realigner to the merged alignment. To use this program, the reference genome should be indexed with SAMtools, and Picard is used to create a sequence dictionary (*see* **Note 23**):

```
samtools faidx NewRefGenome.fa
(*)java -jar picard.jar CreateSequenceDictionary \
R=NewRefGenome.fa O=NewRefGenome.dict
```

8. The GATK realignment is conducted in two steps, the output file of the first program indicates the intervals that need to be realigned:

```
java -jar GenomeAnalysisTK.jar \
   -T RealignerTargetCreator \
   -R NewRefGenome.fa \
   -I Merged.bam \
   -o IndelRealigner.intervals
java -jar GenomeAnalysisTK.jar \
   -T IndelRealigner \
   -R NewRefGenome.fa \
   -I Merged.bam \
   -targetIntervals $WD/IndelRealigner.intervals \
   -o Merged_Realigned.bam
```

9. With the final alignment file the average coverage can be calculated as follows:

```
samtools depth Merged_Realigned.bam |\
awk '{sum+=$3} END {print "Average = ",sum/NR}'
```

3.5 Identification of Variants and Generation of the Final Consensus Sequence

1. Once the final alignment is made, variant calling is the next step. The aim is to detect variants in the aligned reads, i.e., SNVs or INDELs, in relation to the reference sequence. This is not a straightforward task since two sources of errors create uncertainty: sequencing errors at base level and incorrect mapping of the reads. Enough read coverage is the basis for a variant caller algorithm to distinguish randomly distributed base errors. Local realignment, as described previously, partially addresses mapping errors. The proposed variant calling method employs first the SAMtools mpileup function to create a text file that stores the possible variants deduced from the alignment. This text file is created in the standard Variant Call Format (VCF) or in its binary counterpart, BCF (*see* **Note 24**). The second step is performed with BCFtools:

```
samtools mpileup -go mpileup_output.bcf \
-f NewRefGenome.fa Merged_Realigned.bam
bcftools call -vmO v -o\VariantCalls.vcfmpileup_
output.bcf
```

2. Optionally, some statistics of the results can be obtained:

```
bcftools stats -F NewRefGenome.fa -s -
VariantCalls.vcf \
> VariantCalls.stats
```

3. Finally, many of the lines of the VCF file correspond to false positive calls or unlikely variants that must be filtered out. A filter process is shown where the quality value of 150 is used as the threshold. With this command, variants of low quality are

marked with the value "LowQual" in the FILTER column; for the rest of the lines this column will contain the value "PASS":

```
bcftools filter -O v -o VariantCalls_filtered.vcf \
-s LowQual -i'%QUAL>150' VariantCallsOnly.vcf
```

4. To obtain a table with only the reliable variants:

```
grep PASS VariantCalls_filtered.vcf >VariantsTable.txt
```

5. The data of the filtered variants ("VariantsTable.txt") can be exported to a spreadsheet for manual inspection. Gap5 can then be used to view the read alignment and to decide if the filter applied was correct. Eventually, the set of confident variants will be established. Finally, there are several ways to obtain the consensus sequence, one is using Gap5 and the read alignment via its menu options. Anyway, it should be assured that the final genome sequence of the target strain reflects the variants results.

4 Notes

1. Pyrosequencing technology was released by 454 Life Science, being the first available commercial platform for NGS, allowing the generation of 25 Mb in only one 4-h run. This technology was discontinued in 2013, keeping the technical support until 2016.
2. The polony sequencing technology serves as base for the SOLiD (Supported Oligo Ligation Detection) platform, released by Applied Biosystems.
3. The total cost of the first bacterial genomes could be around $50,000, corresponding approximately a half for the draft genome and the other half to close it [7].
4. Some programs run only under Linux; so, for Windows or MacOS users the easiest solution is to create a virtual machine and install on it a Linux distribution. For these, we recommend the free VirtualBox virtualization application (https://www.virtualbox.org/) to build the virtual machine, and the simple Xubuntu distribution (http://xubuntu.org/). Once installed the guest operating system inside a virtual machine, it is recommended to install the so-called Guest Additions to create a shared folder as the most convenient way of sharing files between the Windows host and the Linux guest systems. Please see the user manual (https://www.virtualbox.org/manual/UserManual.html).
5. Some programs require the installation of additional packages, such as Python. Please see the installation instructions available at the referred software websites.

6. The choice of this program is based on the comparative analysis of [10].
7. It can be easily installed using the software center of Ubuntu/Xubuntu distributions.
8. It consists of three separate repositories of which SAMtools are used for reading/writing/editing/indexing/viewing SAM/BAM files, and BCFtools for reading/writing/editing VCF/BCF files. The Sequence Alignment/Map (SAM) format is a generic alignment format for storing short read alignments against reference sequences. The BAM format is the binary compressed version of a SAM file. The Variant Calling Format (VCF) files and their binary version (BCF) store information related to sequence variation.
9. Mauve can compute and interactively browse whole-genome alignments of two or more genomes as a set of contigs or as finished sequences.
10. It works with Illumina or Ion Torrent reads and is capable of providing hybrid assemblies using PacBio, Oxford Nanopore, and Sanger reads.
11. It consists of three algorithms: BWA-backtrack, BWA-SW, and BWA-MEM. The first algorithm is designed for Illumina sequence reads up to 100 bp, while the other two are for longer sequences ranged from 70 bp to 1 Mbp. Both BWA-MEM and BWA-SW support long-read alignment, but BWA-MEM, the latest developed, is the recommended one, since it is faster and more accurate.
12. In this chapter, a process for the quality trimming of the reads is not described, since it can be done by the sequencing service.
13. Coverage is calculated as the amount of bases generated by the sequencer run divided by genome size. According to [27] an average mapped depth of 50× is necessary to identify most of the SNPs and INDELs in human projects.
14. Files in FASTA and GFF formats can be downloaded directly from NCBI (http://www.ncbi.nlm.nih.gov/genome/) and EBI (http://bacteria.ensembl.org/) websites. GFF format is required for QUAST program.
15. The contigs generated by a draft de novo assembly can be used to search the data banks and find the most similar complete genome. In our case, for sequencing the genome of *Mycobacterium neoaurum* NRRL B-3805, the *M. neoaurum* VKM Ac-1815D genome [28] was used as the reference [29].
16. For this and other steps not all the option values of the programs are specified; so default values are applied mostly.

17. It is highly recommended to inspect the above alignment with a viewer, like Gap5.
18. Once the regions with a poor coverage are determined, they are amplified by PCR and sequenced by Sanger method. Since the aim of this chapter is centered on the bioinformatics analysis, the protocol for Sanger sequencing is not included.
19. Please see the program manual, available at the program website, for a detailed description of the output; especially useful is the Icarus contig alignment viewer.
20. The produced contigs can be evaluated doing BLAST searches against the nucleotide databases.
21. Several common DNA editing programs can be used for this task, although the description and use of these programs are out of the scope of this chapter.
22. SAM file specifications are at https://samtools.github.io/hts-specs/
23. The FASTA file for GATK needs to be written with the UTF-8 character encoding and Unix-type ends of line.
24. The details of the VCF format, its header contents, and the columns that form the tabular data can be found at https://vcftools.github.io/specs.html

Acknowledgments

This work was fully supported by a grant of the European Union program ERA-IB [MySterI (EIB.12.010)] through the APCIN call of the Spanish Ministry of Economy and Competitiveness (MINECO, Spain) (PCIN-2013-024-C02-01). The authors want to thank the European Union program ERA-IB; the Spanish Ministry of Economy and Competitiveness (MINECO, Spain) and the MySterI Consortium (INBIOTEC, Pharmins Ltd., University of York, SINTEF, Technische Universität Dortmund and Gadea Biopharma S.L.). We thank J. Merino, B. Martín and A. Casenave for their excellent technical assistance.

References

1. Fleischmann RD, Adams MD, White O et al (1995) Whole-genome random sequencing and assembly of *Haemophilus influenzae* Rd. Science 269:496–512
2. Binnewies TT, Motro Y, Hallin PF et al (2006) Ten years of bacterial genome sequencing: comparative-genomics-based discoveries. Funct Integr Genomics 6:165–185
3. Margulies M, Egholm M, Altman WE et al (2005) Genome sequencing in microfabricated high-density picolitre reactors. Nature 437:376–380
4. Shendure J, Porreca GJ, Reppas NB et al (2005) Accurate multiplex polony sequencing of an evolved bacterial genome. Science 309:1728–1732

5. Buermans HPJ, den Dunnen JT (2014) Next generation sequencing technology: advances and applications. Biochim Biophys Acta 1842:1932–1941
6. Goodwin S, McPherson JD, McCombie WR (2016) Coming of age: ten years of next-generation sequencing technologies. Nat Rev Genet 17:333–351
7. Land M, Hauser L, Jun SR et al (2015) Insights from 20 years of bacterial genome sequencing. Funct Integr Genomics 15:141–161
8. Koren S, Phillippy AM (2015) One chromosome, one contig: complete microbial genomes from long-read sequencing and assembly. Curr Opin Microbiol 23:110–120
9. Hoffmann S, Otto C, Kurtz S et al (2009) Fast mapping of short sequences with mismatches, insertions and deletions using index structures. PLoS Comput Biol 5:e1000502
10. Caboche S, Audebert C, Lemoine Y et al (2014) Comparison of mapping algorithms used in high-throughput sequencing: application to Ion Torrent data. BMC Genomics 15:264
11. Bonfield JK, Whitwham A (2010) Gap5—editing the billion fragment sequence assembly. Bioinformatics 26:1699–1703
12. Gurevich A, Saveliev V, Vyahhi N et al (2013) QUAST: quality assessment tool for genome assemblies. Bioinformatics 29:1072–1075
13. Hoffmann S, Otto C, Doose G et al (2014) A multi-split mapping algorithm for circular RNA, splicing, trans-splicing and fusion detection. Genome Biol 15:R34
14. Quinlan AR, Hall IM (2010) BEDTools: a flexible suite of utilities for comparing genomic features. Bioinformatics 26:841–842
15. Li H, Handsaker B, Wysoker A et al (2009) The Sequence Alignment/Map format and SAMtools. Bioinformatics 25:2078–2079
16. Li H (2011) A statistical framework for SNP calling, mutation discovery, association mapping and population genetical parameter estimation from sequencing data. Bioinformatics 27:2987–2993
17. Li H (2011) Improving SNP discovery by base alignment quality. Bioinformatics 27:1157–1158
18. Darling ACE, Mau B, Blattner FR et al (2004) Mauve: multiple alignment of conserved genomic sequence with rearrangements. Genome Res 14:1394–1403
19. Rissman AI, Mau B, Biehl BS et al (2009) Reordering contigs of draft genomes using the Mauve aligner. Bioinformatics 25:2071–2073
20. Bankevich A, Nurk S, Antipov D et al (2012) SPAdes: a new genome assembly algorithm and its applications to single-cell sequencing. J Comput Biol 19:455–477
21. Li H, Durbin R (2009) Fast and accurate short read alignment with Burrows-Wheeler transform. Bioinformatics 25:1754–1760
22. Li H, Durbin R (2010) Fast and accurate long-read alignment with Burrows-Wheeler transform. Bioinformatics 26:589–595
23. McKenna A, Hanna M, Banks E et al (2010) The Genome Analysis Toolkit: a MapReduce framework for analyzing next-generation DNA sequencing data. Genome Res 20:1297–1303
24. DePristo MA, Banks E, Poplin R et al (2011) A framework for variation discovery and genotyping using next-generation DNA sequencing data. Nat Genet 43:491–498
25. Van der Auwera GA, Carneiro MO, Hartl C et al (2013) From FastQ data to high confidence variant calls: the Genome Analysis Toolkit best practices pipeline. Curr Protoc Bioinformatics 43:11.10.1–11.1033
26. Vincent AT, Derome N, Boyle B et al (2016) Next-generation sequencing (NGS) in the microbiological world: how to make the most of your money. J Microbiol Methods. doi:10.1016/j.mimet.2016.02.016
27. Ajay SS, Parker SCJ, Abaan HO et al (2011) Accurate and comprehensive sequencing of personal genomes. Genome Res 21:1498–1505
28. Shtratnikova VY, Bragin EY, Dovbnya DV et al (2014) Complete genome sequence of sterol-transforming *Mycobacterium neoaurum* strain VKM Ac-1815D. Genome Announc 2:12–13
29. Rodríguez-García A, Fernández-Alegre E, Morales A et al (2016) Complete genome sequence of "*Mycobacterium neoaurum*" NRRL B-3805, an androstenedione (AD) producer for industrial biotransformation of sterols. J Biotechnol 224:64–65

Chapter 5

RNA-Seq-Based Comparative Transcriptomics: RNA Preparation and Bioinformatics

Antonio Rodríguez-García, Alberto Sola-Landa, and Carlos Barreiro

Abstract

The major transcriptome analysis is the determination of differentially expressed genes across experimental conditions. For this, the next-generation sequencing of RNA (RNA-seq) is an increasingly cost-effective technology for the analysis of transcriptomes with several advantages over gene expression microarrays, such as its higher sensitivity and accuracy, broader dynamic range, and the ability to detect novel transcripts, including noncoding RNA molecules, at nucleotide-level resolution. Although these advantages, many microbiology laboratories have not yet applied RNA-seq analyses to their investigations. The high cost of the equipment for next-generation sequencing is no longer an issue, since this intermediate part of the analysis can be provided by commercial or central services. Here, we detail a protocol for the first part of the analysis, the RNA extraction, and an introductory protocol to the bioinformatics analysis of the sequencing data that generates the differential expression results.

Key words *Mycobacterium*, B-3805, Next-generation sequencing, RNA extraction, Transcriptomics, RNA-seq, Bioconductor, Differential expression

1 Introduction

The high-throughput RNA sequencing or RNA-seq allows the analysis of the transcriptome with unprecedented possibilities. Although it is possible to sequence RNA directly [1], currently the RNA-seq analyses are carried out with cDNA generated from the retrotranscription of the RNA. The cDNA library preparation will depend on the platform to be used and the type of RNA to be analyzed [2, 3].

Compared to gene expression microarrays, RNA-seq presents several and significant advantages: (1) greater sensitivity and accuracy; (2) broader dynamic range (*see* **Note 1**); (3) it allows the identification of different isoforms and variants with single-nucleotide resolution; and (4) it can be applied to organisms whose genome is not sequenced [4–6]. In addition, RNA-seq allows the

José-Luis Barredo, Ignacio Herráiz (eds.), *Microbial Steroids: Methods and Protocols*, Methods in Molecular Biology vol. 1645, DOI 10.1007/978-1-4939-7183-1_5, © Springer Science+Business Media LLC 2017

realization of multiple analyses other than differential expression. From the beginning, RNA-seq showed to be an exceptional technique for the identification of noncoding RNAs (ncRNA) or small RNAs (sRNA), as reported in the first bacterial RNA-seq analysis [7] (*see* **Note 2**). ncRNAs were initially identified as RNA expressed from intergenic regions [8] and most of them exert its action as regulators at post-transcriptional level, mainly negatively, and in association with the RNA chaperone Hfq [9–11]. However, these ncRNA differ considerably in length, sequence, and secondary structure, which results in a very complex bioinformatics prediction and identification. RNA-seq of Hfq-bound transcripts showed that sRNA are not only expressed from intergenic regions but also that an important percentage are derived from the 3′ untranslated regions (UTR) of mRNA, expressed from its own promoter or after mRNA processing [12]. In the last years, other mechanisms of regulation mediated by sRNA have been described [10, 13–15], showing the importance of these RNA, most of them being very difficult to identify by techniques other than RNA-seq.

RNA-seq makes possible to localize massively the transcription star sites (TSS), as demonstrated by Sharma and coworkers by differential RNA-seq (dRNA-seq) in *Helicobacter pylori* for the first time [16] and widely employed with other bacteria [17]. It is also possible to determine the gene organization in operons [18] or even to identify RNA modifications in mRNA and ncRNA [19], among other applications [2–4, 20].

In this chapter, we detail the method used routinely in our lab for RNA extraction, a protocol that can be applied not only for RNA-seq but also for other techniques needing high-quality RNA such as qRT-PCR or microarrays. Next, we describe a simple bioinformatics analysis to obtain the results of differential expression from RNA-seq data, with the aim of being an introduction to this type of analysis.

2 Materials

The process presented here comprises a number of steps divided into two phases: first, the steps aimed to isolate high-quality RNA from culture samples that are carried out at the bench; second, the bioinformatics analysis that employs the FASTQ files provided by the sequencing service and that is aimed to reveal the differentially expressed genes.

Culture samples are used for RNA isolation. Working with RNA requires a special care to avoid the degradation by RNases, ubiquitous proteins highly resistant and persistent. It is indispensable to wear globes all along the process, changing them as often as possible. All the solutions (including MilliQ water) and consumable materials, like tips and tubes, should be acquired certified for this use or, alternatively, autoclaved twice. Bench surface and small equipment like pipettes or grids can be cleaned with RNaseZap® (Ambion) or similar. This solution is applied with a paper towel to the surfaces and rinsed with abundant water.

2.1 Buffers, Solutions, and Reagents

1. TE buffer: 10 mM Tris–HCl pH 8.0, 1 mM EDTA pH 8.0.
2. Lysozyme solution: Lysozyme (15 mg/mL) in TE buffer (*see* **Note 3**).
3. Lysis buffer ML (Macherey-Nagel): lysis buffer containing guanidinium thiocyanate (*see* **Note 4**).
4. Acid phenol for molecular biology applications (e.g., Aquaphenol from Qbiogene) or double-distilled phenol saturated with ultra-pure water.
5. Chloroform:isoamyl alcohol (24:1) solution (abbreviated as CIA).
6. Absolute ethanol and 75% ethanol aqueous solution (v/v).
7. 3 M NaCl solution.
8. RNAprotect Bacteria Reagent (Qiagen).
9. TURBO DNA-free kit (Ambion).
10. RNase-free water.

2.2 Laboratory Equipment for RNA Extraction

1. Microcentrifuge.
2. FastPrep-24 instrument and Lysing Matrix B 2-mL tubes (MP Biomedicals) for cell disruption and homogenization.
3. Phase Lock Gel (PLG) Heavy 2 mL tubes (5 PRIME), three tubes per sample (*see* **Note 5**).
4. Thermocycler.
5. NanoDrop spectrophotometer or equivalent.
6. Vortex.

2.3 Bioinformatics Hardware

The programs indicated here, fed with the data of a common bacterial RNA-seq project, can be run on a desktop computer. A minimum of 8 GB of RAM (random access memory) and more than 100 GB of free disk space are recommended.

2.4 Software

Several bioinformatics programs are available for each step of the analysis, however, for some steps are only available in the form of Linux command line programs. For Windows and Mac OS users, running a Linux distribution as a virtual machine is a convenient solution (*see* **Note 6**). The programs selected for the differential expression analysis of the RNA-seq data are the following:

1. The short read aligner BBMap can be downloaded from https://sourceforge.net/projects/bbmap/. It is a Linux command line program.

2. Samtools [21–23] is a suite of Linux command line programs widely used for working with high-throughput sequencing data. Available at http://www.htslib.org/.
3. FastQC is used for quality control of high-throughput sequence data. It is a Java-based program, so a Java Runtime Environment should be installed. Available at http://www.bioinformatics.babraham.ac.uk/projects/fastqc/.
4. The "R" software for statistical computing is available from https://www.r-project.org/. There are versions for Windows, Mac OS, and Linux systems. Once the last version is installed, the Bioconductor packages that perform the analysis are downloaded and installed through the following typed commands:

```
source("https://bioconductor.org/biocLite.R")
biocLite()
biocLite("DESeq2")
biocLite("GenomicFeatures")
biocLite("Rsamtools")
biocLite("GenomicAlignments")
```

If asked to update already installed packages ("Update all/some/none? [a/s/n]:"), allow it typing an "a."

2.5 Input Files

In this introductory protocol we do not cover the process of high-throughput sequencing, usually outsourced. The most used platform for RNA-seq is Illumina. The service provider will require RNA samples of enough quality and concentration (*see* **Note 7**) and will conduct the steps of rRNA removal, library preparation, cDNA sequencing, demultiplexing, and quality trimming of the reads. A guide of the sequencing depth for transcriptomics has been recently published [24]. To illustrate the following steps, two files containing the sequencing reads in FASTQ format are used. These files were generated from a strand-specific library fed to an Illumina NextSeq 500 sequencer that provided reads at a length of 75 bp.

The genome data of the bacterial strain, both the sequence and the annotation, are required. A file containing the sequence in FASTA format is used for mapping the reads, while a GTF (General Transfer Format) file with the genome annotation is used to obtain the table of read counts per gene. To avoid format compatibility issues, it is recommended to prepare a tab-delimited text file with the structure shown (*see* Table 1).

3 Methods

The first and crucial step to assure a successful transcriptomics analysis is the experimental design. In the following procedure, the aim is to interrogate the genes involved in the bioconversion of

Table 1
Selected rows extracted from the annotation file used as an example of a valid GTF (General Transfer Format) file

1	2	3	4	5	6	7	8	9
Myc_B3805_Chromosome	protein_coding	gene	142	1206	.	+	.	gene_id "MyAD_00005";
Myc_B3805_Chromosome	protein_coding	transcript	142	1206	.	+	.	gene_id "MyAD_00005"; transcript_id "MyAD_00005T";
Myc_B3805_Chromosome	pseudogene	gene	3543	4343	.	+	.	gene_id "MyAD_00025";
Myc_B3805_Chromosome	pseudogene	transcript	3543	4343	.	+	.	gene_id "MyAD_00025"; transcript_id "MyAD_00025T";
Myc_B3805_Chromosome	tRNA	gene	27813	27903	.	-	.	gene_id "MyAD_00130";
Myc_B3805_Chromosome	tRNA	transcript	27813	27903	.	-	.	gene_id "MyAD_00130"; transcript_id "MyAD_00130T";
Myc_B3805_Chromosome	rRNA	gene	3584422	3584534	.	-	.	gene_id "MyAD_16805";
Myc_B3805_Chromosome	rRNA	transcript	3584422	3584534	.	-	.	gene_id "MyAD_16805"; transcript_id "MyAD_16805T";
Myc_B3805_Chromosome	misc_RNA	gene	3811712	3812109	.	-	.	gene_id "MyAD_25325";
Myc_B3805_Chromosome	misc_RNA	transcript	3811712	3812109	.	-	.	gene_id "MyAD_25325"; transcript_id "MyAD_25325T";

The GTF format consists of one line per feature, each containing nine columns of data. Fields must be tab-separated and must contain a value; "empty" columns should be denoted with a point ("."). The fields are: (1) the name of the chromosome sequence, it must be the same that the specified in the FASTA file; (2) the data source (database or project name) or the program that generated the feature, we use this field to indicate the gene type; (3) feature type, for simplicity, only "gene" and "transcript" values are used; (4) the start position of the feature, with sequence numbering starting at 1; (5) the end position of the feature, with sequence numbering starting at 1; (6) unused, set to the empty value; (7) strand were the feature is, a plus symbol for the forward or minus symbol for the reverse strand; (8) unused; set to the empty value; (9) this field can serve for attributes as a list of semicolon-separated tag-value pairs; we use this field to indicate the gene identifier

phytosterols to steroid precursors by the "*Mycobacterium neoaurum*" strain B-3805.The availability of the genome sequence of the strain [25] facilitates the bioinformatics process. We use a simple comparative experiment of two conditions, control vs. treatment, in order to reveal the genes induced by the presence in the medium of phytosterols. The control condition is the basal culture condition (defined medium with glycerol as the carbon source), meanwhile the treatment condition only differs in the medium composition: half of the glycerol content of the basal medium is replaced by phytosterols. No biological replicates are included in the presented process. Biological replicates (at least three replicates) are usually recommended to ensure the reliability of the transcriptomics results and are usually required to perform tests that provide statistical significance in the form of *p*-values. However, replication increases the cost of the experiment and sometimes it is not possible to define *a priori* the best experimental conditions. The knowledge gathered from an exploratory assay is the most confident way to decide which conditions are more informative and adequate to do replicates. Therefore, the presented process is intended to perform this exploratory analysis. The DESeq2 approach [26] for the comparative transcriptomics is chosen here since it is capable of estimating *p*-values without the use of replicates (*see* ref. [27] for a comparison of available programs). Once the experimental conditions are validated, the process with added replicates will be straightforward. The following sections cover the steps carried out to obtain total RNA preparations of enough quantity and quality, and to obtain the results of the differential expression analysis.

3.1 Sample Harvest and RNA Stabilization

1. Collect 1 volume of culture sample (*see* **Note 8**) and immediately transfer it to a tube containing 2 volumes of RNAprotect Bacteria Reagent (*see* **Note 9**).
2. Mix as soon as possible with vortex for 5 s.
3. Incubate for 5 min at room temperature (*see* **Note 10**).
4. Centrifuge at top speed for 15 min (10 mL tubes) or 5 min (1.5–2.0 mL tubes).
5. Remove the supernatant and follow with RNA extraction or freeze the pellet until use (*see* **Note 11**).

3.2 RNA Extraction

In contrast with common column-based kits, this purification protocol effectively recovers all sizes of RNA molecules, which is mandatory for the analysis of ncRNAs.

1. Thaw the samples at room temperature during about 10 min.
2. Add 125 μL of lysozyme solution, suspend the cells by pipetting, and transfer to a new tube.

3. Incubate during exactly 10 min at room temperature (*see* **Note 12**).
4. Add 450 μL of lysis buffer ML and mix by pipetting.
5. Transfer to a Lysing Matrix B tube, add 360 μL of acid phenol and shake longitudinally. Keep on ice for a few minutes.
6. Select a speed of 6.5 m/s in the FastPrep instrument and agitate during 30 s.
7. Keep the tubes on ice during 1 min.
8. Repeat **steps 6** and 7.
9. Centrifuge at room temperature and maximum speed for 1 min to pellet the matrix and cell debris.
10. Transfer the lysate to a PLG tube and add 0.5 volumes of CIA (*see* **Note 5**).
11. Shake strongly the tubes longitudinally until a uniform whitish suspension is obtained. Keep on shaking during 30–60 s.
12. Centrifuge at maximum speed during 5 min.
13. Collect the supernatant in a new PLG tube. Add 0.5 volumes of CIA and 0.5 volumes of acid phenol.
14. Repeat **steps 11** and **12**.
15. Collect the supernatant in a new PLG tube and add 1 volume of CIA.
16. Repeat **steps 11** and **12**.
17. Transfer the supernatant to a new (non-PLG) tube. Add 2 volumes of absolute ethanol and 0.1 volumes of 3 M NaCl.
18. Let the RNA precipitate overnight at −20 °C.
19. Centrifuge at maximum speed during 1 h at 4 °C. Discard the supernatant and wash the pellet with 200 μL of 75% ethanol.
20. Centrifuge for 10 min and discard thoroughly the supernatant with the help of fine-tip pipette.
21. Let air-dry the pellet for a few minutes and dilute it in 50 μL of RNase-free water.

3.3 Removal of Contaminant DNA

1. Determine the RNA concentration by absorbance or fluorescence methods.
2. Add 10 μg of RNA to a 0.2 mL tube and prepare the following reaction mix (50 μL of final volume) (*see* **Note 13**):

10 μg of RNA ($x \leq 44$)	x μL
MilliQ water	$(44 - x)$ μL
10× DNase buffer	5.0 μL
DNase I (Ambion)	1.0 μL

3. Mix by repeatedly pipetting up and down, and incubate at 37 °C for 45 min (*see* **Note 14**).
4. Add 1 μL more of DNase I, mix well and incubate at 37 °C for 45 min.
5. Add 10 μL of Inactivation Reagent (included in the TURBO DNase kit). This reagent must be well resuspended before use by flicking or vortexing the tube.
6. Incubate at room temperature for 5 min. It is important to maintain the reagent suspended by flicking the tube three times during the incubation.
7. Centrifuge at maximum speed during 1.5 min and transfer carefully the supernatant to a new tube. Avoid the contact of the tip with the pellet in order to avoid reagent transfer.
8. Store the cleaned RNA solution frozen, preferably at −20 °C for weeks or at −80 °C for months or years.

3.4 Alignment of the Sequencing Reads to the Genome Sequence

Sequencing services commonly provides the reads, i.e., the sequences of nucleotides, in the form of FASTQ files, one for each sample. The bioinformatics procedure begins with these files and consists of a series of steps. Each step can be described as input file(s) that are processed by means of a program that creates output file(s). The output file(s) will be, in turn, the input file(s) of the next step. In case of Linux command line programs, the program is executed by means of a line of text that is typed or copy-pasted in a terminal window. Depending on the installation process, to execute some programs the working directory of the terminal may be anywhere in the directory tree, but for others it must be in the same directory where the program file is (*see* **Note 15**). In the following commands, this requirement is indicated by "(*)" at the beginning of the line.

1. The FASTA file of the strain genome sequence ("Genome.fna") must be indicated to the mapper program. This step is done by typing the first command line in a terminal window. The second and third lines create the read alignments of the control (RNA16) and the treatment (RNA06) samples (please note that the character "\" in the command serves to type a single command as separate lines). The input FASTQ files (*see* **Note 16**) can be either compressed (file extension ".fastq.gz") or not (file extension ".fastq"):

```
(*)./bbmap.sh ref=Genome.fna
(*)./bbmap.sh ./bbmap.sh in=RNA16.fastq /
outm=Mapped_RNA16.samslow=t ambiguous=random
(*)./bbmap.sh ./bbmap.sh in=RNA06.fastq /
outm=Mapped_RNA06.samslow=t ambiguous=random
```

2. The Sequence Alignment/Map (SAM) format is a generic alignment format for storing short read alignments against reference sequences. The BAM format is the binary compressed version of a SAM file to safe disk space. First, it is convenient to inspect the alignment SAM files with FastQC (the program manual guides the interpretation of the output graphs). Second, samtools is used to convert the SAM to BAM format, index the BAM files, and sort the alignments by leftmost coordinates (indexing and sorting are usually required by downstream programs). The following commands do these three steps (only shown the set of commands for the control sample):

```
samtools view -bS Mapped_RNA16.sam> Mapped_RNA16.bam
samtools sort Mapped_RNA16.bam Mapped_RNA16_Sorted
samtools index Mapped_RNA16_Sorted.bam
```

3.5 Differential Expression Analysis with DESeq2

The differential expression analysis is conducted with packages run in the R environment. Input BAM files must be located in the working directory; this directory is "C:/Transcriptomics/Exp1"(Windows system) in the following commands.

1. Open an R session and inside a terminal, run the following commands (comments are preceded by "#" and can be copy-pasted in the terminal as well):

```
library("DESeq2")          #Loading the required packages.
library("Rsamtools")
library("GenomicFeatures")
biocLite("GenomicAlignments")
wdir <- "C:/Transcriptomics/Exp1"
list.files(dir) #To list the files of the working directory.
filenames <- file.path(wdir,c("Mapped_RNA16_Sorted.bam",
"Mapped_RNA06_Sorted.bam"))
bamfiles <- BamFileList(filenames, yieldSize=2000000)
```

2. Prepare a tab-delimited table of the genome annotation with the layout shown in Table 1 and save the file as "Annotation.gtf" in the working directory. To load the annotation, type:

```
gtffile <- file.path(wdir,"Annotation.gtf")
(txdb <- makeTxDbFromGFF(gtffile, format="gtf"))
(genes <- transcriptsBy(txdb, "gene"))
```

3. Before performing the DESeq2 statistical analysis, read counts data are computed from the alignment files. Count data is a matrix of integers: for each gene (in rows) the integers tell the total of reads that map on that gen in each sample (in columns). The output is stored as an object of class

RangedSummarizedExperiment that is named "se" in the following commands:

```
se <- summarizeOverlaps(features=genes, reads=bamfiles,
                        mode="Union",
                        singleEnd=TRUE,
                        ignore.strand=FALSE,
                        fragments=FALSE)
se          #To see the structure of the object "se".
head(assay(se))    #This shows the first rows of the read
counts table.
colSums(assay(se))     #This shows the reads totals.
summary(assay(se))     #This shows the counts statistics.
```

4. The information that must contain the object "se" includes the count data and the information about the samples and the experiment design. Information about the samples is prepared as a tab-delimited file ("SamplesTable.txt") with at least three columns. Each row describes one sample. In the example below, the first row contains the field names, the first column is the sample name, the second column is the name of the alignment file, and the third column assigns each sample to an experimental condition:

	BAMfile	Condition
RNA16	Mapped_RNA16_Sorted.bam	Gol
RNA06	Mapped_RNA06_Sorted.bam	PhyGol

5. The following two command lines add this information to the "se" object:

```
read.delim("SamplesTable.txt", row.names=1) ->
SamplesTable
(colData(se) <- DataFrame(SamplesTable))
```

6. Then, the statistical results are obtained with these commands:

```
dds <- DESeqDataSet(se, ~ Condition)
dds$Condition <- relevel(dds$Condition, "Gol") #To
specify the reference level.
dds <- DESeq(dds)
Results<- results(dds)
```

7. To explore the content of the Results object:

```
Results
resOrd <- Results[order(Results$pvalue),]
head(resOrd)
summary(Results)
plotMA(Results, main="DESeq2: Phy+Gol vs Gol",
ylim=c(-4,4))
mcols(Results)$description#This command will show
the description of the columns of the results table.
```

8. To create a tab-delimited file containing the results, type the command:

```
write.table(as.data.frame(Results),file="Results_
DESeq2.txt",
sep="\t", quote=FALSE)
```

The easiest way to work with the differential expression results is to import the table of the values (file "Results_DESeq2.txt") to a spreadsheet. The spreadsheet functions will serve to easily filter and reorder the rows. Each row corresponds to a gene; the main columns are the gene identifier, the log fold change (header "log-2FoldChange"), the *p*-value and the adjusted *p*-value for multiple testing (header "padj").

4 Notes

1. Gene expression is quantified as reads per kilobase of transcript per million reads mapped (RPKM) or similar measurements (transcripts per million, TPM), or even the number of times every base is transcribed can be counted. Therefore, the expression levels cannot be saturated as light-detection systems can, an important fact for high level expression genes. On the other hand, its high sensitivity allows the study of low abundant transcripts with sufficient sequencing depth [4].
2. The first RNA-seq analyses were realized with the yeasts *Saccharomyces cerevisiae* and *Schizosaccharomyces pombe*, published in June, 2008 [28, 29]. The first bacterial RNA-seq analysis was published in January, 2009 [7], only 7 months later. In this work, the authors identified 13 ncRNA analyzing the differential expression of intergenic regions. All these 13 ncRNA were not included among the 213 putative ncRNA computationally predicted in a previous work.
3. The spatula used to take the lysozyme, thoroughly cleaned, must be flamed before use to eliminate the possible presence of RNases. The solution is aliquoted and stored at −20 °C. No more than two freeze-thaw cycles are recommended.
4. This lysis buffer was chosen because it is sold independently and not only as part of the NucleoSpin miRNA kit (Macherey-Nagel).
5. PLG tubes must be previously prepacked by centrifugation (30 s at maximum speed). The pre-package can be done during Subheading 3.2, **step 9**.
6. We recommend the free VirtualBox software (https://www.virtualbox.org/) to build the virtual machine, and the simple Xubuntu distribution (http://xubuntu.org/). Once installed the guest operating system inside a virtual machine, it is rec-

ommended to install the so-called Guest Additions to create a shared folder as the most convenient way of sharing files between the Windows/Mac OS host and the Linux guest systems.

7. Usually, it is required a minimum of 5 μg of total RNA at 100 ng/μL or higher concentration. The quality of the RNA preparation depends on its purity and on the integrity of the RNA molecules (absence of RNase-dependent degradation). The purity can be checked easily with a spectrophotometer since a pure preparation shows an A_{260}/A_{280} ratio of 2.1. The standard for the integrity control is the RIN (RNA Integrity Number) provided by the Agilent 2100 Bioanlyzer instrument. The usual requirement for the RNA samples being sequenced is a RIN ≥ 8.
8. The needed volume of the culture sample will depend on the growth phase and the culture density. In our experience, an average of 3.5 μg of RNA are extracted from 1 OD_{600} of cells. *See* also **Note** 7 for the requirements of the RNA-seq procedure.
9. In bacteria, RNA biosynthesis and degradation are continuous processes. To minimize the degradation of the RNA after taking the samples, it is necessary to flash-freeze the samples or to add an agent that inhibits the RNase activity, such as the RNAprotect reagent. To reduce as much as possible the time the samples are untreated, it is preferable—disposing an array of tubes containing the necessary amount of the reagent—to add the culture sample to the reagent tube instead of adding the reagent to the sample tube.
10. A temperature higher than 15 °C is necessary to avoid the precipitation of the RNAprotect components.
11. A maximum of 2 weeks at −20 °C or 4 weeks at −80 °C are the recommendations of Qiagen for the storage of the samples. However, we have obtained RNA preparations of the same high quality, compared with 1-month storage, from frozen samples (−20 °C) stored for 1 year.
12. This step is critical since more incubation time might result in appreciable degradation while less time can imply a lower yield.
13. TURBO DNA-free kit (Ambion) is our preferred DNase treatment. Although 0.5 μL of DNase I and 30 min are the standard conditions recommended by the manufacturer, the indicated conditions ensure the complete removal of DNA, at least for RNA-seq applications.
14. Because of the small reaction volume, 0.2 mL tubes and a thermocycler are used for the incubation.
15. To conduct easily the indicated steps, it is highly recommended to follow a Linux command line tutorial. The steps carried out

in the R environment are executed also with command lines, but with a different language and the input and output data are not saved as files in the disk, but as "objects" inside the software. An excellent and free introductory guide to R is found at https://cran.r-project.org/doc/manuals/R-intro.pdf

16. The FASTQ format is widely used to store sequencing reads. It is a text file that contains for each read four lines. The first line always starts with "@" and specifies the name of the read (the software of the Illumina sequencer generates a systematic identifier); the second line contains the sequence letters, the third line begins with "+" which is optionally followed by the same sequence identifier of the first line; the fourth line encodes the quality values (the Phred score) for the sequence of the second line: both lines must contain the same number of characters.

Acknowledgments

This work was supported by a grant of the European Union program ERA-IB [MySterI (EIB.12.010)] through the APCIN call of the Spanish Ministry of Economy and Competitiveness (MINECO, Spain) (PCIN-2013-024-C02-01). The authors want to thank the European Union program ERA-IB; the Spanish Ministry of Economy and Competitiveness (MINECO, Spain) and the MySterI Consortium (INBIOTEC, Pharmins Ltd., University of York, SINTEF, Technische Universität Dortmund and Gadea Biopharma S.L.). We thank J. Merino, B. Martín, and A. Casenave for their excellent technical assistance and the degree students of the group A. Morales and P. González.

References

1. Ozsolak F, Platt AR, Jones DR et al (2009) Direct RNA sequencing. Nature 461:814–818
2. Hrdlickova R, Toloue M, Tian B (2016) RNA-Seq methods for transcriptome analysis. Wiley Interdiscip Rev RNA. doi:10.1002/wrna.1364
3. Creecy JP, Conway T (2015) Quantitative bacterial transcriptomics with RNA-seq. Curr Opin Microbiol 23:133–140
4. Croucher NJ, Thomson NR (2010) Studying bacterial transcriptomes using RNA-seq. Curr Opin Microbiol 13:619–624
5. Zhao S, Fung-Leung WP, Bittner A et al (2014) Comparison of RNA-Seq and microarray in transcriptome profiling of activated T cells. PLoS One 9:e78644
6. Wang Z, Gerstein M, Snyder M (2009) RNA-seq: a revolutionary tool for transcriptomics. Nat Rev Genet 10:57–63
7. Yoder-Himes DR, Chain PSG, Zhu Y et al (2009) Mapping the *Burkholderia cenocepacia* niche response via high-throughput sequencing. Proc Natl Acad Sci U S A 106:3976–3981
8. Mizuno T, Chou MY, Inouye M (1984) A unique mechanism regulating gene expression: translational inhibition by a complementary RNA transcript (micRNA). Proc Natl Acad Sci U S A 81:1966–1970
9. Valentin-Hansen P, Eriksen M, Udesen C (2004) The bacterial Sm-like protein Hfq: a key player in RNA transactions. Mol Microbiol 51:1525–1533
10. de Lay N, Schu DJ, Gottesman S (2013) Bacterial small RNA-based negative regulation: Hfq and its accomplices. J Biol Chem 288:7996–8003

11. Miyakoshi M, Chao Y, Vogel J (2015) Regulatory small RNAs from the 3′ regions of bacterial mRNAs. Curr Opin Microbiol 24:132–139
12. Chao Y, Papenfort K, Reinhardt R et al (2012) An atlas of Hfq-bound transcripts reveals 3′ UTRs as a genomic reservoir of regulatory small RNAs. EMBO J 31:4005–4019
13. Miyakoshi M, Chao Y, Vogel J (2015) Cross talk between ABC transporter mRNAs via a target mRNA-derived sponge of the GcvB small RNA. EMBO J 34:1478–1492
14. Feng L, Rutherford ST, Papenfort K et al (2015) A Qrr noncoding RNA deploys four different regulatory mechanisms to optimize quorum-sensing dynamics. Cell 160:228–240
15. Papenfort K, Sun Y, Miyakoshi M et al (2013) Small RNA-mediated activation of sugar phosphatase mRNA regulates glucose homeostasis. Cell 153:426–437
16. Sharma CM, Hoffmann S, Darfeuille F et al (2010) The primary transcriptome of the major human pathogen *Helicobacter pylori*. Nature 464:250–255
17. Sharma CM, Vogel J (2014) Differential RNA-seq: the approach behind and the biological insight gained. Curr Opin Microbiol 19:97–105
18. Conway T, Creecy JP, Maddox SM et al (2014) Unprecedented high-resolution view of bacterial operon architecture revealed by RNA sequencing. MBio 5:e01442–e01414
19. Marbaniang CN, Vogel J (2016) Emerging roles of RNA modifications in bacteria. Curr Opin Microbiol 30:50–57
20. Bashiardes S, Zilberman-Schapira G, Elinav E (2016) Use of metatranscriptomics in microbiome research. Bioinform Biol Insights 10:19–25
21. Li H, Handsaker B, Wysoker A et al (2009) The sequence alignment/map format and SAMtools. Bioinformatics 25:2078–2079
22. Li H (2011) A statistical framework for SNP calling, mutation discovery, association mapping and population genetical parameter estimation from sequencing data. Bioinformatics 27:2987–2993
23. Li H (2011) Improving SNP discovery by base alignment quality. Bioinformatics 27:1157–1158
24. Lei R, Ye K, Gu Z et al (2015) Diminishing returns in next-generation sequencing (NGS) transcriptome data. Gene 557:82–87
25. Rodríguez-García A, Fernández-Alegre E, Morales A et al (2016) Complete genome sequence of '*Mycobacterium neoaurum*' NRRL B-3805, an androstenedione (AD) producer for industrial biotransformation of sterols. J Biotechnol 224:64–65
26. Love MI, Huber W, Anders S (2014) Moderated estimation of fold change and dispersion for RNA-seq data with DESeq2. Genome Biol 15:550
27. Seyednasrollah F, Laiho A, Elo LL (2015) Comparison of software packages for detecting differential expression in RNA-seq studies. Brief Bioinform 16:59–70
28. Nagalakshmi U, Wang Z, Waern K et al (2008) The transcriptional landscape of the yeast genome defined by RNA sequencing. Science 320:1344–1349
29. Wilhelm BT, Marguerat S, Watt S et al (2008) Dynamic repertoire of a eukaryotic transcriptome surveyed at single-nucleotide resolution. Nature 453:1239–1243

Chapter 6

Intra- and Extra-cellular Proteome Analyses of Steroid-Producer Mycobacteria

Carlos Barreiro, Alejandro Morales, Inés Vázquez-Iglesias, and Alberto Sola-Landa

Abstract

The importance of the pathogenic mycobacteria has mainly focused the omic analyses on different aspects of their clinical significance. In contrast, those industrially relevant mycobacteria have received less attention, even though the steroids market sales in 2011, in example, were estimated in \$8 billion.

The extra-cellular proteome, due to its relevance in the sterols processing and uptake; as well as the intra-cellular proteome, because of its role in steroids bioconversion, are the core of the present chapter. As a proof of concept, the obtaining methods for both sub-proteomes of *Mycobacterium neoaurum* NRRL B-3805, a relevant industrial strain involved in steroids production, have been developed. Thus, procedures and relevant key points of these proteomes analyses are fully described.

Key words Proteome, Secretome, Steroid, 2D-DIGE, *Mycobacterium*, B-3805, Sterol, Phytosterols, Cholesterol

1 Introduction

Mycobacterium genus can be divided, as its close-related genus *Corynebacterium*, under an effective, but informal, classification in: (1) "bad guys," these with clinical significance (*Mycobacterium tuberculosis*, *Mycobacterium leprae*) and (2) "good guys," which are involved in biotechnological applications as steroids production (*Mycobacterium neoaurum*) [1–3]. Additionally, other classifications based on growth rate, tuberculous, or nontuberculous mycobacteria, etc. can also be applied [4].

These Actinobacteria included in the recently created new order of Corynebacteriales are located in the family Mycobacteriaceae. Nowadays, the *Mycobacterium* genus contains 178 species and 13 subspecies as described by Jean P. Euzéby on the website "List of prokaryotic names with standing in nomenclature (LPSN)" (http://www.bacterio.net/mycobacterium.html)

José-Luis Barredo, Ignacio Herráiz (eds.), *Microbial Steroids: Methods and Protocols*, Methods in Molecular Biology vol. 1645, DOI 10.1007/978-1-4939-7183-1_6, © Springer Science+Business Media LLC 2017

[5, 6]. This fact presents it as a wide genus. Its first genome sequence was released on 1998 [7] and currently more than 400 genome entries appear in the NCBI database (http://www.ncbi.nlm.nih.gov/genome). This opened the so-called post-genomic era, mainly focused on omics analysis as Proteomics [8] or Transcriptomics (microarrays [9], RNAseq [10]) of different *Mycobacterium* species.

World Health Organization (WHO) estimates 1.5 million of tuberculosis death in 2013, in addition to 9 million of new cases, which contrasts with the successful story of leprosy control programme in the last 20 years [11–13]. Due to the relevance of these ancient diseases the proteome analyses in *Mycobacterium* genus have been largely focused on the pathogenic species. The attempt to define the protein-based gene expression analysis generates the term proteome, which was the starting point for Proteomics, a new analytic science [14–18]. This science applied to mycobacterial diseases has achieved different goals, such as: (1) mycobacterial strains characterization [19]; (2) pathogenic strains characterization [20–23]; (3) immunogenic targets definition [24, 25]; (4) immune response attenuation [12]; (5) extra-cellular proteome (secretome) analysis [26–28]; (6) 2-D PAGE database generation [29]; and also (7) non-pathogenic model strains have been analyzed as *Mycobacterium smegmatis* [30].

In contrast, the omic analysis based on industrial-relevant mycobacteria has been less tackled, even though the steroid production by *Mycobacterium* strains is an old fashion from the late 1960s [31]. Thus, in 2011 the total revenue from steroid sales was estimated in $8 billion, where just corticosteroids reached $900 million [32, 33]. In case of the steroid production, three are the main concerns standing in the way of the industrial process that also difficult the omics approach: (1) low solubility of sterols as phytosterols ($\approx$0.1 mM), which are the raw materials for steroids production; (2) steroids toxicity for the producer bacteria (end-product inhibition), as well as, (3) the development of microorganisms for the production *a la carte* of high-added-value steroids [34]. These facts somehow affect the proteins sample collection due to the traditional use of, in example, two-phase cultures [aqueous and oil (soya, rape, sunflower) phases], which is critical for secretome analysis. Anyway, some interesting studies applied to the steroid producer mycobacteria have been recently published [10, 35], in addition to the increased number of genome sequences published, which are the key step for protein identification [3, 36].

Three are the sub-proteomes typically analyzed from one cell: intra-cellular, membrane, and extra-cellular (secretome). Last one is the most complex in terms of protein purification, due to the presence of phytosterols or oil derivatives disturbing the protein extraction. The analysis of the extra-cellular proteome is based on some preliminary concepts: (1) use of defined media, if possible, to avoid two-phase culture broth, which make extremely complex the process; (2) the lowest amount of sterol precursors (e.g., cholesterol, phytosterols) compatible with the steroids production and detection (proteins should be active); (3) proper growth phase (e.g., lag phase) to avoid culture lysis, which could enrich the intra-cellular proteins present in the secretome (degradome), to avoid their identification as false moonlighting proteins [18, 37]. These moonlighting proteins comprise an interesting group of multifunctional proteins, which presents two functions in a single polypeptide chain. Nowadays, more than 300 proteins have been described as moonlighting, which develop both functions at the same time or at different times conditioned to environmental changes [38, 39]. Several examples of moonlighting proteins are described at the MoonProt database (http://www.moonlightingproteins.org/).

In contrast, the intra-cellular environment presents a high protein concentration, which allow, if needed, the sample precipitation by using larger volumes of precipitants as acetone-based solutions. This fact typically decreases the number of low soluble proteins precipitated when TCA (trichloroacetic acid) is used, which complicates the subsequent dissolution in the urea-based buffers. In a similar way as described above, the lower amount of sterol precursors, the better for protein purification. Excess of phytosterols in the culture media results in white clumps that precipitate in parallel to the mycobacterial cells when they are collected. Thus, the use of defined media and the lowest amount of sterol precursors compatible with the steroids production and detection are convenient.

The extra-cellular proteome, due to its relevance in the sterols processing and uptake; as well as the intra-cellular proteome, because of its role in steroids bioconversion, are the core of the present chapter. As a proof of concept, both sub-proteomes of *Mycobacterium neoaurum* NRRL B-3805, a relevant industrial strain involved in steroids production, have been used to develop the protein extraction methods described.

M. neoaurum NRRL B-3805 (CECT3018; DSM 2967; NCIMB 11678) was initially obtained after two rounds of ultraviolet irradiation from a soil isolated mycobacteria as part of a mutagenesis program for efficient strains in sterols bioconversion ([40]; patent no. US3759791). This strain is a suitable AD (4-androstene-3,17-dione) producer and low producer of the by-product ADD (androsta-1,4-diene-3,17-dione). Recently, its complete genome sequence has been published, which eases the omic analysis [36].

2 Materials

2.1 Bacterial Growth Media

1. SINTEF's defined cultivation medium for shake flasks (described in this book by Josefsen and coworkers) (per liter in distilled water): Glycerol (99.5%) 10.0 g, PIPES [Piperazine-*N*,*N'*-bis(2-ethanesulfonic acid)] 4.0 g, NH_4Cl 2.5 g,

$MgSO_4{\cdot}7H_2O$ 0.50 g, K_2HPO_4 0.50 g, NaH_2PO_4 0.30 g, aspartic acid 0.10 g, SINTEF's trace mineral solution 4.0 mL, pH 7.0–7.2. As a control condition, this defined medium with glycerol as a carbon source (5.7 g/L, 87% purity) was selected. However, sterols biotransformation analyses were carried out by addition of 7.0 g/L of cholesterol or phytosterols, respectively, to the control medium, which also includes glycerol (5.7 g/L, 87% purity) as a carbon source.

2. SINTEF's trace mineral solution (described in this book by Josefsen and coworkers) (per liter): $FeCl_3$ 3.0 g, $Na_2EDTA{\cdot}2H_2O$ 500 mg, $MnCl_2{\cdot}4H_2O$ 400 mg, $ZnCl_2$ 300 mg, $CuCl_2{\cdot}2H_2O$ 250 mg, $CoCl_2{\cdot}6H_2O$ 20 mg, $Na_2MoO_4{\cdot}2H_2O$ 10 mg, conc. HCl (37%) 25 mL, and distilled water (or similar quality) 975 mL.

2.2 Bacterial Strain

Mycobacterium neoaurum NRRL B-3805 (CECT3018; DSM 2967; NCIMB 11678): mutant strain derived from *Mycobacterium* sp. NRRL B-3683 by ultraviolet irradiation, which lacked the ability to 1-dehydrogenate steroids. The ancestor of *Mycobacterium* sp. NRRL B-3683 was a soil isolated bacterium able to bioconvert stigmast-4-en-3-one to androsta-1,4-diene-3,17-dione also irradiated with ultraviolet light [36, 40].

2.3 Biological and Chemical Materials

1. Acetic acid (Sigma-Aldrich).
2. Acetone (Sigma-Aldrich).
3. Acrylamide/bis-acrylamide solution (ProtoGel (30%), National Diagnostics).
4. Ammonium persulfate (GE Healthcare).
5. Ammonium sulfate (Sigma-Aldrich).
6. Benzonase (DNase and RNase, Merck).
7. Bradford reagent (Bio-Rad Protein Assay Dye Reagent Concentrate, BioRad).
8. Bromophenol Blue (GE Healthcare).
9. BSA (Bovine Serum Albumin, BioRad).
10. CHAPS (3-[(3-Cholamidopropyl)dimethylammonio]-1-propanesulfonate, GE Healthcare).
11. cOmplete™ Protease Inhibitor Cocktail (Roche).
12. Coomassie Brilliant Blue G-250 (Sigma-Aldrich).
13. DTT (Dithiothreitol, GE Healthcare).
14. Ethanol (absolute alcohol, Sigma-Aldrich).
15. Glycerol (87%, AnalaR NORMAPUR).
16. Glycine (GE Healthcare).
17. Hydrochloric acid (HCl, Scharlau).

18. IPG buffer 3–11 NL (GE Healthcare).
19. Lysing Matrix B (MPbio).
20. Methanol (Sigma-Aldrich).
21. NaCl (Sodium chloride, Sigma-Aldrich).
22. Phosphoric acid (Sigma-Aldrich).
23. Precision Plus Protein Standards (All Blue, Bio-Rad).
24. SDS (Sodium dodecyl sulphate, GE Healthcare).
25. TCA (Trichloroacetic acid, Sigma-Aldrich).
26. TEMED (Tetramethylethylenediamine, GE Healthcare).
27. Tris-base [Tris(hydroxymethyl)aminomethane, Sigma-Aldrich].
28. Urea (GE Healthcare).

2.4 Buffers and Solutions

As a general prevention, the Proteomics solutions should be filtered through 0.45 μm filter to avoid contaminants and interfering particles.

1. BSA master solution: bovine serum albumin in distilled water (0.25 mg/mL).
2. Breaking buffer [same than washing buffer (Tris–HCl 50 mM, pH 7.2) supplemented with protease inhibitor (e.g., cOmplete. Roche)].
3. Colloidal Coomassie solution (blue silver): 0.12% dye (Coomassie Brilliant Blue G-250), 10% ammonium sulfate, 10% phosphoric acid, 20% methanol in distilled water [41].
4. Denaturation buffer: 350 mM Tris–HCl (pH 6.8), Glycerol 30% (v/v), SDS 10% (v/v), 0.6 M DTT, 0.02% Bromophenol blue.
5. Fixation solution: 7% methanol, 7% acetic acid in distilled water.
6. NaCl 0.9%: sodium chloride in distilled water.
7. RB (rehydration) buffer: 8 M Urea, 2% CHAPS in water. Filter through 0.45 μm. Save −20 °C. Prior to use supplement with 1% IPG buffer 3–11 NL (nonlinear) and 60 mM DTT.
8. Running solution (10× Laemmli SDS electrophoresis buffer): 250 mM Tris base, 2 M Glycine, 1% (w/v) SDS.
9. SDS solution: SDS in distilled water (10%). Filter through 0.45 μm.
10. TCA precipitation solution: 20% TCA, 0.14% DTT in acetone.
11. Washing buffer (Tris–HCl 50 mM, pH 7.2).
12. Washing solution: 0.07% DTT in acetone.

2.5 Equipment

1. DALTsix Gel Caster (GE Healthcare).
2. Ettan DALTsix Electrophoresis Unit (GE Healthcare).

3. FastPrep®-24 Instrument (MPbio).
4. 0.45 μm filter membrane and SteriCup device (0.22 μm filter) (Millipore).
5. Micropipette (PIPETMAN P2, P20, P200, P1000, Gilson).
6. Microtubes centrifuge (mod. 5415D, Eppendorf).
7. Mini-PROTEAN® 3 Cell (BioRad).
8. Power supply (PowerPac™ Basic Power Supply, BioRad).
9. Shaking platform (Promax 1020, Heidolph).
10. Spectrophotometer (Double Beam Spectrophotometer U-2900, Hitachi).
11. Thermomixer (Eppendorf ThermoMixer® C, Eppendorf).
12. Tubes and microtubes [50 and 1.5 mL (no used Eppendorf Protein LoBind Tubes), Falcon and Eppendorf, respectively].
13. Tubes centrifuge (mod. Allegra 21R Refrigerated, Beckman Coulter).
14. Vortex shaker.

3 Methods

3.1 Protein Collection: Secretome

The secretome, extra-cellular proteome (or less commonly named as exoproteome) is the rich group of molecules secreted from living cells, which in a broader sense also includes molecules shed from the living cells surface and the cellular machinery involved in the protein secretion [42–44]. The protocol for mycobacterial extra-cellular protein obtaining is as follows (*see* Fig. 1):

1. Grow up the bacteria in 100 mL of (preferable defined) liquid medium, nearly 48 h to ensure the exponential phase (if needed).
2. Harvest 25 mL of culture in a 50 mL tube by centrifugation 10 min at 4 °C and 5000 × *g*.
3. Filter the supernatant [0.22 μm; e.g., SteriCup device] to avoid mycobacterial cells.
4. Add 25 mL of TCA precipitation solution per centrifuge tube with 25 mL of supernatant (*see* **Note 1**). Final TCA percentage is 10%.
5. Keep at −20 °C, overnight (*see* **Note 2**).
6. Centrifuge for 10 min at 4500 × *g* and discard the supernatant. A white (to yellowish) pellet should be observed.
7. Wash the pellet with 10 mL of washing solution. Resuspend the pellet by vortexing (*see* **Note 3**).
8. Centrifuge for 10 min at 4500 × *g*. Discard the supernatant.

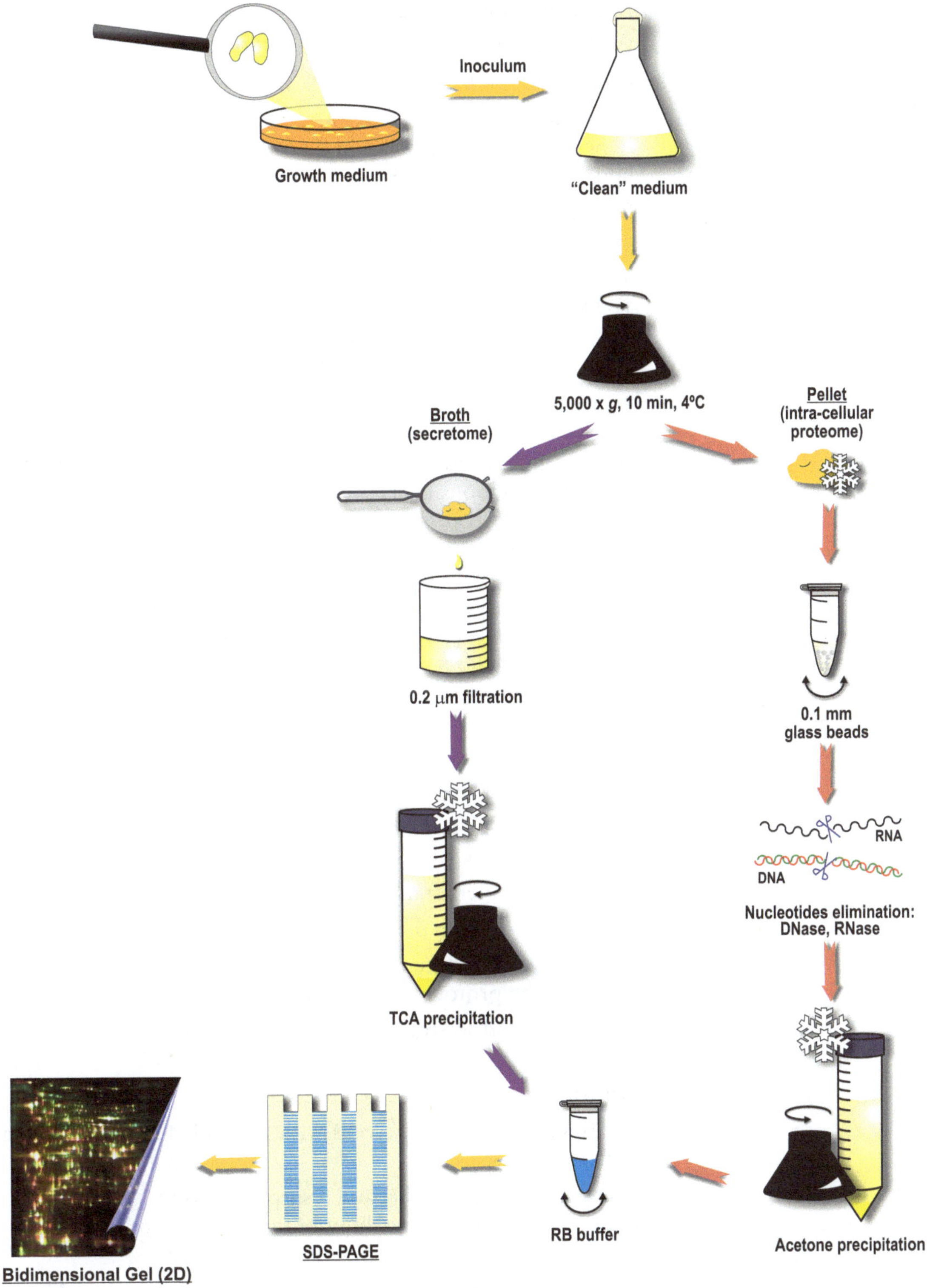

Fig. 1 General overview of the protein extraction methods applied to the analysis of the extra-cellular (*purple arrows*) and intra-cellular (*pink arrows*) proteomes of *M. neoaurum* NRRL B-3805

9. Repeat the washing step.
10. Wash with 10 mL of 80% acetone in distilled water (*see* **Note 4**).
11. Centrifuge for 10 min at 4500 × *g*. Discard the supernatant.
12. Centrifuge for 1 min at 4500 × *g*. Carefully and completely remove the remaining liquid with a micropipette. Let air dry up to 2 min (*see* **Note 5**).
13. Resuspend the protein pellet in 100–200 μL (in function of pellet volume) of RB buffer (*see* **Note 6**).
14. Shake the samples at least 1 h, 20–25 °C (e.g., thermomixer). Transfer the protein solution to 1.5 mL microtube. Centrifuge for 1 min at top speed to discard insoluble proteins (*see* **Note 7**).
15. Transfer the supernatant to a new 1.5 mL microtube.
16. Preserve the samples at −20 °C.

3.2 Protein Collection: Intra-cellular Proteome

The intra-cellular proteome present a snapshot of the microorganism cellular metabolism. Due to the initial harvest steps, the amount of obtained proteins, usually, is higher than the secreted ones, then the precipitation process can be carried out by using higher precipitant volumes (e.g., acetone).

When the proteome analysis consists in the comparison of different nutrient effects (e.g., glycerol vs. sterols), it is recommended to collect the samples in the lag phase to avoid the analysis of proteins from lysis phase. The protocol for mycobacterial intra-cellular protein extraction is as follows (*see* Fig. 1):

1. Grow up the bacteria in 100 mL of (defined) liquid medium, nearly 48 h to ensure the exponential phase (if needed).
2. Collect 5–10 mL of culture by centrifugation for 10 min at 4 °C and 5000 × *g*.
3. Discard the supernatant.
4. Resuspend and wash the cells with NaCl 0.9%, 4 °C (*see* **Note 8**).
5. Centrifuge to collect the cellular pellet. Keep at −20 °C or continue the protocol (*see* **Notes 9** and **10**).
6. Resuspend and wash the cellular pellet with precooled (4 °C) NaCl 0.9%.
7. Centrifuge to collect the cellular pellet.
8. Resuspend and wash the cells with precooled (4 °C) washing buffer (*see* **Note 11**).
9. Harvest the cellular pellet and resuspend in 0.8–1 mL of breaking buffer [same than washing buffer supplemented with protease inhibitor (e.g., cOmplete)] (*see* **Note 12**).
10. Transfer to a FastPrep tube (lysing matrix B) (*see* **Note 13**).
11. Shake three times for 30 s in the Fastprep instrument at top speed (6.5 m/s). Place 1 min on ice among every shacking step (*see* **Note 14**).

12. Centrifuge for 30 min, 12,000 × *g*, 4 °C to discard cell debris.
13. Collect the supernatant (600–700 μL if the initial buffer volume was 800 μL).
14. Add 2 μL of Benzonase (commercial DNase and RNase mixture), 30 min, 37 °C (*see* **Note 15**).
15. Centrifuge for 30 min, 12,000 × *g*, 4 °C.
16. Collect the supernatant. Save at −20 °C or precipitate (*see* **Note 16**).
17. Add 4–9 volumes of cold acetone (−20 °C). Keep at −20 °C overnight (*see* **Note 17**).
18. Centrifuge for 10 min, 12,000 × *g*, 4 °C. A white protein pellet should be observed. An optional washing step with 80% acetone (5 mL) can be included.
19. Discard the supernatant. Air-dry 1 h in a fume hood (acetone smell must disappear).
20. Resuspend in 100–500 μL (in function of pellet volume) of RB buffer.
21. Shake the samples for at least 1 h at R/T (room temperature) by using a thermomixer (20–25 °C) (*see* **Note 7**).
22. Centrifuge for 1 min at 16,100 × *g* (*see* **Note 18**).
23. Transfer the supernatant to a new 1.5 mL microtube.
24. Conserve the protein samples at −20 °C.

3.3 Protein Quantitation by Bradford Method

Different methods and improvements for protein quantitation have been described along the history based on: dyes as Coomassie (Bradford [45]); or copper depending as the well-known "biuret reaction" (Lowry [46]; BCA (bicinchoninic acid) [47]). These are the base for the commercial kits currently available from different suppliers: (1) Pierce 660 nm protein assay (Thermo); (2) Bradford Protein Assay (BioRad); (3) 2-D Quant Kit (GE Healthcare) (see: Thermo Scientific Pierce Protein Assay Technical Handbook: https://tools.thermofisher.com/content/sfs/brochures/1602063-Protein-Assay-Handbook.pdf). The protocol for protein quantitation presented in this chapter is the Bradford method.

1. Prepare the BSA master solution.
2. Following Table 1 distribute the master solution in 1.5 mL microtubes and add distillate water (up to 800 μL) in duplicate.
3. In parallel, prepare the proteins samples as follows: (1) dilution 1 (1:100) 5 μL of proteins:495 μL of distilled water; (2) dilution 2 (1:2000) 40 μL of dilution 1:760 μL distilled water (*see* **Note 19**).
4. Add 200 μL of Bradford reagent. Mix and keep 5 min in the dark.

Table 1
Volumes to reach the standard amounts required to perform the Bradford calibration curve

Proteins (μg)	0	0.5	1.25	2.5	3.75	5.0	6.25	7.5	8.75	10
Master sol. (μL)	0	2	5	10	15	20	25	30	35	40
Distilled water (μL)	800	798	795	790	785	780	775	770	765	760
Bradford reagent	200	200	200	200	200	200	200	200	200	200

5. Mix again and measure the standards and samples absorbance spectrophotometrically at 595 nm.

3.4 Samples Integrity Validation

Some Actinobacteria as *Streptomyces* species can produce colored antibiotics (e.g., actinorhodin, prodigiosins), which remain along the whole protein extraction process. As a result, these and other metabolites can affect the proteins quantitation (under- or over-estimate). Besides, the protein integrity should be evaluated to detect possible protein degradation or heterogeneity among biological replicates, which should be highly similar. Thus, protein visualization in a mono-dimensional (1D) and subsequently in a small 2D gel [48, 49] is strongly recommended to check these parameters and validate the extraction procedures.

1. Clean Mini-PROTEAN® 3 Cell including glass plates, spacers, gel caster, and equipment very thoroughly (*see* **Note 20**). Rinse with distilled water and absolute ethanol. If fluorescence labeling is considered, replace ethanol by some drops of 10% SDS solution (*see* **Note 21**).
2. Place the glass plates (spacers also if no joined to the glasses) in the gel caster following the supplier's instructions (*see* **Note 22**). A 0.75-thickness gel is enough for loading and visualization of 20 μg of proteins per well. When small 2D gels are used, 1 mm is the recommended thickness.
3. Mix the reagents of the running solution (Table 2), with the exception of ammonium persulfate (*see* **Note 23**). Mixture degasification by means of a kitasato and a vacuum pump is always recommended. It is mandatory for large gels as the Ettan Dalt format (21 × 26 cm) (*see* **Note 24**).
4. Immediately before to pour the polyacrylamide solution add the ammonium persulfate (*see* **Note 25**). Shake briefly and pour the solution to the gels. Keep space for the stacking solution and the comb if a 1D gel is being performed. Save a few milliliters as polymerization reporter.
5. Immediately after pouring the solution overlay each gel with a layer of water-saturated 1-butanol, 2-propanol or spray 0.1% SDS/water to minimize gel exposure to oxygen and to create

Table 2
Composition for 10 mL volume of the 1D gel running section of two small gels (mini-protean format)

Reagent	7.5%	12%	15%
Acrylamide/bis-acrylamide 30% (*see* **Note 32**)	2.5 mL	4.0 mL	5.0 mL
Tris–HCl 1.5 M, pH 8.8	2.5 mL	2.5 mL	2.5 mL
SDS 20% (w/v)	50 μL	50 μL	50 μL
Ammonium persulfate 10% (w/v)	50 μL	50 μL	50 μL
TEMED (*see* **Note 33**)	5 μL	5 μL	5 μL
Distilled water	4.9 mL	3.4 mL	2.4 mL

Three acrylamide concentrations are shown in function of the molecular masses of the proteins to be separated (7.5%: 250–50 kDa; 12%: 100–20 kDa; 15%: 60–10 kDa)

Table 3
Composition for 10 mL volume of the stacking section of two 1D gels

Reagent	Volume
Acrylamide/bis-acrylamide 30%	1.3 mL
Tris–HCl 0.5 M, pH 6.8	2.5 mL
SDS 20% (w/v)	50 μL
Ammonium persulfate 10% (w/v)	50 μL
TEMED	10 μL
Distilled water	6.09 mL

a flat gel surface (*see* **Note 26**). For 1D small gels 30–60 min of polymerization are enough. When bigger gels are used overnight polymerization is recommended (*see* **Note 27**).

6. Prepare the stacking solution (Table 3) if needed [the strip of the isoelectrofocusing (IEF, first dimension) acts as staking gel when 2D gels are used].
7. After 30 min to 1 h [in function of gel size and the solidification of the remaining solution out of the gel (reporter)] discard the covering solution (butanol, isopropanol or 0.1% SDS) and add the stacking gel or add running solution (10× Laemmli SDS electrophoresis buffer) if a 2D gel is done and save at 4 °C for a few days.
8. Place the comb for 1D gels over the running gel and fill the gaps with stacking gel solution. Avoid bubbles formation by continuous addition of the solution until the solidification begins. Full polymerization takes 15–20 min.

9. Mix 20 μg per protein sample and 1/10 of denaturation buffer. Boil 5 min. Place on ice, spin down the samples, and load into the gel wells located inside the running chamber (covered by running buffer). When 2D gels are used place the IEF strip laying over the running gel.
10. Apply 100–150 V continuous current along 1 h for small gels [mini-protean format (BioRad)]. When Ettan Dalt format (GE Healthcare) gels are used, 3 W per gel, 45 min and 20 W per gel during 3.5 h are recommended.
11. Run is completed when the bromophenol blue line (from denaturation buffer) reaches the bottom of the gel. Remove the gels from the running chamber, disassemble the glass plates, and proceed to stain. 20 μg of total proteins are easily visualized by using colloidal Coomassie staining [50, 51] (*see* **Note 28**).
12. Fix the proteins (*see* **Note 29**) with fixation solution. 30 min in a shaking platform.
13. Discard the fixation solution. Wash three times 10 min with distilled water.
14. Cover the gels with colloidal Coomassie [41]. Keep overnight (*see* **Note 30**).
15. Save colloidal Coomassie (up to 6 months in the dark). Wash the gel with distilled water until background becomes transparent (1–2 h) (*see* **Note 31**).
16. Test the protein integrity by visual inspection, verify the similarities among biological replicates of the same condition, and validate the Bradford quantitation.

After these protein extraction steps and the subsequent quantification and methods validation (*see* Fig. 2), the protein samples are ready for proteomic analyses (e.g., 2D-DIGE).

4 Notes

1. TCA (trichloroacetic acid) is easily dissolved in acetone at room temperature. The solution should be kept at −20 °C before use. Traditionally, acetone precipitation is recommended in a ratio 4–9 per sample milliliter, which is unapproachable when secretomes are analyzed due to the large volumes used. Thus, TCA precipitation in a vol/vol ratio is preferred.
2. If the microorganism presents high secretion capability protein precipitation can be detected almost immediately and precipitation time can be decreased. On the contrary, if white lumps are not easily observed, keep at −20 °C overnight.
3. TCA is an extremely acidic compound, then washing steps with acetone and subsequently with 80% acetone improve

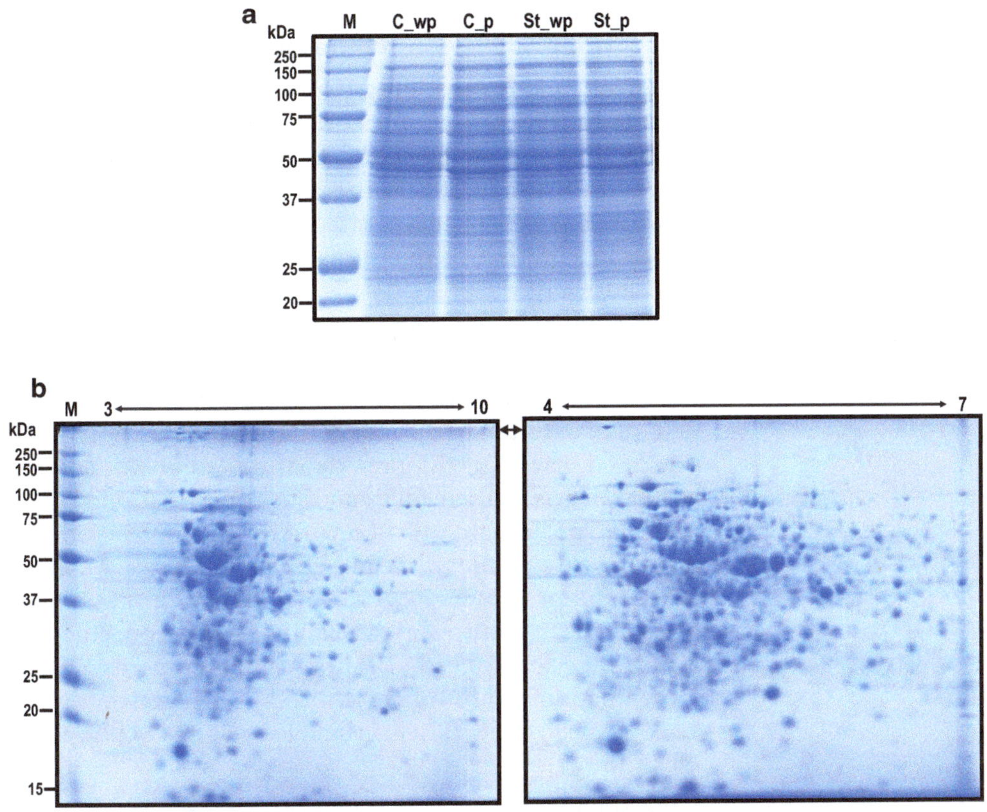

Fig. 2 Polyacrylamide gels for protein extraction methods validation of intra-cellular proteins. (**a**) Monodimensional gel (1D) comparing samples from control (*C*) versus sterols (*St*) supplemented conditions. Besides, the effect of the acetone precipitation was also tested (wp: without precipitation; p: precipitated). (**b**) Bidimensional (2D) gels using a pI range 3–10 NL (non-lineal) (*left*) and 4–7 (*right*). The loaded sample was the same without (*left*) and with (*right*) acetone precipitation. Note the big-size proteins lost (*double arrowhead middle upper part*) when precipitation is carried out, as well as, the dirtiest aspect of the proteins without precipitation in the central section (75–25 kDa) of the gel (*left*). *M* Precision Plus Protein Standards (Bio-Rad)

TCA traces elimination. It is capital in any case to avoid protein acidification, but it is mandatory if DIGE labeling is considered. DIGE optimal labeling conditions are pH 8.5 and 5–10 μg/μL of protein when minimal labeling is carried out. However, if TCA continues in the sample the NaOH amount needed to reach the optimal pH will dilute the sample behind the recommended limits. Thus, a careful removal of the washing solutions also helps in the later pH equilibration.

4. Protein resuspension along the washing steps is not a "real resuspension." It means to detach the pellet from the bottom of the tube, but the proteins continue insoluble in acetone. Then, this "resuspension" is a kind of white (preferable) to yellowish lumps floating in the solution. If proteins became totally

or suddenly soluble in cold acetone solutions something is going wrong (e.g., water solution instead of acetone was used).

5. When acetone is used, long drying period (up to 1 h) can be accepted without disturbing the protein integrity and solubilization. In contrast, when TCA is used the over-dry is a real risk, which deeply affects the subsequent protein solubilization. TCA is a more aggressive precipitant than acetone, which results in a partial precipitation of some protein hardly soluble. If additionally, the protein pellet is over-dried the final number of insoluble proteins could be increased.
6. Concentrated chaotropes as urea (or thiourea and urea combinations) in addition of one (or more) nonionic or zwitterionic detergents are used to enhance the protein solubility. On the one hand, urea and thiourea denature (solubilize and unfold) most proteins to their fully random conformation. Thus, their ionizable groups are exposed to solution. On the other hand, zwitterionic detergents ensure complete sample solubilization preventing protein aggregation through hydrophobic interactions [see: http://www.gelifesciences.com/file_source/GELS/Service%20and%20Support/Documents%20and%20Downloads/Handbooks/pdfs/2-D%20Electrophoresis.pdf; "2-D Electrophoresis: Principles and Methods" (GE Healthcare)].
7. Urea solutions (high concentration), if rise above 30–37 °C, begin to form cyanates (isocyanate), which are detrimental for the proteins by carbamylation, resulting in artifactual "charge trains" [see: "Urea Solution 8 M; Cat. No: U4883" (SIGMA) and "2-D Electrophoresis: Principles and Methods" (GE Healthcare)]. Thus, the protein solubilization should be done by means of shaking periods without any kind of temperature support.
8. Washing steps are capital for intra-cellular proteome analysis in order to avoid growth media interferences. The protein cleanliness influences in the subsequent isoelectrofocusing (IEF) process. High salt concentrations (>50–100 mM) disturb the first dimension.
9. When several samples from different conditions, which interfere in the growth speed, are collected, it is usual to develop the process at different times along the day. Thus, one washing step and a fast frozen process are recommended. Later, when all the samples are saved at −20 °C, a second washing step helps to thaw the samples all together and start the disruption process at the same time.
10. Frozen proteins at −20 °C can be kept for long periods of time. Up to 1 year could be possible. Acetone precipitation solution is also a good preservation solution for long periods of time.

11. Washing and breaking buffers present the same composition to adapt (e.g., osmotic conditions) the cells to the breaking buffer in the last washing step. Note the low salt content (50 mM) to avoid the interference in the IEF. In addition, the pH of the buffer (7.2–7.4) is similar to the typically described as intra-cellular.
12. Proteases may be liberated or activated upon cell disruption, which induce proteolysis. This fact greatly complicates analysis of the 2-D resulting gels. Protein manipulation at low temperatures decreases the proteases activity. In addition, the presence of Tris, sodium carbonate, or basic carrier ampholyte mixtures in the breaking buffers can often inhibit the proteolysis. Anyway, the use of cocktails of protease inhibitors is highly recommended to inhibit those recalcitrant proteases. Several inhibitors are commercially available. Along this work the inhibitor used was: cOmplete™ Protease Inhibitor Cocktail.
13. Matrix lysis is composed by glass beads (silica spheres). A 0.1 mm diameter is quite convenient for the protein extraction in Actinobacteria.
14. Due to the violent movement of the Fastprep apparatus, it is highly recommended to keep the samples on ice among breaking steps. It avoids the protein degradation by the system overheating.
15. Nucleic acids interfere the 2D gels development because of: (1) cause smear, (2) increase sample viscosity, (3) stick to the proteins disturbing focusing, (4) if silver staining is used, the nucleic acids are also detected [for detailed description see the manual from GE Healthcare: "2-D Electrophoresis. Principles and Methods"]. Thus, one step of nucleic acids degradation is highly recommended. Benzonase from Merck, a cocktail of DNase plus RNase, allows an effective viscosity reduction and removal of nucleic acids.
16. If possible, it is better to avoid protein precipitation since every additional step means protein lost. But it is highly recommended to test both options (with and without precipitation) when a new procedure is updated. Sometimes, the number of proteins lost is recompensed by the better resolution of the 2D gels. Usually, these lost proteins correspond to high molecular mass spots. The validation of the protocol by 1D gel to test protein integrity and later by mini 2D gels [7-cm strips (IEF) in a mini-protean format gel (SDS PAGE: 8 × 6 cm)] is enough to detect protein lost and resolution variations.
17. In contrast to the secretome (*see* **Note 1**), the intra-cellular proteome can be precipitated by addition of acetone. The cells are concentrated in a small volume after centrifugation and the final breaking product is also small. Acetone precipitation

allows an easy protein pellet resuspension in contrast to the TCA, which always presents insoluble proteins.

18. After protein resuspension a short centrifuge spin is recommended to collect those insoluble particles. Usually, none particles are visible after acetone precipitation.
19. The yield of the described methods for protein extraction ranges from 2 to 15 μg/μL. Thus, the suggested dilutions allow a proper measurement in the linear section of the standard calibration curve. Based on the experience, concentrations below 1 μg/μL do not generate fully trustable measures by Bradford method.
20. It is recommended to work with SDS-PAGE using the cautions described for mass spectrometry, even though the final application of the gel is just for visualization. Also for 1D gel only high-quality chemicals must be used. Every insignificant mistake in an 8 × 6 cm gel is strongly magnified in a 2D(-DIGE) gel (21 × 26 cm). In addition, current mass spectrometers are extremely sensitive showing up every contamination, which affects the final result and subsequent protein/modification identifications. Keratin stemming from hair, pet hair, clothes wool, and skin (typical nude finger pointing out to the protein band in the gel along the destaining process) are the most frequently contaminant sources. Thus, it is a recommended practice to filter each monomer and buffer solution through a membrane filter before use. Powder-free gloves and tissue are mandatory.
21. When the room temperature is higher than 20–25 °C, the gel polymerization, as well as the previous mixture, should be done using precooled reagents (e.g., water, acrylamide, tris buffers except the SDS, which became solid at low temperatures) in a cold room or a refrigerator. In contrast to the agarose gels, the polymerization of acrylamide run slow at low temperatures, which allow better polymerization. High temperarture of room, reagents, or glass plates cause extremely fast and irregular acrylamide polymerization.
22. When the handling of multiple gels is required, the use of glass plates carrying the spacers is a good option.
23. Ammonium persulfate is recommended to be prepared immediately prior to use (see: 2-D Electrophoresis. Principles and Methods), but some laboratory traditions support the stock conserved at −20 °C for long periods of time.
24. Oxygen can cause incomplete gel polymerization, then the oxygen removing from the gel environment by 5–10 min of degas using a kitasato and a vacuum pump is highly recommended.

25. Polyacrylamide polymerization process starts with ammonium persulfate as catalyst, whereas TEMED is the tertiary amino groups provider to release the radicals. Once the gel reagents mixture is completed, the shaken done and the degasification process (if considered) is finished, the ammonium persulfate can be added. Then, just around 10 min are available before the polymerization process shows its effects [52]. Proceed as fast as possible in a safety compatible way.
26. Westermeier and coworkers [52] supported the 0.1% SDS spraying as the best option to reach straight gel surfaces. They explain it due to "very small detergent solution droplets glide down on the glass surface very quickly and will not mix with the monomer solution," because they are very small and have a low density' in contrast to pipetting water-saturated butanol or isopropanol, since they have a high affinity to the vinyl spacer material and can cause bent surfaces. Anyway, the alcoholic solutions are worldwide used with successful results.
27. The polyacrylamide gel polymerization has a fast solidification step, which is well known, but a slow silent polymerization also occurs along several hours to complete the final matrix formation. Incomplete gel polymerization cause high background in the mass spectrogram. Thus, the gels should be used not before than 2 h of polymerization or overnight saved in the refrigerator [52].
28. From the Coomassie discovery as wool stain in textile industry to the current protein stain use, different adaptations have been carried out to reach the colloidal Coomassie dye employed nowadays (see complete methods review on "Coomassie Blue staining of polyacrylamide gels" in Westermeier, 2006 [53]).
29. Proteins denaturalization and precipitation in the gel matrix is the aim of the fixation. It prevents protein diffusion, which enhances the band resolution, as well as, removes buffer components as SDS (see details at: https://www.nationaldiagnostics.com/electrophoresis/article/protein-fixation-gels). In order to diminish the ester formation in the fixation buffer, by means of the well-known chemical reactions acid plus alcohol is equal to ester plus water, both acetic acid and methanol at 7% can be used, which decreases the final ester production.
30. Silver Coomassie is an endpoint staining described by Candiano and coworkers [41]. It is an effective, low reagents consuming and fast staining protocol. The destaining step using water saves reagents because it does not need the traditional ethanol and acetic acid in this solution. Besides, thin gels (e.g., 0.75 mm) after 10 min of staining can be observed to test the proteins isolation. The gel background is totally transparent after water destaining.

31. The higher number of washing steps, the faster gel destain.
32. Bis-acrylamide (*N*,*N*'-methylene-bisacrylamide) allows the polymerization process (copolymerization of acrylamide and bis-acrylamide) mediated by ammonium persulfate and TEMED. It is the responsible of the random crosslink of the elongating polymer chains. This results in the characteristic porosity that depends on the monomer concentrations and polymerization conditions.
33. TEMED is subject to oxidation, which changes its color from transparent to yellowish decreasing the catalytic activity. The practical result is greater amounts required for an adequate polymerization [see Tech note no. 1156 from BioRad: Acrylamide Polymerization—A Practical Approach (http://www.bio-rad.com/webroot/web/pdf/lsr/literature/Bulletin_1156.pdf)]. Thus, the TEMED bottle renovation once per year is strongly suggested.

Acknowledgments

This work was fully supported by a grant of the European Union program ERA-IB [MySterI (EIB.12.010)] through the APCIN call of the Spanish Ministry of Economy and Competitiveness (MINECO, Spain) (PCIN-2013-024-C02-01). The authors want to thank the European Union program ERA-IB; the Spanish Ministry of Economy and Competitiveness (MINECO, Spain); and the MySterI Consortium (INBIOTEC, Pharmins Ltd., University of York, SINTEF, Technische Universität Dortmund and Gadea Biopharma S.L.). We thank J. Merino, B. Martín, and A. Casenave for their excellent technical assistance and the degree students of the group A. Ibáñez and A. del Árbol.

References

1. Bott M (2007) Corynebacteria: the good guys and the bad guys. Microbiol Today 34:74–77
2. Barreiro C, Martín JF (2015) *Corynebacterium* genus, a dual group of clinical and industrial relevant bacteria. In: Barreiro C (ed) New trends in *Corynebacterium glutamicum*: beyond the amino acids. Nova Science Publishers, Inc, New York
3. Shtratnikova VY, Bragin EY, Dovbnya DV et al (2014) Complete genome sequence of sterol-transforming *Mycobacterium neoaurum* strain VKM Ac-1815D. Genome Announc 2:12–13
4. Tortoli E (2014) Microbiological features and clinical relevance of new species of the genus *Mycobacterium*. Clin Microbiol Rev 27:727–752
5. Euzéby JP. List of Prokaryotic names with standing in nomenclature. http://www.bacterio.net/
6. Goodfellow M, Kämpfer P, Busse HJ, et al. (2012) Bergey's manual of systematic bacteriology. The actinobacteria, part B, vol 5. Springer, New York.
7. Cole ST, Brosch R, Parkhill J et al (1998) Deciphering the biology of *Mycobacterium tuberculosis* from the complete genome sequence. Nature 393:537–544
8. Mattow J, Siejak F, Hagens K et al (2009) Two-dimensional gel electrophoresis-based proteomics of mycobacteria. Methods Mol Biol (Clifton, NJ) 465:111–142

9. Uhía I, Galán B, Kendall SL et al (2012) Cholesterol metabolism in *Mycobacterium smegmatis*. Environ Microbiol Rep 4:168–182
10. Liu M, Zhu ZT, Tao XY et al (2016) RNA-Seq analysis uncovers non-coding small RNA system of *Mycobacterium neoaurum* in the metabolism of sterols to accumulate steroid intermediates. Microb Cell Fact 15:64
11. World Health Organisation (WHO) (2014) Global tuberculosis report 2014
12. Korb VC, Chuturgoon AA, Moodley D (2016) *Mycobacterium tuberculosis*: manipulator of protective immunity. Int J Mol Sci 17:131
13. Kumar B (2015) World Leprosy Day 2015: Renewing commitment for a leprosy free world! Indian J Med Res 141(1):1–4
14. Wasinger VC, Cordwell SJ, Cerpa-Poljak A et al (1995) Progress with gene-product mapping of the Mollicutes: *Mycoplasma genitalium*. Electrophoresis 16:1090–1094
15. Wilkins MR, Sanchez JC, Gooley A et al (1996) Progress with proteome projects: why all proteins expressed by a genome should be identified and how to do it. Biotechnol Genet Eng Rev 13:19–50
16. Wilkins MR, Pasquali C, Appel RD et al (1996) From proteins to proteomes: large scale protein identification by two-dimensional electrophoresis and amino acid analysis. Biotechnology 14:61–65
17. James P (1997) Protein identification in the post-genome era: the rapid rise of proteomics. Q Rev Biophys 30:279–331
18. Barreiro C (2015) Methods in proteomics applied to *Corynebacterium glutamicum*. In: Barreiro C (ed) New trends in *Corynebacterium glutamicum*: beyond the amino acids. Nova Science Publishers, Inc, New York
19. Bisht D, Singhal N, Sharma P et al (2006) Analysis of mycobacterial strains by two-dimensional gel electrophoresis. J Commun Dis 38:255–262
20. Betts JC, Dodson P, Quan S et al (2000) Comparison of the proteome of *Mycobacterium tuberculosis* strain H37Rv with clinical isolate CDC 1551. Microbiology 146:3205–3216
21. Jungblut PR, Schaible UE, Mollenkopf HJ et al (1999) Comparative proteome analysis of *Mycobacterium tuberculosis* and *Mycobacterium bovis* BCG strains: towards functional genomics of microbial pathogens. Mol Microbiol 33:1103–1117
22. Kruh NA, Troudt J, Izzo A et al (2010) Portrait of a pathogen: the *Mycobacterium tuberculosis* proteome *in vivo*. PLoS One 5:e13938
23. Schmidt F, Donahoe S, Hagens K et al (2004) Complementary analysis of the *Mycobacterium tuberculosis* proteome by two-dimensional electrophoresis and isotope-coded affinity tag technology. Mol Cell Proteomics 3:24–42
24. Gupta MK, Subramanian V, Yadav JS (2009) Immunoproteomic identification of secretory and subcellular protein antigens and functional evaluation of the secretome fraction of *Mycobacterium immunogenum* a newly recognized species of the *Mycobacterium chelonae–Mycobacterium abscessus* group. J Proteome Res 8:2319–2330
25. Kruh-Garcia NA, Murray M, Prucha JG et al (2014) Antigen 85 variation across lineages of *Mycobacterium tuberculosis*—implications for vaccine and biomarker success. J Proteomics 97:141–150
26. Mattow J, Schaible UE, Schmidt F et al (2003) Comparative proteome analysis of culture supernatant proteins from virulent *Mycobacterium tuberculosis* H37Rv and attenuated *M. bovis* BCG Copenhagen. Electrophoresis 24:3405–3420
27. Ranganathan S, Garg G (2009) Secretome: clues into pathogen infection and clinical applications. Genome Med 1:113
28. Yadav JS, Gupta M (2012) Secretome differences between the taxonomically related but clinically differing mycobacterial species *Mycobacterium abscessus* and *M. chelonae*. J Integr OMICS 2(2):64–79
29. Mollenkopf HJ, Jungblut PR, Raupach B et al (1999) A dynamic two-dimensional polyacrylamide gel electrophoresis database: the mycobacterial proteome via Internet. Electrophoresis 20:2172–2180
30. He Z, De Buck J (2010) Cell wall proteome analysis of *Mycobacterium smegmatis* strain MC2 155. BMC Microbiol 10:121
31. Sih CJ, Wang KC (1965) A new route to estrone from sterol. J Am Chem Soc 87:1387–1388
32. Al Jasem Y, Khan M, Taha A et al (2014) Preparation of steroidal hormones with an emphasis on transformations of phytosterols and cholesterol—a review. Mediterr J Chem 3:796–830
33. Donova MV, Egorova OV (2012) Microbial steroid transformations: current state and prospects. Appl Microbiol Biotechnol 94:1423–1447
34. Wang F, Yao K, Wei D (2011) From soybean phytosterols to steroid hormones. In: El-Shemy H (ed) Soybean and health. InTech, Rijeka, Croatia, pp 231–252

35. Bragin EY, Shtratnikova VY, Dovbnya DV et al (2013) Comparative analysis of genes encoding key steroid core oxidation enzymes in fast-growing *Mycobacterium* spp. strains. J Steroid Biochem Mol Biol 138:41–53
36. Rodríguez-García A, Fernández-Alegre E, Morales A et al (2016) Complete genome sequence of "*Mycobacterium neoaurum*" NRRL B-3805, an androstenedione (AD) producer for industrial biotransformation of sterols. J Biotechnol 224:64–65
37. Barreiro C, Martín JF, García-Estrada C (2012) Proteomics methodology applied to the analysis of filamentous fungi—new trends for an impressive diverse group of organisms. In: Prasain JK (ed) Tandem mass spectrometry—applications and principles. InTech, Rijeka, Croatia
38. Jeffery CJ (2014) An introduction to protein moonlighting. Biochem Soc Trans 42:1679–1683
39. Jeffery CJ (2009) Moonlighting proteins—an update. Mol Biosyst 5:345–350
40. Marsheck WJ, Kraychy S, Muir RD (1972) Microbial degradation of sterols. Appl Microbiol 23:72–77
41. Candiano G, Bruschi M, Musante L et al (2004) Blue silver: a very sensitive colloidal Coomassie G-250 staining for proteome analysis. Electrophoresis 25(9):1327–1333
42. Greenbaum D, Luscombe NM, Jansen R et al (2001) Interrelating different types of genomic data, from proteome to secretome: 'oming in on function. Genome Res 11:1463–1468
43. Makridakis M, Roubelakis MG, Vlahou A (2013) Stem cells: insights into the secretome. Biochim Biophys Acta 1834:2380–2384
44. Tjalsma H, Bolhuis A, Jongbloed JD et al (2000) Signal peptide-dependent protein transport in *Bacillus subtilis*: a genome-based survey of the secretome. Microbiol Mol Biol Rev 64:515–547
45. Bradford MM (1976) A rapid and sensitive method for the quantitation of microgram quantities of protein utilizing the principle of protein-dye binding. Anal Biochem 72:248–254
46. Lowry OH, Rosebrough NJ, Farr AL et al (1951) Protein measurement with the Folin phenol reagent. J Biol Chem 193:265–275
47. Smith PK, Krohn RI, Hermanson GT et al (1985) Measurement of protein using bicinchoninic acid. Anal Biochem 150:76–85
48. Laemmli UK (1970) Cleavage of structural proteins during the assembly of the head of bacteriophage T4. Nature 227:680–685
49. O'Farrell PH (1975) High resolution two-dimensional electrophoresis of proteins. J Biol Chem 250:4007–4021
50. Fazekas de St Groth S, Webster RG, Datyner A (1963) Two new staining procedures for quantitative estimation of proteins on electrophoretic strips. Biochim Biophys Acta 71:377–391
51. Meyer TS, Lamberts BL (1965) Use of coomassie brilliant blue R250 for the electrophoresis of microgram quantities of parotid saliva proteins on acrylamide-gel strips. Biochim Biophys Acta 107:144–145
52. Westermeier R, Naven T, Hopker HR (2008) Proteomics in practice: a guide to successful experimental design, 2nd Completely Revised edn. WILEY-VCH Verlag GmbH & Co. KGaA, Weinheim
53. Westermeier R (2006) Sensitive, quantitative, and fast modifications for Coomassie Blue staining of polyacrylamide gels. Proteomics 6 Suppl 2:61–64

Chapter 7

Genetic Techniques for Manipulation of the Phytosterol Biotransformation Strain *Mycobacterium neoaurum* NRRL B-3805

Jessica K. Loraine and Margaret C.M. Smith

Abstract

Mycobacterium neoaurum is a saprophytic, soil-dwelling bacterium. The strain NRRL B-3805 converts phytosterols to androst-4-ene-3,17-dione (androstenedione; AD), a precursor of multiple C19 steroids of importance to industry. NRRL B-3805 itself is able to convert AD to other steroid products, including testosterone (Ts) and androst-1,4-diene-3,17-dione (androstadienedione; ADD). However to improve this strain for industrial use, genetic modification is a priority. In this chapter, we describe a range of genetic techniques that can be used for *M. neoaurum* NRRL B-3805. Methods for transformation, expression, and gene knockouts are presented as well as plasmid maintenance and stability.

Key words *Mycobacterium*, Steroids, Transformation, Plasmids, Mutagenesis

1 Introduction

As knowledge of the roles and uses of steroids in healthcare emerged in the 1950s, the demand for steroid drugs increased. Microbial strains were identified that could convert sterols to high value steroids and therefore a route to industrial scale production of steroids through bioconversion [1]. However, the industrial application of microbial sterol biotransformation was not widely adopted due to problems with the solubility of starting materials, microbial strain development, and low product yields [2, 3]. Recent advances in molecular biology and genetics of industrial microbes, as well as the availability of cheap phytosterols from agricultural waste, have led to increased interest in this area [4].

Mycobacterial sterol catabolism has been widely investigated in several Actinobacteria (others include *Rhodococcus* and *Nocardia* species) that are able to catabolize sterols and produce steroidal intermediates [4–7]. Mycobacteria are highly efficient producers of androst-4-ene-3,17-dione (androstenedione; AD) and

José-Luis Barredo, Ignacio Herráiz (eds.), *Microbial Steroids: Methods and Protocols*, Methods in Molecular Biology vol. 1645, DOI 10.1007/978-1-4939-7183-1_7, © Springer Science+Business Media LLC 2017

androst-1,4-diene-3,17-dione (androstadienedione; ADD) as they selectively degrade the 17-alkyl side chain of sterols, leaving the steroid nucleus intact. A screening programme identified a soil *Mycobacterium* that could convert a plant sterol, stig-mast-4-en-3--one, to ADD. This strain was subjected to UV mutagenesis to produce NRRL B-3805 that had enhanced biotransformation activity and produced, as its major product, the steroid AD, a more useful precursor for several valuable C19 steroids than ADD [8]. *Mycobacterium neoaurum* NRRL B-3805 is now considered to be the benchmark bioconversion strain for the production of AD from plant sterols [9]. Clearly, a desirable goal is to further enhance the biotransformation capabilities of this strain by genetic manipulation.

1.1 Growth Media for M. neoaurum NRRL B-3805

The growth medium for *M. neoaurum* has been described by others [10]. Previous studies have used a variety of carbon sources for mycobacterial biotransformation of phytosterols, including glycerol, glucose and the dissolving agent (such as Tween 80) [11–13]. Increasing the concentration of glucose in the biotransformation media was found to affect oxidation and reduction reactions; Liu and Lo demonstrated that the presence of glucose favors reduction and can inhibit oxidation [13]. Using testosterone (Ts) as the starting material they showed that an increasing glucose concentration reduced the biotransformation of Ts but, once glucose is consumed, Ts is oxidized to AD [13]. Glycerol is an attractive carbon source for industrial biotransformation as it is a by-product from biodiesel production; however, Marques et al. found that *M. neoaurum* NRRL B-3805 needed to be adapted to use waste glycerol as a carbon source [11]. In our studies we tested a variety of solvents for the addition of sterol to media, settling on the use of Tween80/dimethyl formamide (DMF). We also tested whether supplementing Tween80 with addition of carbon sources, glucose, and glycerol would affect bioconversion.

1.2 Overview of Genetic Techniques for Strain Development

Mycobacterium smegmatis is the most genetically amenable species in the *Mycobacterium* genus. Our approach was to examine the extent to which we could transfer the existing vectors, promoters, and mutagenesis technologies to *M. neoaurum* NRRL B-3805. The first stage was to be able to grow NRRL B-3805 and to analyze the biotransformation at laboratory scale (*see* Subheading 3.1). We then profiled innate antibiotic resistance in NRRL B-3805 so that we could choose the most appropriate antibiotic resistance markers for selection of transformants (*see* Subheadings 3.3–3.5). The stability of freely replicating (high and low copy variants of pAL5000) and integrating plasmids could then be determined (*see* Subheading 3.3).

Strain improvement requires reliable promoters and we have used the constitutive *hsp60* promoter from *M. bovis*. However, inducible promoters have a number of advantages including the ability to switch on genes that might be detrimental to growth only when required during the biotransformation stage of fermentation. The original inducible expression system developed for mycobacteria utilized the acetamide-dependent regulon from *M. smegmatis* [14, 15]. More recently, researchers have used a nitrile responsive promoter, *nitA*, regulated by the product of *nitR* [16] and a tetracycline-inducible expression system. The tetracycline-inducible system permitted conditional mutant generation and led to higher gene expression than the acetamide system [17]. Each system has advantages and disadvantages but a common theme is the need for a regulator (e.g., *tetR, nitR*) or a resistance gene (e.g., thiostrepton resistance to allow use of the *tipA* promoter) or a large promoter region (e.g., acetamide regulated promoter) being present on the vector leading to unwieldy plasmid sizes (*see* ref. 18 for discussion).

Here, we describe the use of plasmids encoding a controllable promoter, originally created for use in *Streptomyces,* and new to mycobacteria. The plasmids contain an inducible synthetic, riboswitch expression system utilizing the constitutive *ermE** promoter and the *gusA* reporter gene [19]. The synthetic riboswitches are theophylline-dependent, allowing conditional dose-dependent gene expression [19]. Only a short sequence of approximately 85 nucleotides, placed between a transcriptional start site and the start codon, is required for regulation [19, 20]. Addition of the theophylline ligand, which binds to the aptamer, causes a structural rearrangement opening up the Shine-Delgarno (SD) sequence and start codon [19, 20].

Possibly, the most important technique that we describe in this chapter is the targeted mutagenesis technology. Recently, recombineering technology has been developed for *M. smegmatis* [21] and we attempted to transfer this to *M. neoaurum* NRRL B-3805. While we were able to generate knockout mutations easily in *M. smegmatis*, the recombineering failed with NRRL B-3805. Instead, we use homologous recombination as a method for unmarked mutant generation in combination with counter selection using *sacB* from *Bacillus subtilis* [22].

2 Materials

2.1 Media Optimization for Laboratory Scale Biotransformation

1. *Mycobacterium neoaurum* strain NRRL B-3805.
2. M3-Tw medium: dissolve 0.5 g of KH_2PO_4, 0.5 g of K_2HPO_4, 1.5 g of $(NH_4)_2HPO_4$, 0.005 g of $FeSO_4{\cdot}7H_2O$, 0.002 g of $ZnSO_4{\cdot}7H_2O$, 0.2 g of $MgSO_4{\cdot}7H_2O$, 5 g of yeast extract and 5 g of glycerol and make up to 1 L with distilled (d)H_2O, adjust pH to 7.2 and autoclave. Supplement with 0.05% (v/v) filter sterilized Tween 80.

3. Bioconversion medium: In 400 mL dH_2O dissolve 0.5 g of urea, 3 g of $(NH_4)_2HPO_4$, 4 g of KH_2PO_4, and 1 g of K_2HPO_4; in 50 mL dH_2O dissolve 0.2 g of $MgSO_4{\cdot}7H_2O$, 0.01 g of $FeSO_4{\cdot}7H_2O$ and 0.002 g of $ZnSO_4{\cdot}7H_2O$. Combine solutions, adjust pH to 7.0, and make up to 1 L with dH_2O before autoclaving. Glycerol or glucose should be added to a final concentration of 0.5% (w/v), prior to sterilization (if required).
4. Cholesterol stock solution: for supplementation of bioconversion media; prepared by adding cholesterol (100 g/L) to a 1:1 solution of Tween 80 and DMF and heated to 80 °C until dissolved.
5. Bioconversion medium with added cholesterol: prepared by adding 0.01 volume of the cholesterol stock solution to sterile bioconversion media to give a final concentration of 1 g/L cholesterol. In growth experiments without cholesterol, add solvent only (Tween 80/DMF, 0.01 volume).
6. Spectrophotometer to read at 600 nm.
7. 1 mL spectrophotometer cuvettes.
8. Thin-layer chromatography (TLC) tank.
9. TLC plates: X25 ALUGRAM sheets SIL G/UV254 (Fisher).
10. Filter paper.
11. Chloroform.
12. Diethyl ether.
13. Oven, able to reach 110 °C.
14. Developer spray: 1 M H_2SO_4.
15. Fume Hood.

2.2 Measurement of Antibiotic Minimum Inhibitory Concentrations

1. M3-Tw medium.
2. 50 mL sterile Falcon tubes.
3. Apramycin sulfate stock solution 20 mg/mL prepared in sterile double distilled $(dd)H_2O$.
4. Gentamicin sulfate stock solution 10 mg/mL prepared in sterile double distilled $(dd)H_2O$.
5. Streptomycin sulfate stock solution 20 mg/mL prepared in sterile double distilled $(dd)H_2O$.
6. Kanamycin stock solution 50 mg/mL prepared in sterile double distilled $(dd)H_2O$.
7. Rifampicin stock solution 50 mg/mL prepared in DMSO.
8. Hygromycin B stock solution 50 mg/mL (Thermo fisher scientific).
9. 96-well plates.

Table 1
Plasmids for genetic techniques in *M. neoaurum* NRRL B-3805

Plasmid	Description	Reference
pJL001	pSMT3 containing *ermE** promoter, E* riboswitch and *gusA*	J. Loraine, unpublished
pJL002	pSMT3 containing *hsp60* promoter, E* riboswitch and *gusA*	J. Loraine, unpublished
pJL003	pSMT3 containing *sacB* gene	J. Loraine, unpublished
pJL020	pGEM7 containing *sacB* and *hygR* marker	J. Loraine, unpublished
pSMT3	*E. coli/Mycobacterium* shuttle vector, *hygR*, *hsp60* promoter, low copy number	[23]
pSMT3-M	*E. coli/Mycobacterium* shuttle vector, *hygR*, *hsp60* promoter, high copy number	[24]
pGEM7	*E. coli* cloning vector, *ampR*	Promega

2.3 Electroporation, Plasmid Maintenance, and Stability

1. M3-Tw medium.
2. M3 medium: dissolve 0.5 g of KH_2PO_4, 0.5 g of K_2HPO_4, 1.5 g of $(NH_4)_2HPO_4$, 0.005 g of $FeSO_4{\cdot}7H_2O$, 0.002 g of $ZnSO_4{\cdot}7H_2O$, 0.2 g of $MgSO_4{\cdot}7H_2O$, 5 g of yeast extract and 5 g of glycerol and make up to 1 L with distilled (d)H_2O, adjust pH to 7.2 and autoclave.
3. Glycerol 10% (w/v), sterilized by autoclaving.
4. Electroporation apparatus.
5. Electroporation cuvettes with 2 mm gap electrodes.
6. Purified plasmid DNA (e.g., pSMT3 and pSMT3-M; Table 1) prepared using a kit (e.g., Qiaprep® miniprep).
7. Lysozyme solution (10 mg/mL) dissolved in cell resuspension buffer.
8. Cell resuspension buffer: 50 mM Tris pH 8.0, 10 mM EDTA, and 100 μg/mL RNAse A.
9. M3 agar plates: M3 media containing 18 g/L agar and autoclaved.
10. M3-Tw-hyg: M3-Tw media supplemented 50 μg/mL hygromycin.
11. M3-hyg agar plates: M3 agar supplemented with 100 μg/mL hygromycin.

2.4 Promoter Activity and Reporter Gene Expression

1. Purified plasmid DNA of pJL001 and pJL002 (Table 1).
2. Electroporation apparatus.
3. Electrocompetent B-3805 cells.

4. M3-hyg plates.
5. M3-Tw-Hyg medium.
6. Reporter plates: M3-hyg agar containing 2 mM theophylline (from a stock solution dissolved in 0.1 M NaOH) and 80 μg/mL X-glucuronide (dissolved in DMSO).

2.5 Generation of Knockout Mutants by Homologous Recombination and Counter Selection

1. Purified plasmid DNA pJL003 (Table 1).
2. Purified plasmid DNA pJL020 (Table 1).
3. Primers to amplify 1 kb upstream and 1 kb downstream of the gene to be targeted.
4. Electroporation apparatus.
5. Electrocompetent NRRL B-3805 cells.
6. Recovery medium: M3-Tw.
7. M3-hyg plates.
8. 50 mL Falcon tubes.
9. M3-Tw medium.
10. M3-Tw-hyg medium.
11. M3-suc plates: M3 agar, cooled to approximately 55 °C and 5% (v/v) sucrose added from a filter sterilized 50% (w/v) stock solution.
12. DIG-High Prime DNA Labeling and Detection Starter Kit II (Roche).

2.6 Purification of Genomic DNA from *M. neoaurum* NRRL B-3805

1. 50 mL Falcon tubes.
2. M3-Tw medium.
3. Tris-EDTA (TE) buffer: 10 mM Tris-Cl (pH 8.0) and 1 mM EDTA, sterilized by autoclaving.
4. Chloroform/methanol (2:1 v/v).
5. 1 M Tris-base, pH 9.0, sterilized by autoclaving.
6. Lysozyme (10 mg/mL) stock solution in dH_2O, filter sterilized.
7. DNase-free RNase (500 μg/mL).
8. SDS solution in dH_2O 10% (w/v).
9. Proteinase K stock solution (10 mg/mL) in dH_2O.
10. Phenol/chloroform/isoamyl alcohol (25:24:1).
11. 3 M sodium acetate, pH 5.2, filter sterilized.
12. Isopropanol.
13. Ice-cold 70% (v/v) ethanol.
14. DNase-free dH_2O.
15. Bench top and high-speed centrifuges.
16. 1.5 mL tubes.
17. Heat block.

3 Methods

3.1 Media Optimization for Laboratory Scale Biotransformation

1. Inoculate 5 mL cultures with single colonies of *M. neoaurum* B-3805 in M3-Tw medium. Grow until stationary phase at 30 °C with 180 rpm shaking (approximately 48 h).
2. Centrifuge (12,000 × *g*) the stationary phase culture to pellet the cells. Remove the culture supernatant and resuspend cells in 10 mL bioconversion medium (*see* **Note 1**).
3. Prepare duplicate 10 mL tubes of bioconversion media containing glycerol or glucose or without any additional carbon source and add cholesterol solution to one set and solvent to the other.
4. Inoculate each tube with 10% (v/v) washed stationary phase *M. neoaurum* B-3805 from **step 2**. Grow as in **step 1**.
5. Take 800 μL aliquots for OD measurements (600 nm) at multiple time points (Fig. 1a) (*see* **Note 2**).
6. For TLC analysis take 1 mL samples and extract with an equal volume of chloroform (*see* **Note 3**).
7. Recover the organic phase (it will be the lower phase).

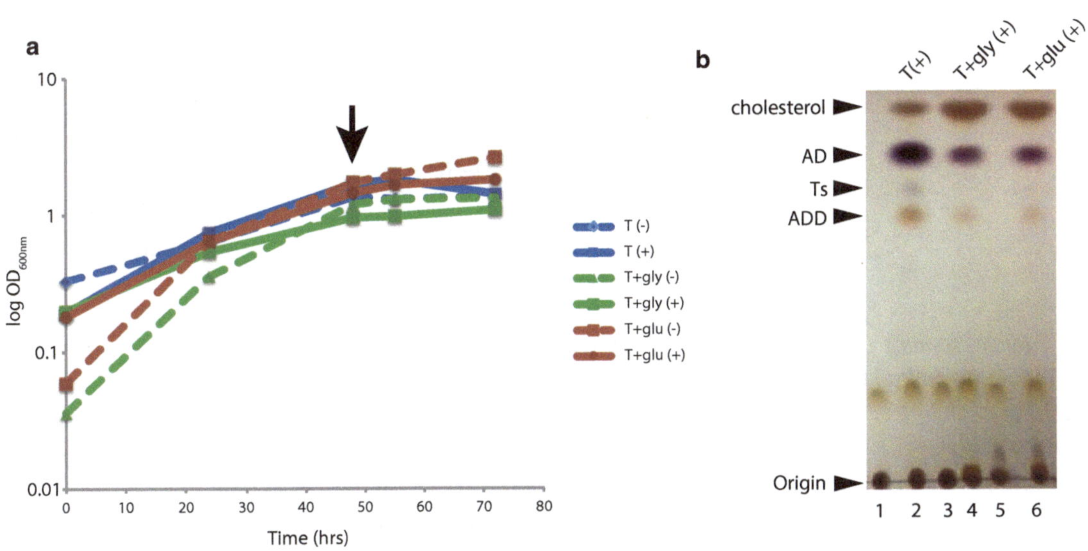

Fig. 1 Investigating sterol bioconversion in the presence of an added carbon source. (**a**) Growth curves for *M. neoaurum* NRRL B-3805 in bioconversion medium with and without cholesterol (indicated by +/–) and either solvent (T) or solvent with glycerol (T + gly) or solvent with glucose (T + glu). Samples were taken for TLC analysis at 48 h (*arrow*). (**b**) TLC of extracted steroids from the 48 h time point from (**a**). Cholesterol, androstenedione (AD), androstenedienedione (ADD), and testosterone (Ts) are indicated. Lanes contained extracts of cultures from (**a**) without added cholesterol (*lanes 1*, *3*, and *5* from cultures T(–), T + gly(–), and T + glu(–), respectively) and with added cholesterol (*lanes 2*, *4*, and *6* from cultures T(+), T + gly(+), and T + glu(+), respectively). The TLC plate was sprayed with 1 M H_2SO_4 and baked at 110 °C for 15 min. Biotransformation with Tween80 and no addition carbon source provided the better conditions for steroid production

Table 2
Antibiotic concentrations for selection with *M. neoaurum* NRRL B-3805

Antibiotic	MIC (μg/mL)	Concentration used in B3805 (μg/mL)	Concentration used in *E. coli* (μg/mL)
Kanamycin	0.5–1	50	50
Streptomycin	0.2–0.5	20	–
Gentamycin	1–2	10	10
Rifampicin	10–20	–	–
Hygromycin	10–20	50–150	100
Apramycin	0.2–0.5	30	30

8. Saturate the TLC tank by placing a piece of filter paper in the solvent system and letting it stand prior to inserting the TLC plate.
9. Spot 2 μL aliquots up to a total volume of 30 μL of the recovered organic phase onto a TLC plate, letting each aliquot air dry before spotting the next. Allow the loaded TLC plate to air dry and then place it in a tank containing diethyl ether (*see* **Note 4**).
10. Remove the TLC plate from the tank once the solvent front has almost reached the top of the plate; using fresh solvent this takes approximately 60–90 min.
11. Air dry the plate and carefully spray with developer spray, remove from the fume hood, and place in an oven at 110 °C for 15 min to develop (Fig. 1b) (*see* **Note 5**).

3.2 Measurement of Antibiotic Minimum Inhibitory Concentrations

1. Take a fresh stationary phase culture of *M. neoaurum* B-3805 and use a 50 μL inoculum to grow 5 mL cultures of *M. neoaurum* B-3805 in M3-Tw medium. Grow at 30 °C with 180 rpm shaking overnight. Standardize the OD (600 nm) of all cultures to 1 and prepare 10^{-1} and 10^{-2} dilutions.
2. Prepare 96-well plates with 100 μL M3-Tw medium per well. Supplement columns with different concentrations of antibiotic (*see* **Note 6**).
3. Inoculate rows with 10 μL of 10^{-1} or 10^{-2} dilutions of inoculum. Incubate the plate for 72 h, at 30 °C with 180 rpm shaking.
4. Inspect the plates for growth versus no growth. The MIC is the lowest antibiotic concentration at which no growth was observed (Table 2).

3.3 Electroporation and Plasmid Stability in *M. neoaurum* NRRL B-3805

1. This method is adapted from Goude and Parish [25]. Grow 5 mL precultures of *M. neoaurum* NRRL B-3805 in M3-Tw at 30 °C with 180 rpm shaking until stationary phase (approximately 48 h).
2. Inoculate 50 mL cultures with 0.5 mL stationary phase preculture. Incubate cultures at 30 °C with 180 rpm shaking until OD (600 nm) = 0.8–1 (approx. 16–24 h). Place cells on ice for 1.5 h.
3. Harvest bacteria by centrifugation (3000 × *g*, 10 min, 4 °C). Wash cells three times with ice-cold sterile 10% glycerol. For the first wash use 25% of the original culture volume, second wash use 10% of the volume, and final wash use 5%.
4. Resuspend in 1/10–1/100 of original volume in cold 10% glycerol. Store at −80 °C or use immediately for electroporation.
5. For electroporation, use 2 mm cuvettes (*see* **Note 7**) with 100 μL aliquots of electrocompetent cells mixed well with no more than 5 μL of DNA (0.5–5 μg; *see* **Note 8**). Immediately electroporate cells at 1000 Ω and 2500 V.
6. Remove cells to 1 mL M3-Tw medium and incubate at 30 °C for 3 h with 180 rpm shaking for recovery.
7. Plate out the cells onto M3 agar plates containing the appropriate antibiotic selection. For suicide plasmid or non-replicating plasmid selection, centrifuge the cells, resuspend in 100 μL M3 medium, and inoculate the entire transformation mix onto selective plates.
8. Determine the transformation efficiencies using a series of plasmid concentrations (1, 10, and 100 ng). Typical transformation efficiencies for B3805 are 10^4–10^5 transformants/μg DNA.
9. As some plasmids are prone to deletions in mycobacteria, plasmid structural stability should be examined for each construct. Grow transformants of NRRL B-3805 to stationary phase in M3-Tw medium (supplemented with appropriate antibiotic) at 30 °C with 180 rpm shaking (approximately 48 h) prior to plasmid purification using a commercially available plasmid purification kit (e.g., Qiagen®). This is performed according to manufacturer's instructions with the additional step for lysozyme (10 mg/mL) treatment (4 h, 37 °C; *see* **Note 9**).
10. Use purified plasmid DNA extracted from NRRL B-3805 to transform *E. coli* DH5α. Prepare plasmid minipreps from *E. coli* transformants and compare restriction fragment lengths to the original plasmid construct.
11. Determine plasmid maintenance, e.g., by the low and high copy number mycobacterial shuttle vectors pSMT3 and

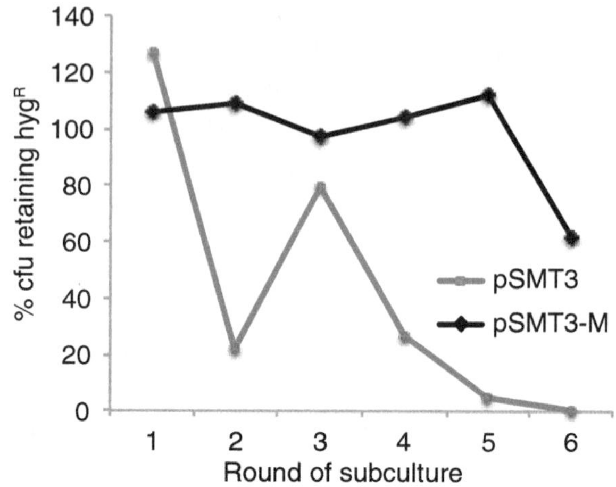

Fig. 2 Plasmid segregational instability in *M. neoaurum* NRRL B-3805 in the absence of selection. Initial transformants were cultured and grown on agar without selection, then subcultured to stationary phase and the % of cfu/mL retaining hygromycin resistance determined by plating on agar with and without hygromycin. Subculture was repeated for six rounds. The low copy number version of pSMT3 was lost rapidly in the absence of selection, whereas the high copy number version, pSMT3-M, was more stable

pSMT3-M, respectively, by growing transformants first in the presence of selection to stationary phase (in M3-Tw-hyg media, incubated at 30 °C with 180 rpm shaking for approximately 48 h).

12. Subculture (1/100 dilution) into 5 mL M3-Tw in the absence of hygromycin and into 5 mL M3-Tw-hyg media to stationary phase (48 h, approximately).
13. Prepare serial dilutions and plate 100 μL on M3 and M3 + hyg plates to count colony forming units (cfu) per mL.
14. Pick single colonies from M3 plates (without selection) and repeat **steps 12–14**. Calculate percentage of hygromycin-resistant cfu/mL to total cfu/mL (plated without hygromycin) after each round. Plasmid loss after several rounds of growth without selection (an effect exaggerated in the low copy plasmid; Fig. 2) indicates that antibiotic selection is a requirement throughout *M. neoaurum* B-3805 biotransformation.

3.4 Promoter Activity and Reporter Gene Expression

The plasmids pJL001 and pJL002 (Table 1) were used to test the activity of a promoter, *ermE** (originally from *Streptomyces*), located upstream of a synthetic, theophylline inducible riboswitch, E*, and a reporter gene, *gusA* [19]. Plasmid pJL001 contains the *ermE** promoter, E* and *gusA* from the pGusT_ermE_E* plasmid [19] inserted into the low copy mycobacterial shuttle vector pSMT3, replacing the *hsp60* promoter (Fig. 3a). pJL002 contains the *hsp60* promoter reading into E* and *gusA* in pSMT3.

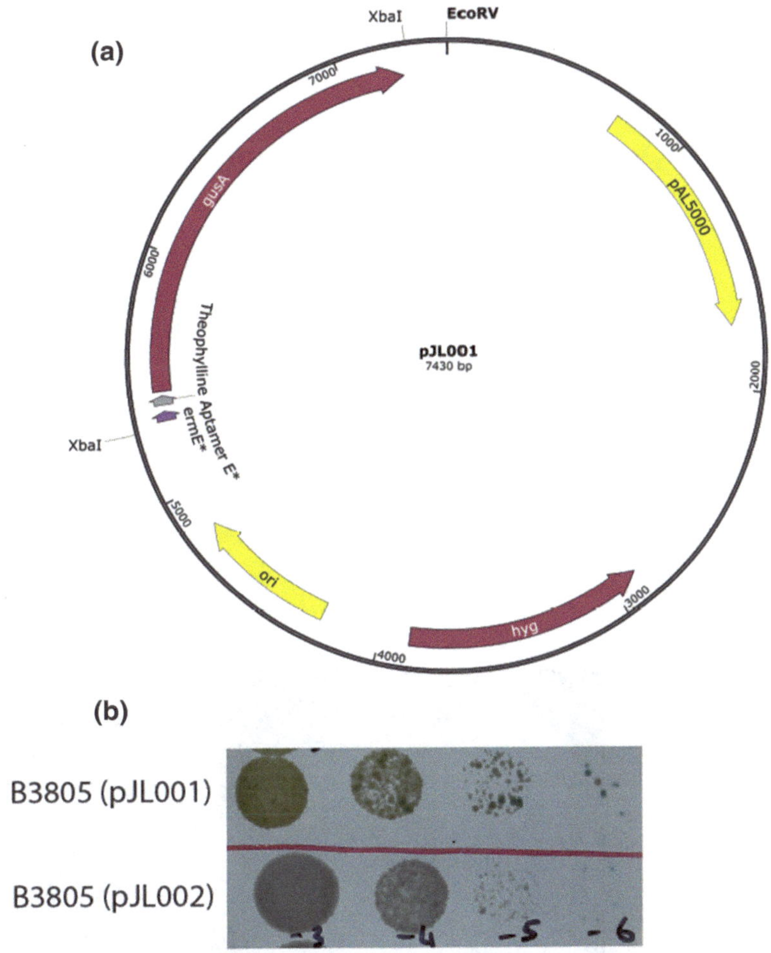

Fig. 3 Plasmid pJL001 was used to test whether the *ermE*_E*riboswitch-*gusA* reporter cassette was functional in *M. neoaurum* NRRL-B3805. (**a**) Map of the plasmid pJL001 containing *ermE_E*_gusA*. pJL002 has the *hsp60* promoter (in place of the *ermE* promoter) reading into *E*_gusA*. (**b**) Plate containing dilutions of *M. neoaurum* NRRL-B3805 (pJL001) or (pJL002) spotted onto medium containing theophylline and X-Gluc. The *blue staining* is indicative of expression of the *gusA* gene in NRRL-B3805 (pJL001), indicating that the *ermE** promoter is active in the presence of theophylline

1. Electroporate competent *M. neoaurum* NRRL B-3805 cells with either pJL001 or pJL002, and select for transformants on M3-hyg plates.
2. Screen transformants for a plasmid encoded gene by colony polymerase chain reaction (PCR) (*see* **Note 10**).
3. Grow positive transformants of NRRL B-3805 (pJL001) and NRRL B-3805 (pJL002) in 5 mL of M3-Tw-hyg medium at 30 °C with 180 rpm shaking (for 48 h).

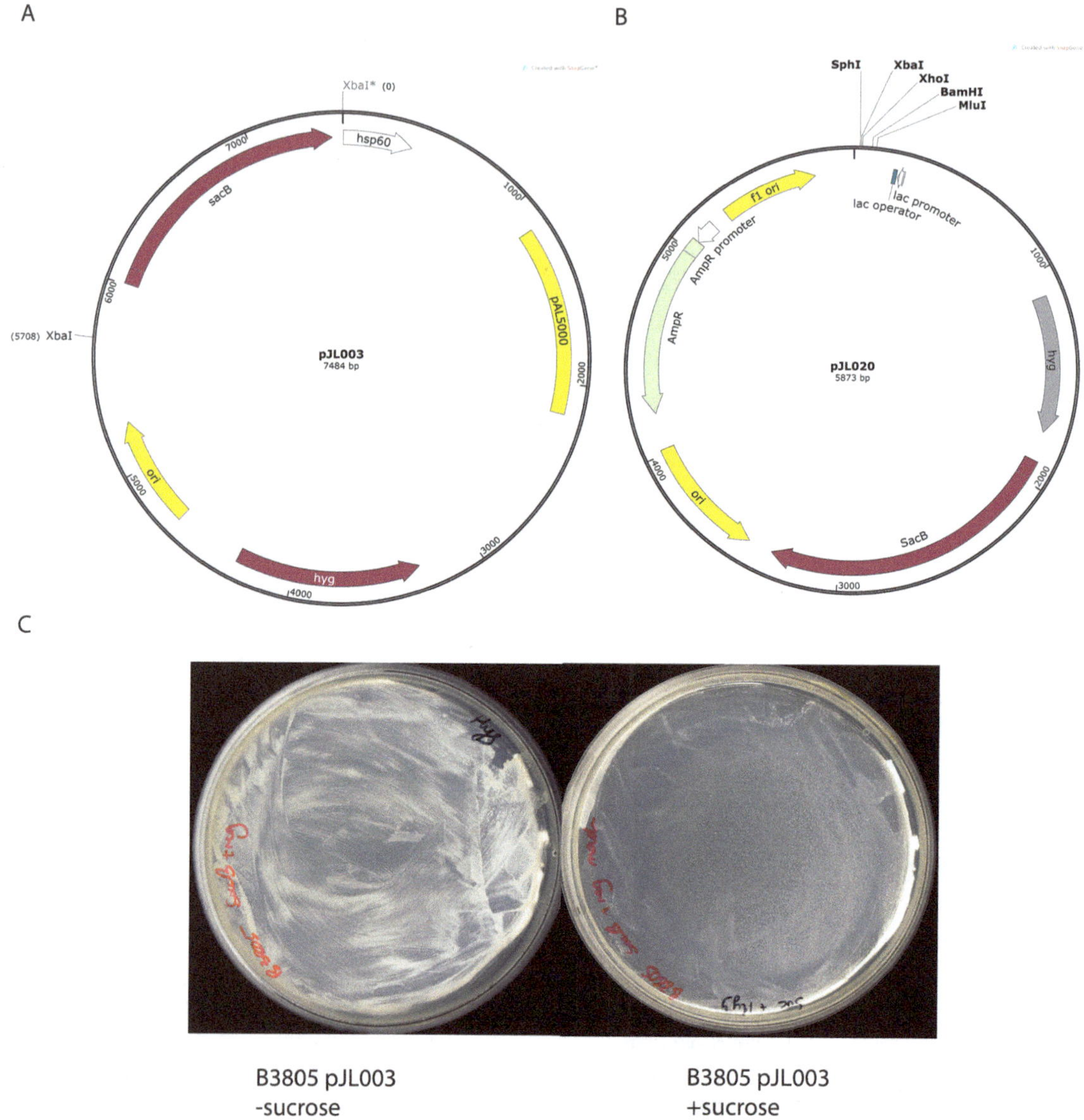

Fig. 4 Use of the *sacB* gene as a counterselectable marker in *M. neoaurum* NRRL B-3805. (**a**) Plasmid pJL003 contains the *sacB* gene from *B. subtilis* in pSMT3. (**b**) Plasmid pJL020 is a vector designed for the generation of markerless knockout mutants in *M. neoaurum* NRRL B-3805. (**c**) *M. neoaurum* NRRL B-3805 (pJL003) fails to grow on plates containing sucrose (+), while the same culture grows well on medium without (–) sucrose indicating that the *sacB* can be used as a reliable counter selectable marker

4. Prepare serial dilutions of cultures and spot 10 μL aliquots on reporter plates. Incubate plates at 30 °C for up to 5 days (Fig. 3b).

3.5 Generation of Knockout Mutants by Homologous Recombination and Counter Selection

Plasmid pJL003 contains the *sacB* gene (from *Bacillus subtilis*) inserted in pSMT3 (at the *XbaI* site; Fig. 4a) and is used to test the culture conditions for counterselection. pJL020 was constructed to provide a vector for the generation of unmarked gene knockouts (Fig. 4b). pJL020 is derived from *E. coli* vector pGEM®-7Zf(−), contains *hygR* and the counter selectable marker *sacB* but no origin of replication for *Mycobacterium*. As expected, *M. neoaurum* NRRL B-3805 (pJL003) fails to grow on medium containing sucrose (Fig. 4c).

1. To construct a plasmid that can be used to create a markerless deletion of a chromosomal gene of interest, amplify (using PCR) regions of DNA from upstream and downstream of the target gene using genomic DNA as the template (*see* Subheading 3.6) and primers designed to contain short overlapping identical ends (*see* **Note 11**).
2. Perform an overlap extension PCR using purified PCR products of the upstream and downstream regions (*see* **Note 12**).
3. Clone the PCR product containing the joined upstream and downstream regions of the gene of interest from **step 2** into linearized pJL020 (e.g., *BamHI*, Fig. 4b).
4. Electroporate *M. neoaurum* NRRL B-3805 (*see* Subheading 3.3) with 1–5 μg of purified plasmid DNA (*see* **Note 13**).
5. Transfer to the recovery medium and incubate at 30 °C with 180 rpm shaking for 3 h. Plate out the entire transformation mix onto M3-hyg plates and incubate at 30 °C to isolate clones that have undergone a single crossover (SCO).
6. Grow up transformants in M3-Tw media with no antibiotics and plate out serial dilutions (10^0 to 10^{-2}) onto M3 plates and incubate until large colonies appear.
7. From the M3 plates pick approximately 100 single colonies and plate onto M3-suc plates. Incubate for up to 7 days. The sucrose-resistant colonies are potential double cross overs (DCOs) containing the desired gene deletion (*see* **Note 14**).
8. Screen the potential DCOs by colony PCR (*see* **Note 10**) and Southern blotting (DIG-High Prime DNA Labeling and Detection Starter Kit II).

3.6 Purification of Genomic DNA from M. neoaurum B-3805

1. This method is adapted from Belisle et al. [26]. Inoculate 5 mL of M3-Tw medium with *M. neoaurum* B-3805 and incubate at 30 °C with 180 rpm (for 48 h).
2. In a 50 mL Falcon inoculate 20 mL M3-Tw medium with 200 μL of the stationary phase preculture. Grow as in **step 1**.
3. Harvest the cells by centrifugation at 3000 × *g* for 20 min.
4. Discard the supernatant and resuspend the pellet in 2 mL TE buffer.
5. Repeat the centrifugation and discard the supernatant again, freeze the pellet at −80 °C for a minimum of 4 h or overnight.

6. Thaw cells and resuspend in 1 mL TE buffer.
7. Add an equal volume of chloroform/methanol (2:1) and mix at room temperature for 1 h.
8. Centrifuge the suspension at 3000 × *g* for 20 min to separate the phases, carefully discard both phases leaving the bacterial pellet behind.
9. Place the uncapped tube at 55 °C for up to 20 min to remove residual solvent.
10. Add 1 mL TE buffer and vortex vigorously until resuspended.
11. Add 0.1 volume of 1 M Tris-base (pH 9.0), 0.01 volume lysozyme stock solution and incubate overnight (12–16 h) at 37 °C.
12. Add 10 μL DNase-free RNase solution, mix and incubate at 37 °C for 30 min.
13. Add 0.1 volume 10% SDS solution and 0.01 volume of Proteinase K stock solution to the lysate. Mix by inversion and incubate at 55 °C for 3 h.
14. Add an equal volume of phenol/chloroform/isoamyl alcohol (25:24:1) and mix gently for 30 min before centrifugation at 12,000 × *g* for 30 min at room temperature.
15. Transfer the aqueous layer to a new tube and remove residual phenol by adding an equal volume of chloroform, mix for 10 min and centrifuge as before. Transfer the aqueous layer to a new tube.
16. To precipitate the DNA add 0.1 volume 3 M sodium acetate (pH 5.2) and 1 volume isopropanol to the aqueous layer and mix by inversion. Place at 4 °C for a minimum of 1 h.
17. Centrifuge the solution at 12,000 × *g* for 30 min at 4 °C to pellet the DNA. Remove the supernatant and wash the pellet with ice-cold 70% ethanol.
18. Centrifuge as before and remove the ethanol. Allow the pellet to air dry before dissolving the pellet in 300–500 μL TE or DNase-free dH_2O.

4 Notes

1. Removal of the supernatant prevents transfer of M3 media to the biotransformation culture.
2. As the cultures become denser, take 200 μL samples and dilute with 600 μL of the appropriate medium.
3. Vortex until homogenous mixture produced. Centrifuge in a bench top centrifuge to collect cell debris. Perform **steps 6–11** in a fume hood.

4. Solvent may be reused for additional runs.
5. Sterols and steroids should become visible. Some compounds are visible without acid treatment under UV light.
6. The antibiotics kanamycin, streptomycin, rifampicin, hygromycin, and apramycin MICs were analyzed at the following concentrations (μg/mL): 0, 0.1, 0.15, 0.2, 0.5, 1, 2, 5, 10, 20, 30, and 50, while gentamycin concentrations (μg/mL) were 0, 0.02, 0.03, 0.04, 0.1, 0.2, 0.4, 1, 2, 6, and 10.
7. Precool the electroporation cuvettes by incubation on ice for 30 min.
8. Higher concentrations of DNA are required for experiments using suicide vectors for gene knockouts (*see* Subheading 3.5).
9. Lysozyme treatment should be performed in resuspension buffer.
10. Use a pipette tip to resuspend cells in 100 μL distilled water and heat at 95 °C for 5 min, spin down and use supernatant for PCR reactions.
11. Use flanking regions of 1 kb in length to generate the substrates for homologous recombination events upstream and downstream of the gene to be deleted. Overlap PCR will join the upstream and downstream flanking regions to create a single PCR product that can be cloned into pJL020. The primers for each flanking region should have a minimum of 18 bp of identical sequence (the overlap) to mediate joining.
12. Perform two rounds of PCR, the first round should be approximately 10 cycles and should contain no primers. The second round should be performed with primers that will amplify the joined product and should be approximately 20 cycles.
13. UV treatment of DNA with 100 mJ/cm^2 may increase homologous recombination.
14. *sacB* spontaneous mutations arise with a similar frequency to desired deletion generated through the second recombination event, therefore screening a large number of colonies is necessary.

Acknowledgments

We are grateful to Beatrix Suess for donating the plasmid pGUS_ermE_E* and to the Biotechnology and Biological Sciences Research Council, UK, for funding (project BB/L003619).

References

1. Bhatti HN, Khera RA (2012) Biological transformations of steroidal compounds: a review. Steroids 77(12):1267–1290
2. Wei W et al (2010) Inactivation and augmentation of the primary 3-ketosteroid-{delta}1-dehydrogenase in Mycobacterium neoaurum NwIB-01: biotransformation of soybean phytosterols to 4-androstene-3,17-dione or 1,4-androstadiene-3,17-dione. Appl Environ Microbiol 76(13):4578–4582
3. Perez C et al (2006) A very efficient bioconversion of soybean phytosterols mixtures to androstanes by mycobacteria. J Ind Microbiol Biotechnol 33(8):719–723
4. Van der Geize R et al (2007) A gene cluster encoding cholesterol catabolism in a soil actinomycete provides insight into Mycobacterium tuberculosis survival in macrophages. Proc Natl Acad Sci U S A 104(6):1947–1952
5. Shao M et al (2015) Enhanced production of androst-1,4-diene-3,17-dione by mycobacterium neoaurum JC-12 using three-stage fermentation strategy. PLoS One 10(9):e0137658
6. Fernandes P et al (2003) Microbial conversion of steroid compounds: recent developments. Enzym Microb Technol 32(6):688–705
7. Bergstrand LH et al (2016) Delineation of steroid-degrading microorganisms through comparative genomic analysis. MBio 7(2): e00166
8. Marsheck WJ, Kraychy S, Muir RD (1972) Microbial degradation of sterols. Appl Microbiol 23(1):72–77
9. Rodríguez-García A et al (2016) Complete genome sequence of 'Mycobacterium neoaurum' NRRL B-3805, an androstenedione (AD) producer for industrial biotransformation of sterols. J Biotechnol 224:64–65
10. Bragin EY et al (2013) Comparative analysis of genes encoding key steroid core oxidation enzymes in fast-growing Mycobacterium spp. strains. J Steroid Biochem Mol Biol 138:41–53
11. Marques MPC et al (2010) Steroid bioconversion: towards green processes. Food Bioprod Process 88(C1):12–20
12. Llanes N et al (1995) Glucose and lactose effect on ad and add bioconversion by Mycobacterium Sp. Biotechnol Lett 17(11):1237–1240
13. Liu WH, Lo CK (1997) Production of testosterone from cholesterol using a single-step microbial transformation of Mycobacterium sp. J Ind Microbiol Biotechnol 19(4):269–272
14. Parish T et al (1997) Regulation of the inducible acetamidase gene of Mycobacterium smegmatis. Microbiology 143:2267–2276
15. Brown AC, Parish T (2006) Instability of the acetamide-inducible expression vector pJAM2 in Mycobacterium tuberculosis. Plasmid 55(1): 81–86
16. Pandey AK et al (2009) Nitrile-inducible gene expression in mycobacteria. Tuberculosis 89(1):12–16
17. Carroll P, Muttucumaru DGN, Parish T (2005) Use of a tetracycline-inducible system for conditional expression in Mycobacterium tuberculosis and Mycobacterium smegmatis. Appl Environ Microbiol 71(6):3077–3084
18. Sassetti CM (2008) Inducible expression systems for mycobacteria. In: Parish T, Brown AC (eds) Mycobacteria protocols. Humana Press, New York, pp 255–264
19. Rudolph MM, Vockenhuber MP, Suess B (2013) Synthetic riboswitches for the conditional control of gene expression in Streptomyces coelicolor. Microbiology 159:1416–1422
20. Suess B et al (2004) A theophylline responsive riboswitch based on helix slipping controls gene expression in vivo. Nucleic Acids Res 32(4):1610–1614
21. van Kessel JC, Hatfull GF (2008) Efficient point mutagenesis in mycobacteria using single-stranded DNA recombineering: characterization of antimycobacterial drug targets. Mol Microbiol 67(5):1094–1107
22. Pelicic V, Reyrat JM, Gicquel B (1996) Generation of unmarked directed mutations in mycobacteria, using sucrose counter-selectable suicide vectors. Mol Microbiol 20(5):919–925
23. Carroll P et al (2010) Sensitive detection of gene expression in mycobacteria under replicating and non-replicating conditions using optimized far-red reporters. PLoS One 5(3):e9823
24. Herrmann JL et al (1996) Bacterial glycoproteins: a link between glycosylation and proteolytic cleavage of a 19 kDa antigen from Mycobacterium tuberculosis. EMBO J 15(14):3547–3554
25. Goude R, Parish T (2008) Electroporation of mycobacteria. In: Parish T, Brown AC (eds) Mycobacteria protocols. Humana Press, New York, pp 203–215
26. Belisle JT, Mahaffey SB, Hill PJ (2008) Isolation of Mycobacterium species genomic DNA. In: Parish T, Brown AC (eds) Mycobacteria protocols. Humana Press, New York, pp 1–12

Chapter 8

Identification and Characterization of the Genes and Enzymes Belonging to the Bile Acid Catabolic Pathway in *Pseudomonas*

José M. Luengo and Elías R. Olivera

Abstract

The study of the catabolic potential of microbial species isolated from different habitats has allowed the identification and characterization of bacteria able to assimilate bile acids and other steroids (e.g., testosterone and 4-androsten-3,17-dione). From soil samples, we have isolated several strains belonging to genus *Pseudomonas* that grow efficiently in chemical defined media containing some cyclopentane-perhydro-phenantrene derivatives as carbon sources. Genetic and biochemical studies performed with one of these bacteria (*P. putida* DOC21) allowed the identification of the genes and enzymes belonging to the 9,10-*seco* pathway, the route involved in the aerobic assimilation of steroids. In this manuscript, we describe the most relevant methods required for (1) isolation and characterization of these species; (2) determining the chromosomal location, nucleotide sequence, and functional analysis of the catabolic genes (or gene clusters) encoding the enzymes from this pathway; and (3) the tools employed to establish the role of some of the proteins that participate in this route.

Key words *Pseudomonas*, Catabolism, Cholic acid, Chenodeoxycholic acid, Lithocholic acid, Deoxycholic acid, 4-Androstene-3,17-dione (AD), 1,4-Androstadiene-3,17-dione (ADD), Tn5 mutagenesis, Genome editing, Acyl-CoA synthetase, 3-Ketosteroid dehydrogenase

1 Introduction

Bile acids (BA) are the main end products of cholesterol metabolism in animals. Those compounds, which are produced exclusively in the liver, represent about 90% of the actively metabolized cholesterol in the body; the remaining 10% of the cholesterol is used for steroid hormone biosynthesis. In humans, the final products of liver synthesis (primary bile acids, PBA) are cholic acid (CA) and chenodeoxycholic acid (CDCA). Both CA and CDCA are synthesized from cholesterol through complex processes that involve hydroxylation, saturation of the double bond at C5-C6, epimerization of the 3-hydroxyl group (from β configuration in cholesterol to α configuration in bile acids), and oxidative cleavage of a

José-Luis Barredo, Ignacio Herráiz (eds.), *Microbial Steroids: Methods and Protocols*, Methods in Molecular Biology vol. 1645, DOI 10.1007/978-1-4939-7183-1_8, © Springer Science+Business Media LLC 2017

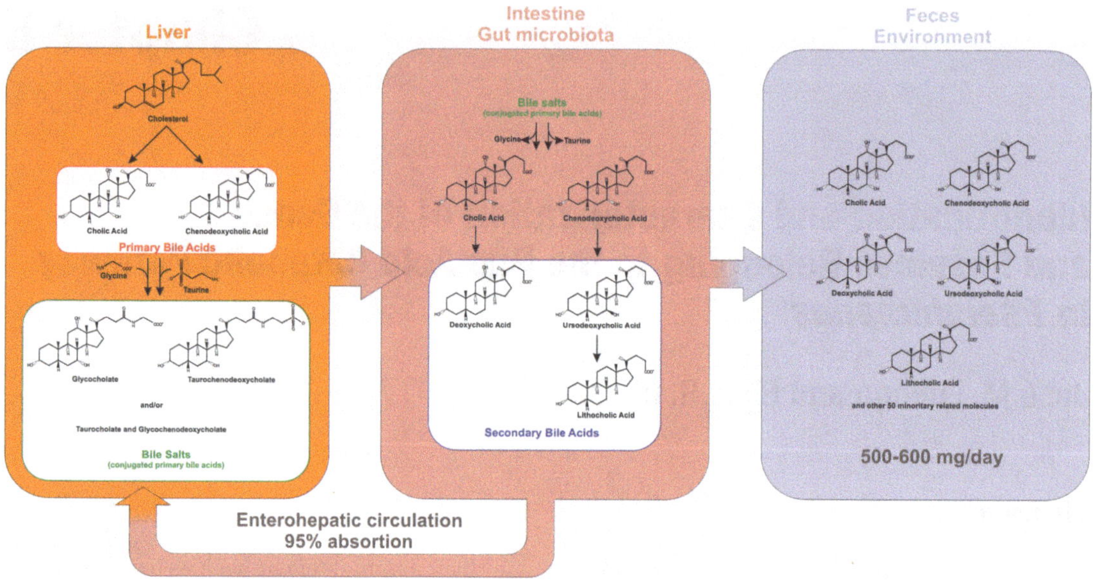

Fig. 1 Schematic representation of the metabolism of bile acids in the human being: Hepatic synthesis from cholesterol of PBA (CA and CDCA), and their glycine and taurine conjugates (BS); enterohepatic circulation recycling (95% of the synthesized BS) and intestinal transformation of BS in SBA (DCA, LCA and UDCA) by gut microbiota; and release on feces (500–600 mg/day) of the most representative bile acids (CA, CDCA, DCA, LCA, and UDCA)

3-carbon unit from the lateral chain at C17 [1, 2] (*see* Fig. 1). In the liver, the final metabolic process consists in the condensation of PBA [3] with glycine or taurine, to generate bile salts (BS) [4] (*see* Fig. 1). The BS stored in the gallbladder are liberated into the duodenum as a result of food ingesta. They act as biological detergents and emulsifiers, thus improving the absorption of fats and lipid soluble vitamins (vitamins A, D, E, and K) along the small intestine tract [2]. CA and CDCA conjugates are mainly reabsorbed from ileum (>95%) and, later, they enter in the portal vein going back to the liver, to be recycled. This process is known as enterohepatic circulation [5]. However, 3–5% of BS reach the colon (400–800 mg/day) and interact with the intestinal microbiota [6]. Then, BS are deconjugated by bacterial hydrolases, releasing free PBA which are converted by gut bacterial enzymes in secondary bile acids (SBA; deoxycholic acid, DCA, lithocholic acid, LCA, and ursodeoxycholic acid, UDCA) (*see* Fig. 1). Part of SBA are reabsorbed through passive diffusion in the upper small intestine and in the colon, being recycled through the enterohepatic circulation. However, approximately, 5% of BA are not reabsorbed from the small intestine. Thus, as much as 600 mg of bile acids/day are secreted into the feces. Although DCA and LCA are the most abundant SBA detected in human feces, more than 50 different secondary bile acids have been found [7]. Therefore, due

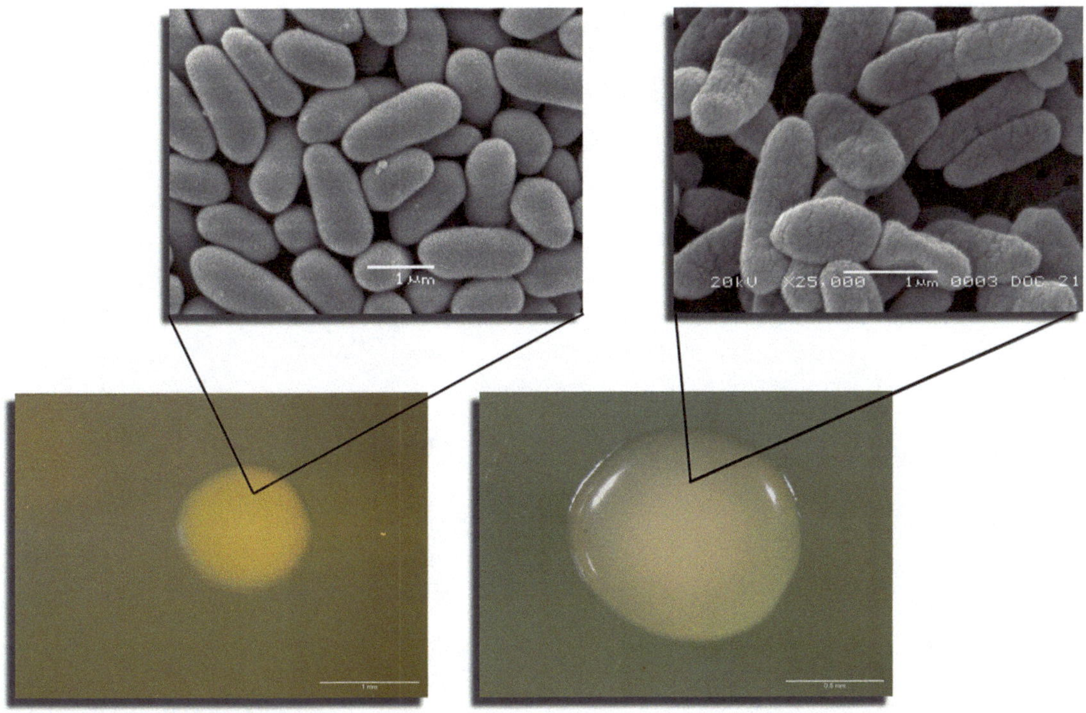

Pseudomonas putida
strain DOC19

Pseudomonas putida
strain DOC21

Fig. 2 Morphological aspect of the colonies (optical microscopy) and cells (scanning electron microscopy) from *P. putida* DOC19 and *P. putida* DOC21, the two strains isolated from soil samples in base to their ability to catabolize bile acids

to the decomposition of biomass and excreta, BA are available to environmental microorganisms. Some of these microbes have developed metabolic systems which allow the assimilation of these compounds.

In our laboratory, we have isolated, from soil samples, two different gram-negative bacteria able to catabolize BA and testosterone, but unable to metabolize sterols. Both strains were identified as *Pseudomonas putida* in base to homology and phylogenetically analysis using partial sequences of 16S rDNA and *rpoB* genes, morphological aspect, and metabolic profiling. These two strains were referred to as *P. putida* DOC19 and *P. putida* DOC21 (*see* Fig. 2) [8].

P. putida strain DOC21 was used as a model to identify and characterize the genes encoding the catabolic enzymes required for degradation of BA. The genetic studies revealed that in the genome of this strain, BA catabolic genes are grouped into four different clusters. The different ORFs were identified in *P. putida* DOC21 in base to random Tn*5* transposon mutagenesis, and homology and synteny comparison with other steroid degradative strains which catabolize testosterone (*Comamonas testosteroni*) [9], cholesterol

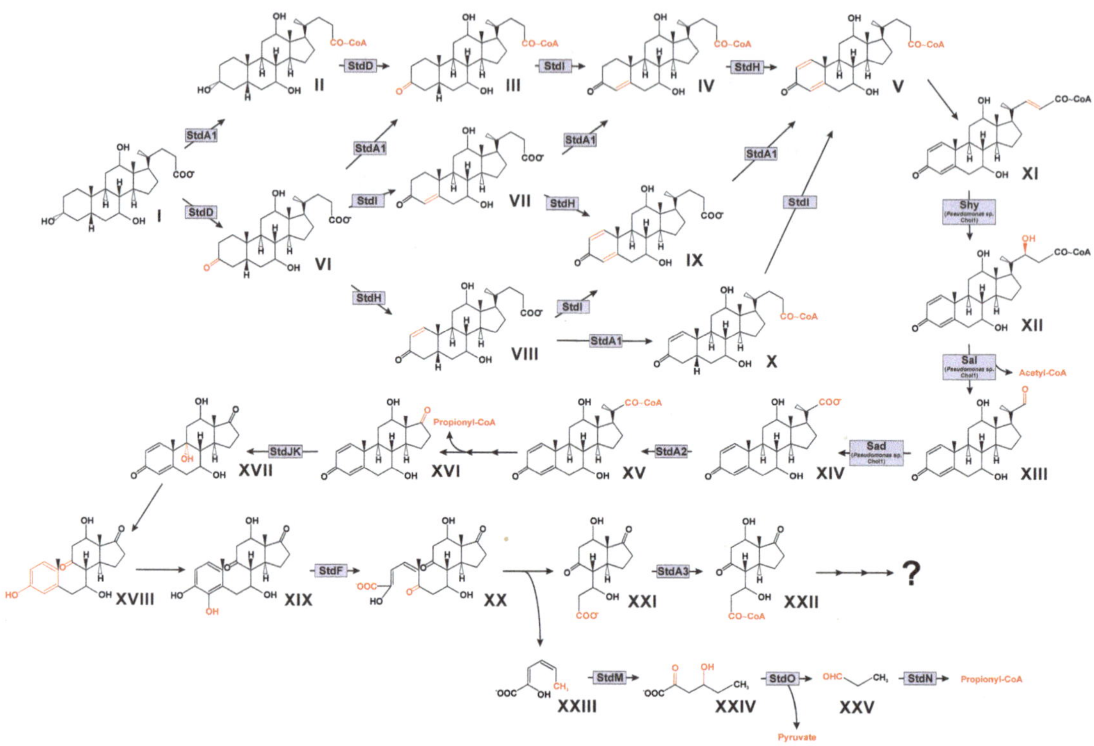

Fig. 3 Catabolic pathway proposed for aerobic assimilation of bile acids and other steroid compounds in *P. putida* DOC21, exemplified with the molecule of cholic acid. (**I**) Cholic acid; (**II**) cholyl-CoA; (**III**) 3-keto-5β-cholyl-CoA; (**IV**) Δ^4–3-keto-cholenoyl-CoA; (**V**) $\Delta^{1,4}$–3-keto-choladienoyl-CoA; (VI) 3-keto-5β-cholanoate; (**VII**) Δ^4–3-keto-cholenoate; (**VIII**) Δ^1–3-keto-5β-cholenoate; (**IX**) $\Delta^{1,4}$–3-keto-choladienoate; (**X**) Δ^1–3-keto-5β-cholenoyl-CoA; (**XI**) (22*E*)7α,12α-dihydroxy-3-oxochola-1,4,22*E*–triene-24-oyl-CoA; (**XII**) 7α,12α,22-trihydroxy-3-oxochola-1,4-diene-24-oyl-CoA; (**XIII**) 7α,12α-dihydroxy-3-oxopregna-1,4-diene-20*S*-carbaldehyde; (**XIV**) 7α,12α-dihydroxy-3-oxopregna-1,4-diene-20-carboxylate; (**XV**) 7α,12α-dihydroxy-3-oxopregna-1,4-diene-20-carboxylyl-CoA; (**XVI**) 7α,12α-dihydroxy-androsta-1,4-diene-3,17-dione; (**XVII**) 7α,9α,12α-trihydroxy-androsta-1,4-diene-3,17-dione; (**XVIII**) 3,7,12-trihydroxy-9,10-secoandrosta-1,3,5(10)-triene-9,17-dione; (**XIX**) 3,4,7,12-tetrahydroxy-9,10-secoandrosta-1,3,5(10)-triene-9,17-dione; (**XX**) 4,5–9,10-diseco-3,7,12-trihydroxy-5,9,17-trioxoandrosta-1(10),2-dien-4-oate; (**XXI**) 3aα-*H*-4α(3′(*R*)-hydroxy-3′-propanoate)-7α-hydroxy-7aβ-methylhexahydro-1,5-indanedione; (**XXII**) 3aα-*H*-4α(3′(*R*)-hydroxy-3′-propanoyl-CoA)-7α-hydroxy-7aβ-methylhexahydro-1,5-indanedione; (**XXIII**) 2-hydroxyhexa-2,4-dienoate; (**XXIV**) 4-hydroxy-2-oxohexanoate; (**XXV**) Propionaldehyde

(*Mycobacterium*, *Rhodococcus*, *Gordonia*) [10–14], or CA (*Comamonas*, *Rhodococcus*, and *Pseudomonas*) [9, 15–17]. The essentiality of these ORFs for BA catabolism was established by deletion of certain genes [8, 17].

Sequence analysis of the genes found in *P. putida* strain DOC21, the loss of function caused by their mutation (either by Tn*5* insertion or gene deletion), and the identification of catabolites [8, 17] allow us to conclude that *P. putida* DOC21 uses a 9,10-*seco* pathway for the assimilation of BA and testosterone, androstanolone, or 4-androstene-3,17-dione (AD) (*see* Fig. 3).

According to the proposed catabolic pathway, the first step corresponds to the oxidation of the hydroxyl function present in C-3 (α-configuration) to a keto group. This reaction is catalyzed by a 3α-hydroxysteroid dehydrogenase (StdD). Later, or at the same time, the carboxy group from side chain in C-17 is activated to a coenzyme A thioester throughout a reaction catalyzed by a bile acid-CoA ligase (StdA1) [17]. Once activated, a β-oxidation-like process takes place. First, a dehydrogenation in C-2/C-3 of the acyl-CoA side chain occurs. The specific enzyme catalyzing this step remains unknown, although some putative candidates have been proposed. In the second step, hydration of the α,β-double bond takes place. The introduction of a hydroxy group at C-3 in *Pseudomonas* sp. Chol1, is catalyzed by the enoyl-CoA hydratase Shy1 [18]. Until this step, the process is identical to a canonical β-oxidation. However, the next enzymatic step does not imply, as expected, the oxidation of the hydroxy function to a keto group (followed for a thiolytic cleavage mediated by the introduction of a CoA molecule), but an aldolytic cleavage of the C-C bond (catalyzed by Sal1), generating an aldehyde and releasing an acetyl-CoA molecule [18, 19]. This aldehyde is oxidized to the corresponding carboxylic acid by a specific aldehyde dehydrogenase (Sad in *Pseudomonas* sp. Chol1), and later, a second acyl-CoA synthetase (StdA2) activates it to a CoA derivative [17]. The subsequent degradation of the remaining side chain putatively occurs through a similar mechanism, yielding a molecule of propionyl-CoA and a steroid derivative with a keto function in C-17. Concomitantly with the elimination of the C-17 side chain, the oxidation of the rings from the steroid nucleus starts with the introduction of double bonds in C-4 (catalyzed by StdI) and in C-1 (reaction driven by StdH) [20]. In sum, after the elimination of the acyl-side chain and the introduction of two unsaturations in the A ring, 7α,12α-dihydroxy-androsta-1,4-diene-3,17-dione is obtained if CA was the BA catabolized. However, the degradation of LCA (lacking the hydroxy groups in 7 and 12) renders at this stage androsta-1,4-diene-3,17-dione (ADD), which is a catabolite of convergence when testosterone and AD degradation occur. Thus, ADD, as well as the hydroxylated derivatives generated from other bile acids, is a convergence point for the degradative pathway of steroids in *Pseudomonas* species. All these metabolites suffer an α-hydroxylation in C-9, catalyzed by a two-component hydroxylase (StdJK) [20], originating an unstable intermediate that suffers a spontaneous opening of the B ring. At the same time, A ring is aromatized, leading to 3,7α,12α-trihydroxy-9,10-secoandrosta-1,3,5(10)-triene-9,17-dione when CA is the BA that is being catabolized. The next step in CA degradation involves a hydroxylation at C-4 (A-ring), a reaction catalyzed by a monooxygenase, and that yields 3,4,7,12-tetrahydroxy-9,10-secoandrosta-1,3,5(10)-triene-9,17-dione. Now, the 3,4-dihydroxylated aromatized A-ring is a substrate of StdF [8], an extradiol dioxygenase homologous to TesB from *C. testosteroni* [21] and to HsaC from *R. jostii* [10] and *M. tuberculosis* [22]. The study of the catabolism of testosterone and cholesterol revealed that the *meta*-cleavage of A-ring is performed by TesB and HsaC. When catabolism of CA is approached in *Pseudomonas*, the reaction is catalyzed by StdF, giving 4,5–9,10-diseco-3,7,12-trihydroxy-5,9,17-trioxoand rosta-1(10)2-diene-4-oic acid, which is further degraded by a hydrolytic cleavage, leading to the products 2-hydroxyhexa-2,4-dienoic acid (HHD) and a fully reduced methylindanone derived from C and D rings [3aα-*H*-4α(3′(*R*)-hydroxy-3'propanoate)-7β-hydroxy-7aβ-methylhexahydro-1,5-indane-dione, 3′7-diOH-HIP]. The catabolism of HHD is carried out in *P. putida* DOC21 by three enzymes: (1) StdM, which catalyzes the hydration of HHD yielding 4-hydroxy-2-oxohexanoic acid; (2) StdO that catalyzes the aldolytic cleavage of this compound giving pyruvate and propionaldehyde; and (3) StdN, a CoA-dependent acylating propionaldehyde dehydrogenase that transforms propionaldehyde in propionyl-CoA (Olivera et al., manuscript in preparation).

On the other hand, the metabolism of 3′7-diOH-HIP derived from CA, as well as other related compounds (originated from other bile acids or testosterone), continues through its CoA activation, which is catalyzed by StdA3 in *P. putida* DOC21 [17]. When cholesterol metabolism was studied in actinobacteria, a similar activity catalyzing the CoA activation of 3aα-*H*-4α(3'propanoate)-7aβ-methylhexahydro-1,5-indane-dione (HIP) has been reported [23]. However, in *C. testosteroni*, although it has been speculated that HIP (and their hydroxylated derivatives) is activated to CoA, the thioesterifications are attributed to a CoA-transferase [24]. Furthermore, in this bacterium, two acyl-CoA dehydrogenases, as well as an enoyl-CoA hydratase, have been proposed to be involved in a β-oxidation acting over the HIP-CoA generated from testosterone [25, 26]. However, at the present moment, the final metabolic fate of this molecule, as well as its hydroxylated derivatives, remains unknown.

2 Materials

2.1 Isolation of Environmental Strains Able to Degrade Cholate [8]

1. *Pseudomonas* chemically defined minimal medium (MM) [27]: KPO_4H_2 (13.6 g/L); $(NH_4)_2SO_4$ (2.0 g/L); $MgSO_4.7H_2O$ (0.25 g/L); $FeSO_4.7H_2O$ (5 mg/L), pH 7.0; add carbon source at the final concentration required in each experiment (usually 5 mM when BA or testosterone was used). In other cases (4-hydroxyphenylacetic acid or other compounds), 10 mM final concentration is used. For solid MM, 2% (w/vol) of agar must be added.

2. Cholic acid (CA, Sigma-Aldrich Co).
3. Erlenmeyer flasks (500 mL).
4. Incubator shakers (INNOVA 4330, New Brunswick Scientific Co., NJ, USA).
5. Applied Biosystems 2720 Thermal Cycler (Thermo Fisher Scientific In., MA, USA).
6. Polymerase Chain Reaction reagents.

2.2 Isolation of Mutants of *P. putida* DOC21 Unable to Catabolize BA by Tn5 Transposon Mutagenesis

1. Luria-Bertani broth (LB): BactoTryptone 10 g/L; Yeast extract 5 g/L, NaCl 10 g/L. For solid LB medium, agar 2% (w/vol) has to be supplied.
2. MM solid media supplied with different carbon sources.
3. Kanamycin, chloramphenicol, and ampicillin.
4. *E. coli* DH10B pGS9 [28].
5. *E. coli* HB101 pRK600 [29].
6. *P. putida* DOC21 [8].
7. 4-Hydroxyphenylacetic acid.
8. Incubator shakers (INNOVA 4330, New Brunswick Scientific Co., NJ, USA).
9. Laboratory heaters (at 30 and 37 °C).
10. Beckman spectrophotometer DU 640 (Beckman Coulter Ltd., UK).
11. Eppendorf 5810R centrifuge (Eppendorf AG, Germany).
12. Eppendorf MiniSpin microcentrifuge (Eppendorf AG, Germany).
13. 50 and 15 mL polypropylene tubes and 1.5–2 mL Eppendorf tubes.
14. Filters, 0.45 μm pore size.
15. Sterile toothpicks.

2.3 Identification of the Tn5 Insertion Point in the Genome of *P. putida* Strain DOC21 Mutants and Location of the Genomic Sequences Surrounding the Insertion Point

1. Plasmid pJQ200KS [30].
2. Plasmid pGEM-T*easy* (Promega Co.).
3. QIAquick PCR purification Kit (QIAGEN Sciences, MD, USA) or QIAquick Gel Extraction Kit (QIAGEN Sciences, MD, USA).
4. Gentra Puregene Yeast/Bacteria kit (QIAGEN Sciences, MD, USA).
5. Applied Biosystems 2720 Thermal Cycler (Thermo Fisher Scientific In., MA, USA).
6. Polymerase Chain Reaction reagents.
7. Restriction endonucleases and T4 DNA ligase.
8. *E. coli* DH10B chemically competent cells.

9. LB broth and solid media.
10. Kanamycin, chloramphenicol, ampicillin, and gentamycin.
11. Incubator shakers (INNOVA 4330, New Brunswick Scientific Co., NJ, USA).
12. Laboratory heaters (at 30 and 37 °C).
13. Beckman spectrophotometer DU 640 (Beckman Coulter Ltd., UK).
14. Eppendorf 5810R centrifuge (Eppendorf AG, Germany).
15. Eppendorf MiniSpin microcentrifuge (Eppendorf AG, Germany).
16. 50 and 15 mL polypropylene tubes and 1.5–2 mL Eppendorf tubes.
17. Filters, 0.45 μm pore size.
18. Sterile toothpicks.
19. Sodium acetate 3 M.
20. Ethanol.
21. QIAprep Spin Miniprep Kit (QIAGEN Sciences, MD, USA).

2.4 Genome Editing Through Deletion of Specific Genes Involved in Bile Acids Catabolism in *P. putida* DOC21

1. Plasmid pJQ200KS [30].
2. Plasmid pGEM-T*easy* (Promega Co.).
3. QIAquick PCR purification Kit (QIAGEN Sciences, MD, USA) or QIAquick Gel Extraction Kit (QIAGEN Sciences, MD, USA).
4. Gentra Puregene Yeast/Bacteria kit (QIAGEN Sciences, MD, USA).
5. Applied Biosystems 2720 Thermal Cycler (Thermo Fisher Scientific In., MA, USA).
6. Polymerase Chain Reaction reagents.
7. Restriction endonucleases and T4 DNA ligase.
8. *E. coli* DH10B chemically competent cells.
9. LB broth and solid media.
10. Sucrose.
11. Kanamycin, chloramphenicol, ampicillin, and gentamycin.
12. Incubator shakers (INNOVA 4330, New Brunswick Scientific Co., NJ, USA).
13. Laboratory heaters (at 30 and 37 °C).
14. Beckman spectrophotometer DU 640 (Beckman Coulter Ltd., UK).
15. Eppendorf 5810R centrifuge (Eppendorf AG, Germany).
16. Eppendorf MiniSpin microcentrifuge (Eppendorf AG, Germany).

17. 15 and 50 mL polypropylene tubes and 1.5–2 mL Eppendorf tubes.
18. Filters, 0.45 μm pore size.
19. Sterile toothpicks.

2.5 Identification of 4-Androstene-3,17-Dione and 1,4-Androstadiene-3,17-Dione as Metabolic Intermediates Using Specific Mutants of *P. putida* DOC21

1. Succinate.
2. LCA (Sigma-Aldrich Co).
3. 500 mL flasks.
4. Rotary shaker.
5. Microfuge.
6. Filters, 0.22 μm pore size.
7. TLC plates (TLC Silica gel60, aluminum sheets 20 × 20 cm, Merck KGaA).
8. Drummond Glass Micropipettes (10 μL).
9. Soft pencil, ruler.
10. Organic Solvents: benzene, acetone.
11. TLC developing chamber.
12. Filter paper.
13. Hair drier.
14. Exhaust hood.
15. Benzene.
16. Acetone.
17. AD (Sigma-Aldrich Co.).
18. ADD (Gadea Pharmaceutical Group).
19. H_2SO_4 solution (30% by vol.).
20. Compressed gas sprayer for TLC.
21. Oven.
22. Gloves.

2.6 Identification of StdA1 and StdA2 from *P. putida* DOC21 as ATP-Dependent Acyl-CoA Synthetases Involved in the Catabolism of the C-17 BA-Acyl Chain

1. Erlenmeyer flasks (2000 mL).
2. Succinate.
3. CA (Sigma-Aldrich Co).
4. Incubator shakers (INNOVA 4330, New Brunswick Scientific Co., NJ, USA).
5. Filters, 0.22 μm pore size.
6. HCl.
7. Ethyl acetate.
8. Separation funnels.
9. Avanti J-30I High Performance Beckman centrifuge equipped with an Avanti JA-10 rotor (Beckman Coulter Ltd., UK).

10. HPLC system consisting of two high-pressure pumps (Shimadzu LC-10ATvp), an autoinjector (Gilson 234), a C18-reversed phase column, and a UV-Visible light diode array detector (Shimadzu SPD-M 10) with a scanning range from 190 to 500 nm.
11. A semipreparative reversed-phase column [250 by 8 mm, Eurosphere II, 100–5 C18 H (Knauer)].
12. K-Na-phosphate buffer (10 mM, pH 7.1).
13. Acetonitrile HPLC grade.
14. 50 mM Tris–HCl (pH 8.0).
15. $NAD(P)^+$.
16. $K_3Fe[CN]_6$.
17. α,β-Hydroxysteroid dehydrogenase from *Pseudomonas testosteroni* (Sigma-Aldrich Co).
18. Spectrophotometer.
19. MOPS buffer (pH 7.8).
20. CoA.
21. ATP.
22. $MgCl_2$.
23. French press (Amico, Richmond Hill, ON, Canada).
24. Sephadex G-25 PD-10 columns (Pharmacia Biotech, Uppsala, Sweden).
25. Bradford protein assay reagents (Bio-Rad, Spain) [31].

2.7 Identification of ATP-Dependent Acyl-CoA Synthetase StdA3 from *P. putida* DOC21 Involved in the Degradation of the C and D Rings of the Steroid Core

1. Erlenmeyer flasks (2000 mL).
2. Incubator shakers (INNOVA 4330, New Brunswick Scientific Co., NJ, USA).
3. Avanti J-30I High Performance Beckman centrifuge equipped with an Avanti JA-10 rotor (Beckman Coulter ltd., UK).
4. Separation funnels.
5. HPLC system consisting of a Waters binary pump1525 HPLC module (Milford, MA, USA), a Luna 3 μm PFP(2) 50 × 4.6 mm column (Phenomenex, Torrance, CA), and a Waters 2996 photodiode array detector.
6. Eluents: ethyl acetate, hexane, and acetic acid; all of them HPLC grade.
7. Bradford protein assay reagents (Bio-Rad, Spain) [31].
8. HEPES buffer (pH 7.3).
9. $MgCl_2$.
10. ATP.
11. CoA.

12. Pyruvate.
13. HCl.
14. Ethyl acetate.

3 Methods

3.1 Isolation of Environmental Strains Able to Degrade Cholate [8]

1. Make suspensions of soil samples (1 g of soil in 200 mL of sterile distilled water) and shake them vigorously for 1 h at 25 °C.
2. Inoculate 500 mL Erlenmeyer flasks containing 100 mL of MM supplied with CA (4 mM) as carbon source with 500 μL of the suspensions obtained in **step 1**.
3. Incubate the flasks on a rotary shaker (250 rpm) for 48 h at 30 °C.
4. After 24 and 48 h of incubation, spread serial dilutions from these cultures into MM plates containing CA 5 mM as the sole carbon source.
5. Incubate the plates for 48–72 h at 30 °C, select isolated colonies, and streak them on new MM plates containing CA (5 mM, MM-CA). Incubate the plates for 48 h at 30 °C.
6. Pick the colonies from these new plates with sterile toothpicks and suspend the cells in 200 μL of sterile MilliQ water. Plate serial dilutions of these cultures on MM-CA, and incubate the plates for 48 h at 30 °C.
7. Streak the new colonies in MM-CA and test the isogenicity of cultures through PCR amplification and sequencing of the amplicons from 16S rDNA (primers 6F: 5′-GGAGAGTTA GATCTTGGCTCAG-3′ and 1510R: 5′-GTGCTGCAGG GTTACCTTGTTACGACT-3′; amplicon size of 1450 bp) and *rpoB* (primers LAPS5: 5′-TGGCCGAGAACCAGTTC CGCGT-3′ and LPAS27: 5′-CGGCTTCGTCCAGCTTG TTCAG-3′; amplicon size of 1240 bp).

3.2 Isolation of Mutants of *P. putida* DOC21 Unable to Catabolize BA by Tn5 Transposon Mutagenesis

Tn*5* transposon mutagenesis has been proven to be a very useful technique in bacterial genetics [8, 32–34]. Tn*5* insertion into a gene leads to its inactivation, being the insertion stable and non-leaky. However, although the insertion of Tn*5* transposon occurs randomly, existence of hot spots in the acceptor genome has been observed. Moreover, the insertion of the Tn*5* into a bacterial genome implies the creation of a new genetic and physical marker that could be readily mapped, allowing the sequencing and identification of the insertion point and surrounding DNA from kanamycin resistant transconjugants. However, sometimes, the insertion of the transposon on a set of cotranscribed genes (an operon) exerts polar effects silencing the expression genes downstream from the insertion point. This effect has been attributed to the presence of transcriptional and translational termination signals into the transposon [35].

1. Stabilize three different cultures in sterile 50 mL polypropylene tubes: (1) *E. coli* DH10B containing pGS9 [28] in a 10 mL aliquot of LB broth supplemented with kanamycin (Km, 25 μg/mL) and chloramphenicol (Cm, 30 μg/mL); (2) *E. coli* HB101 pRK600 [29] in other 10 mL of LB broth supplemented with Cm (30 μg/mL); and (3) *P. putida* DOC21 in 10 mL of LB broth supplied with ampicillin (Ap, 100 μg/mL). Incubate *E. coli* strains at 37 °C, and *P. putida* strain at 30 °C during 8–10 h.
2. Determine the growth of each culture as a function of their absorbance at 540 nm.
3. Collect aliquots of each culture in a unique 15 mL sterile polypropylene tube and mix them in a proportion of 5:1:1 (five times more cells of the acceptor *P. putida* DOC21 than the donor *E. coli* pGS9 and the helper *E. coli* pRK600, in function of the absorbance reached for each individual culture). Use, at least, 1 mL of *E. coli* pGS9 culture, and the share of the other ones. Preserve the surplus of the cultures for future actions (*see* **step 11**).
4. Centrifuge the bacterial mixture of cells at 3220 × *g*, 10 min.
5. Discard the supernatant and suspend the bacteria in 1 mL of fresh, sterile LB broth. Transfer them to a 1.5 mL Eppendorf tube.
6. Centrifuge the microbial mixture at 12,700 × *g*, 2 min.
7. Suspend the bacteria in 1 mL of fresh, sterile LB broth.
8. Repeat **step 6**.
9. Suspend the mixed cultures in 50–80 μL of LB broth.
10. Dispose, using sterile forceps, a sterile Millipore bacterial filter (0.45 μm pore size) on the surface of a plate containing solid LB medium without any antibiotic.
11. Apply the cell suspension from **step 9** onto the sterile filter. At the same time, using 10–20 μL from the rest of the cultures from **step 3**, inoculate LB plates containing Km, Cm, Ap, or combination thereof, at the concentrations indicated in **step 1**. They are used as control plates to verify the absence of contamination. All the plates are incubated at 30 °C for 12 h (*see* **Notes 1** and **2**).
12. Once verified the validity of the strains (through examination of the control plates), place the filter with the mixture of strains onto a 15 mL polypropylene tube containing 3 mL of sterile LB broth. Discard the control plates.

13. Shake the tube until complete suspension of the bacteria.
14. Make serial dilutions from the suspension and seed 100 μL of each dilution onto LB plates supplied with Ap and Km (at the concentrations indicated in **step 1**).
15. Incubate the plates 48 h at 30 °C, until rising of transconjugant colonies (*see* **Notes 3** and **4**). Use those plates where the higher number of isolated colonies arise.
16. Select transconjugants by seeding with a sterile toothpick from the LB plates supplied with Km and Ap to paired MM plates containing (1) 10 mM 4-hydroxyphenylacetic acid (or other steroid nonrelated carbon source, i.e., succinate), Ap and Km, and (2) CA (5 mM) plus Km (not Ap) (*see* **Note 5**). Incubate the plates containing MM for 12–20 h. Those transconjugants that do not grow in MM containing CA as sole carbon source, but that are able to grow in the counterselection MM plates (those containing a non-steroid carbon source, i.e., 4-hydroxyphenylacetic acid) should be considered mutants specifically affected in BA metabolism.

The full process is schematized (*see* Fig. 4).

3.3 Identification of the Tn*5* Insertion Point in the Genome of *P. putida* Strain DOC21 Mutants and Location of the Genomic Sequences Surrounding the Insertion Point

A method for the identification of the surrounding sequences around the insertion point of the Tn*5* has been developed in our laboratory. This method is based on the homologous recombination of a specifically designed suicide plasmid (derivate from pJQ200KS [30]) containing a DNA fragment corresponding to the inverted repeated sequences flanking the Tn*5* transposon (construct pJQ-Tn*5* [34]). Later, the genomic DNA from the recombinants is isolated, and, after restriction of this DNA with specific endonucleases, and ligation of the constructs, the plasmids containing the adjacent sequences are recovered. The transformation of *E. coli* competent cells with the ligated DNA allows the obtaining of colonies containing the plasmid (and thus, the replicative origin contained in it) together with the genomic DNA adjacent to one of the IS*50* from Tn*5*.

3.3.1 Construction of pJQ-Tn5 (See Fig. 5) [34]

1. Amplify by PCR a 1113 bp fragment present in both, the IS*50*L and IS*50*R sequences from Tn*5* transposon, using the oligonucleotide primers Tn*5*-ext. (5′-GGGTGATCCTCGC CGTACTGCC-3′) and Tn*5*-int (5′-GCGGACTGGCTTTCT ACGTGTTC-3′) (*see* **Note 6**).
2. Clone the amplified fragment into pGEM-T*easy* (Promega Co.) according to the manufacturer's instructions.
3. Transform *E. coli* DH10B chemically competent cells. Plate transformed cells in LB plates supplied with Ap (100 μg/mL). Select clones.
4. Culture selected clones in 15 mL sterile polypropylene tubes containing LB broth plus Ap (100 μg/mL).

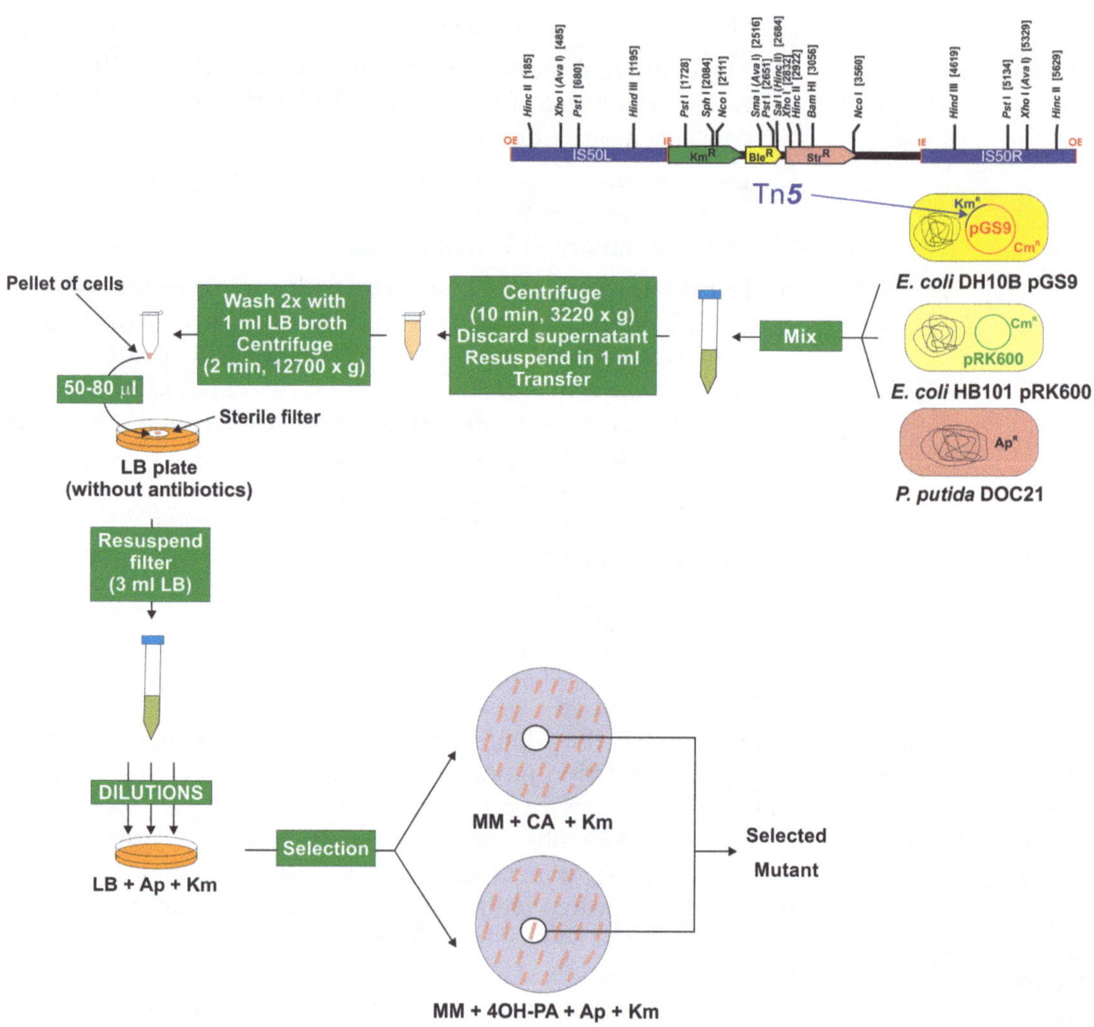

Fig. 4 Schematic representation of the procedure followed for collecting Tn*5* insertional mutants from *P. putida* DOC21

5. Collect the cells by centrifugation at maximum speed in a microcentrifuge for 5 min.
6. Extract the plasmids by using QIAprep Spin Miniprep Kit (QIAGEN).
7. Rescue the cloned fragment through *Apa*I and *Sal*I restriction of the construction.
8. Ligate the fragment into a pJQ200KS [30] previously restricted with *Apa*I and *Sal*I.
9. Transform competent *E. coli* DH10B cells with the ligation reaction. Plate the transformed cells onto LB plates supplied with Gm (30 μg/mL).
10. Select the appropriate clone containing the construct pJQ-Tn5.

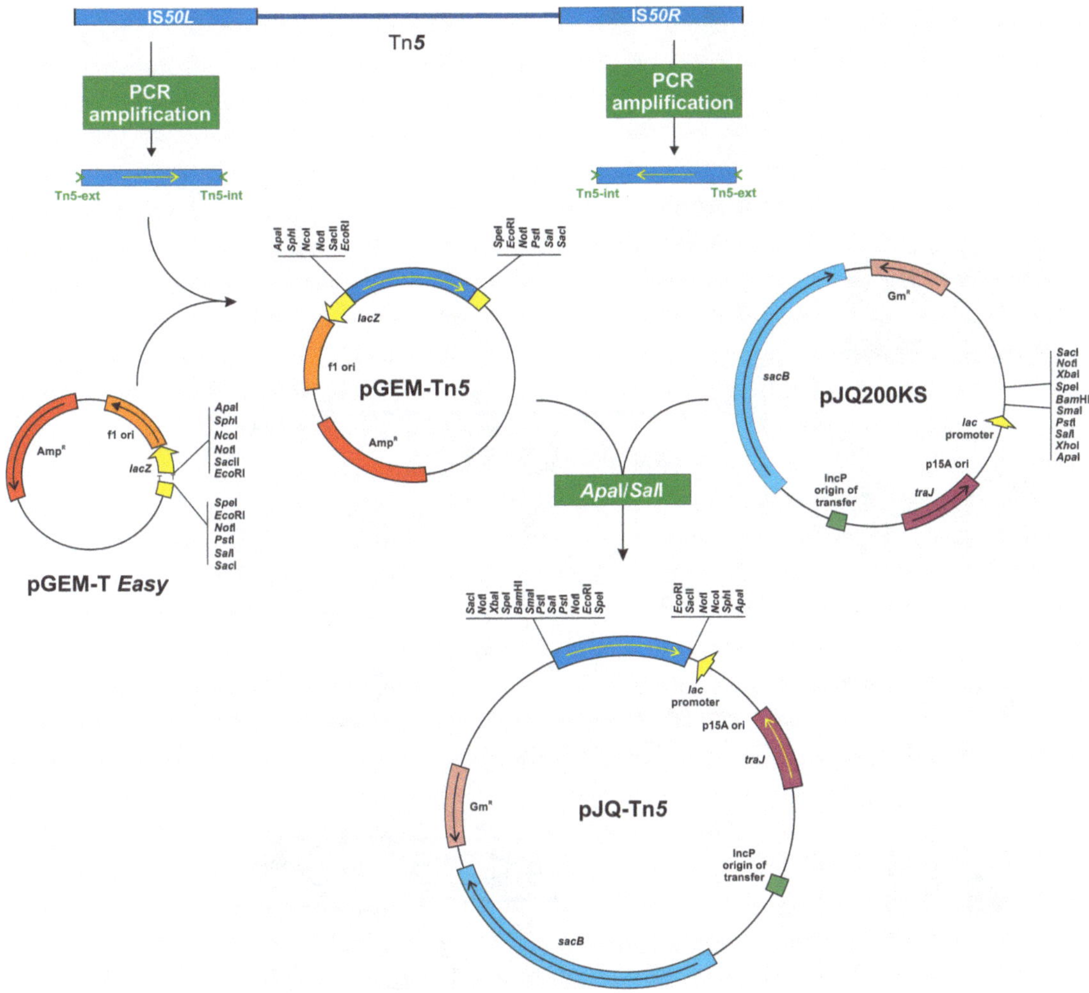

Fig. 5 Schematic representation of the process followed for the collection of the plasmid pJQ-Tn5, needed for the identification of both, the insertion point of Tn*5* transposon in the genome of *P. putida* strain DOC21 and the genomic sequences surrounding this point

3.3.2 Transferring of pJQ-Tn5 into the P. putida DOC21 Tn5 Mutant by Triparental Mating (See Fig. 6)

1. Prepare three different cultures in sterile 50 mL polypropylene tubes: (1) *E. coli* DH10B containing pJQ-Tn5 in 10 mL of LB broth supplemented with gentamycin (Gm, 30 μg/mL); (2) *E. coli* HB101 pRK600 [29] in other 10 mL of LB broth supplemented with Cm (30 μg/mL); and (3) *P. putida* DOC21 mutant in 10 mL of LB broth supplied with Ap (100 μg/mL) and Km (25 μg/mL). Incubate *E. coli* strains and *P. putida* for 8–10 h at 37 °C and 30 °C, respectively.
2. Determine the growth of each culture as a function of their absorbance at 540 nm.
3. Taking into account the absorbance reached by the cultures, collect, in a 15 mL sterile polypropylene tube, aliquots of *E.*

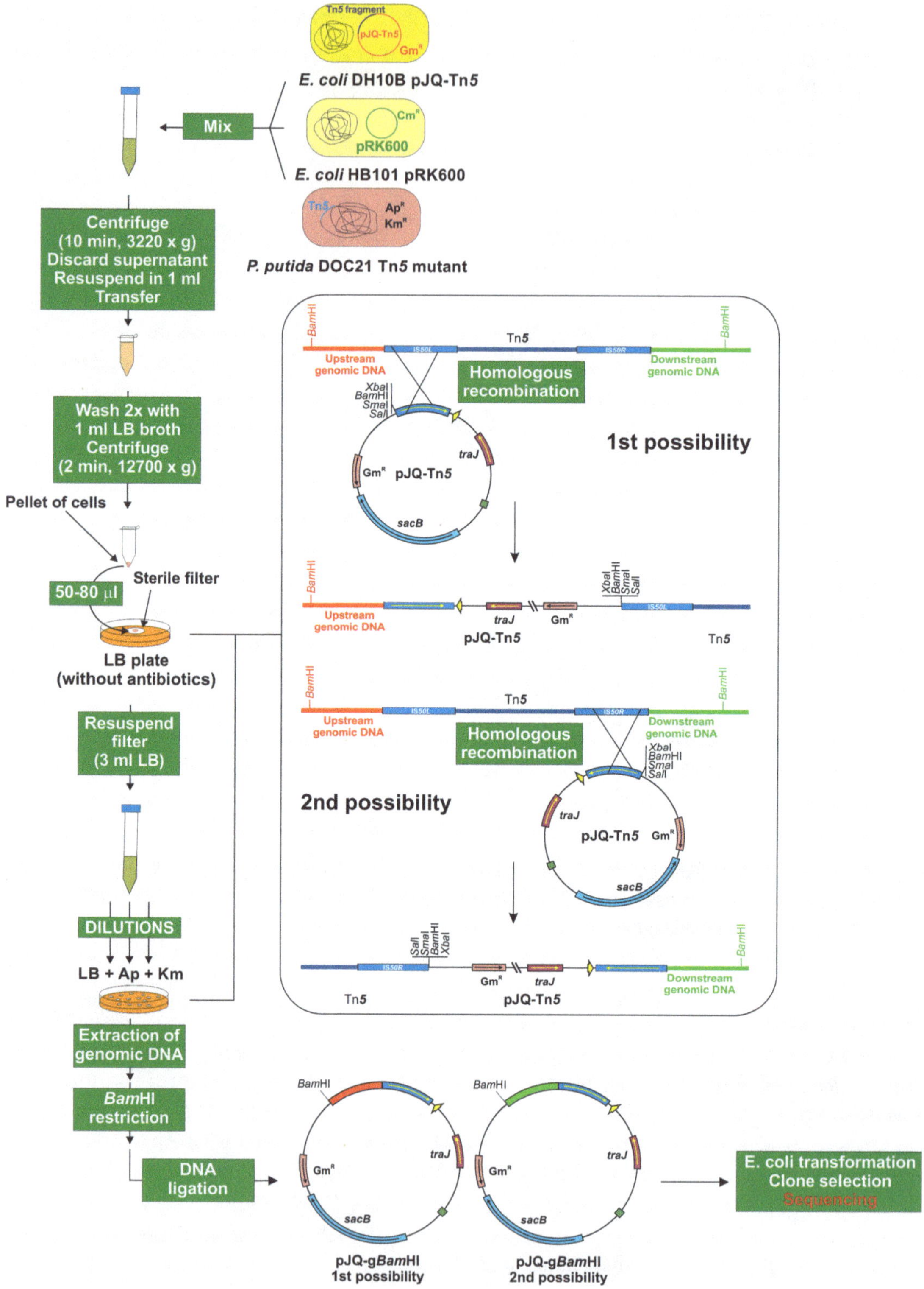

Fig. 6 Schematic representation of the procedure followed to get the identification of: (1) the point at which the transposon Tn*5* is nested in the genome of *P. putida* DOC21 mutants, as well as (2) the genomic surrounding sequences. The knowledge of the Tn*5* insertion points is based on the homologous recombination of pJQ-Tn*5*

coli cultures containing the same amount of cells. Collect in the same tube the amount of *P. putida* DOC21 Tn5-mutant culture corresponding to five times more cells of the acceptor (*P. putida* DOC21) than the donor (*E. coli* pJQ-Tn5) and the helper (*E. coli* pRK600). Use, at least, 1 mL of *E. coli* pJQ-Tn5 culture and the share of the other ones. Preserve the surplus of the cultures to use them later for establishing controls of the process (*see* **step 11**).

4. Centrifuge the bacterial mixture at 3220 × *g*, 5 min.
5. Discard the supernatant and suspend the bacteria in 1 mL of fresh, sterile LB broth without any antibiotic. Transfer them to a 1.5 mL Eppendorf tube.
6. Centrifuge the mixture at 12,700 × *g*, 2 min.
7. Suspend the bacteria in 1 mL of fresh, sterile LB broth.
8. Repeat **step 6**.
9. Suspend the mixed cultures in 50–80 μL of LB broth.
10. Dispose (by using sterile forceps) a sterile Millipore bacterial filter (0.45 μm pore size) on the surface of a plate containing solid LB medium without any antibiotic.
11. Apply the cell suspension from **step 9** onto the sterile filter. At the same time, using 10–20 μL from the rest of the cultures from step 3, inoculate LB plates containing Km, Gm, Cm, Ap, or combinations thereof, at the concentrations indicated in **step 1** (control plates to verify no contamination of the strains). Incubate these plates at 30 °C for 12 h. (*see* **Notes 1** and **2**).
12. Once verified the suitability of the strains (through examination of the control plates), place the filter with the mixture of strains (using sterile forceps) onto a 15 mL polypropylene tube containing 3 mL of sterile LB broth and shake the tube until complete suspension of the bacteria. Discard the control plates.
13. Make serial dilutions from the suspension and seed 200 μL of each dilution onto LB plates supplied with Ap, Gm, and Km (at the concentrations indicated in **step 1**).
14. Incubate the plates at 30 °C for 48 h. or the time required (until rising of transconjugant colonies).
15. Select isolated colonies streaking them on LB plates containing Ap, Gm and Km.

Fig. 6 (continued) over both IS*50* fragments flanking the transposon. In the scheme are indicated the two different recombination possibilities. The use of *Bam*HI restriction of the genomic DNA from the recombinant strains in which pJQ-Tn*5* has been nested is only for didactical. This restriction could be performed with *Bam*HI, *Xba*I, *Sal*I, or *Sac*I, according to the presence of these targets in the genomic DNA

3.3.3 Cloning of the DNA Fragments in which Tn5 has Been Inserted (See Fig. 6)

1. Grow the selected pJQ-Tn*5* recombinant clones (*see* **Note 7**) in sterile 50 mL polypropylene tubes containing 10 mL of LB broth plus Ap, Gm and Km (concentrations indicated in Subheading 3.2, **step 2**, paragraph 1). Incubate the cultures at 30 °C for 8–10 h. with vigorous shaking (250 rpm).
2. Transfer 1 mL from each culture into 1 mL sterile Eppendorf tubes on ice.
3. Centrifuge at 12,700 × *g* for 1 min.
4. Purify the genomic DNA using the Gentra Puregene Yeast/Bacteria kit (QIAGEN) according to the instructions of the manufacturer.
5. Digest the 50 μL of the genomic DNA obtained from **step 4** with the endonucleases *Bam*HI, *Xba*I, *Sma*I, or *Sal*I (*see* **Note 8**).
6. Precipitate the restriction reactions by adding 1/10 volume sodium acetate 3 M, pH 5.2 and 2.5 volumes of cold 100% ethanol. Place at −20 °C for >20 min.
7. Centrifuge at maximum speed on a microfuge for 15–20 min.
8. Decant the supernatant.
9. Wash the DNA pellet with 1 mL 70% ethanol. Spin at maximum speed on a microcentrifuge for 5 min. Decant the supernatant.
10. Air dry the pellet for 5 min.
11. Suspend the pellet in 35 μL of sterile MilliQ water by incubating at 65 °C, 5–10 min. And, at room temperature, for 10–15 additional min.
12. Circularize the DNA fragments: use 17 μL of the DNA from **step 11** in a 20 μL ligation reaction (2 μL of 10× T4 DNA ligase ligation buffer, 1 μL of T4 DNA ligase, and 17 μL of DNA).
13. Use 10 μL from the ligation reaction to transform chemically competent *E. coli* DH10B cells.
14. Seed the transformed bacteria in LB plates containing Gm (30 mg/mL).
15. Select clones.
16. Culture the selected clones in 3 mL of LB broth at 37 °C during 8–12 h.
17. Collect the bacteria by centrifuging in Eppendorf tubes.
18. Extract the plasmids by using QIAprep Spin Miniprep Kit.
19. Check the plasmids through restriction with the same endonuclease used in **step 5** (*see* **Note 9**).
20. Sequence the selected constructions from oligonucleotides M13 forward (−47): 5′ CGCCAGGGTTTTCCCAGTCA CGAC 3′, M13 reverse (−48): 5′ AGCGGATAACAATTTC

ACACAGGA 3′ (universal primers of the plasmid) and Tn*5*–2: 5′ CCGCCGAAGAGAACACAG 3′ (oligonucleotide primer present at 152 bp from the end of IS*50* sequences) (*see* **Note 10**).

3.4 Genome Editing Through Deletion of Specific Genes Involved in Bile Acids Catabolism in *P. putida* DOC21

In order to study the metabolic role of specific genes and enzymes, or to obtain catabolic intermediates belonging to the BA degradative pathway, a procedure of genome editing based on the specific deletion of selected genes is proposed. This technique consists, summarily, in force a double homologous recombination of a suicide plasmid (pJQ200KS [30]) into the genome, using DNA fragments upstream and downstream of the fragment that is desired to delete as homologous recombinational places. The first recombination is triggered by the selective pressure to conserve the Gm resistance through culture of the clones in LB plates containing Gm, and inserts the plasmid into de chromosome; the second recombinational process is induced by culturing the selected clones in LB plates without Gm but sucrose. Plasmid pJQ200KS harbors the *sacB* gene, whose enzymatic product synthesizes high molecular weight, β-(2, 6)-linked levan from sucrose by transfer of fructosyl units in the periplasmic space, having a strong deleterious effect over the cell. To avoid cell death, some of the clones suffer a second homologous recombinational event that eliminates the inner fragment of the target gene as well as the suicide plasmid.

1. Amplify by PCR the fragments (>400 bp) adjacent (upstream and downstream) at the DNA fragment to be deleted. The oligonucleotides in 3′ (for the fragment located upstream) and 5′ (for the fragment located downstream) should include a restriction target sequence not present inside of the amplified fragments nor *Not*I.
2. Purify the amplified fragments from the PCR amplification reaction (use QIAquick PCR purification Kit, QIAGEN) or from an agarose gel (use QIAquick Gel Extraction Kit, QIAGEN).
3. Digest both fragments with the endonuclease whose target has been included in the upstream 3′ primer and the downstream 5’ primer.
4. Using equal amounts of each fragment, ligate them in a T4 DNA ligase reaction.
5. Using the T4 DNA ligation reaction as template DNA, amplify by PCR both fragments ligated using the primer in 5′ from the fragment upstream and the primer in 3′ from the fragment downstream.
6. Ligate the PCR amplicon into pGEM-T*easy* plasmid.
7. Transform *E. coli* DH10B with the constructs and select the appropriate clones in LB plates containing Ap (100 μg/mL).

8. Rescue the fragment from the selected clones using *Not*I endonuclease (cutting to the restriction targets from the multiple cloning sites of pGEM-T*easy* disposed at both sites of the cloned fragment).
9. Ligate the fragment into *Not*I restricted and dephosphorylated (using calf alkaline phosphatase) pJQ200KS.
10. Transform *E. coli* DH10B with the constructs and select the appropriate clones in LB plates containing Gm (30 μg/mL) (The full process is schematized, *see* Fig. 7).
11. Prepare three different cultures in sterile 50 mL polypropylene tubes: (1) *E. coli* DH10B containing the pJQ200KS derivative in 10 mL of LB broth supplemented with gentamycin (Gm, 30 μg/mL); (2) *E. coli* HB101 pRK600 in other 10 mL of LB broth supplemented with Cm (30 μg/mL); and (3) *P. putida* strain DOC21 in 10 mL of LB broth supplied with Ap (100 μg/mL). Incubate the *E. coli* strains at 37 °C and *P. putida* at 30 °C for 8–10 h.
12. Determine the growth of each culture as a function of their absorbance at 540 nm.
13. Taking into account the absorbance reached by the cultures, collect, in a 15 mL sterile polypropylene tube, aliquots of each *E. coli* culture containing the same amount of cells. Add to this tube a volume of the *P. putida* DOC21 culture containing five times more cells than the donor *E. coli* containing the plasmid pJQ200KS and that the helper strain (*E. coli* pRK600). Use, at least, 1 mL of donor *E. coli* pJQ200KS culture, and the share of the other ones. Preserve the surplus of the cultures, they will be used later to establish controls of the process (*see* **step 21**).
14. Centrifuge the bacterial mixture at 3220 × *g*, 5 min.
15. Discard the supernatant and suspend the cells in 1 mL of fresh, sterile LB broth without any antibiotic. Transfer the cells to a 1.5 mL Eppendorf tube.
16. Centrifuge them at 12,700 × *g*, 2 min.
17. Suspend the bacteria in 1 mL of fresh, sterile LB broth.
18. Repeat **step 16**.
19. Suspend the mixed cultures in 50–80 μL of LB broth.
20. Dispose, with sterile forceps, a sterile Millipore bacterial filter (0.45 μm pore size) on the surface of a plate containing solid LB medium without any antibiotic.

Fig. 7 (continued) two fragments. In this particular example, the use of a *BamHI* restriction sequence into the oligonucleotides, those annealing in 3' position of fragment U and in 5' position of fragment D, are for didactical reasons. The endonuclease sequences to be created in the degenarated oligonucleotides should be dependent on the restriction sites present on the fragments to be amplified, avoiding the use of sequences recognized by endonucleases that would be present on them, as well as *NotI*, due to the use of this endonuclease in the process of cloning

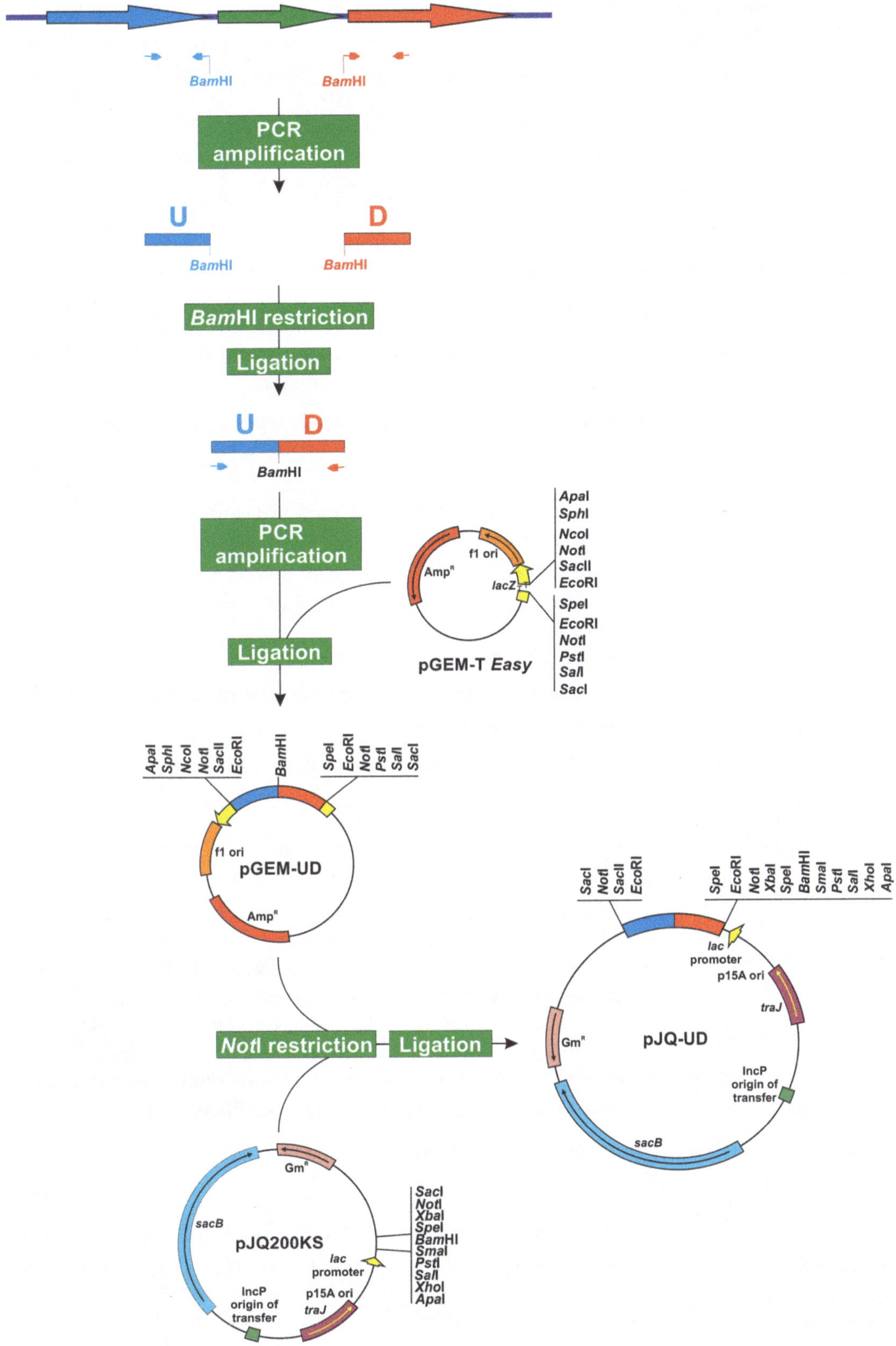

Fig. 7 Schematic representation of the construction of a theoretical plasmid derived from pJQ200KS illustrating the process for obtaining a construction to delete genes in *P. putida* DOC21. U (from "upstream") and D (from "downstream") indicated the two flanking regions used for the deletion of a gene located between these

21. Apply the cell suspension from **step 19** onto the sterile filter. At the same time, using 10–20 μL from the surplus of the cultures from **step 3**, inoculate LB plates containing Gm, Cm, and Ap or combination thereof at the concentrations indicated in **step 1**, and incubate them at 30 °C for 12 h (*see* **Notes 1** and **2**).
22. Once verified the suitability of the strains (through examination of the control plates), place the filter with the mixture of strains (using sterile forceps) onto a 15 mL polypropylene tube containing 3 mL of sterile LB broth and shake the tube until complete suspension of the bacteria. Discard the control plates.
23. Make serial dilutions from the suspension and seed 200 μL of each dilution onto LB plates supplied with Gm (30 μg/mL).
24. Incubate the plates at 30 °C, until rising of transconjugant colonies (about 48 h). This medium only allows the growth of these *P. putida* DOC21 derivatives in which the pJQ200KS-derived plasmid has been inserted onto the chromosome.
25. Pick isolated colonies with a sterile toothpick and suspend each colony on an Eppendorf tube containing 200 μL of sterile MilliQ quality water.
26. Make serial dilutions of the cells and plate them in LB plates supplied with 10% sucrose without Gm.
27. Incubate the plates at 30 °C until colonies arise (about 48 h).
28. Select isolated clones by streaking them in LB plates containing 10% sucrose and LB plates supplied with Gm. The clones that have suffered a deletion should be fully sensitivity to Gm (*see* **Note 11**).
29. Check that the deletion has occurred in the selected clones. Thus, amplify by PCR using specific oligonucleotide primers designed over genomic DNA (upstream and downstream) of the fragments that has been cloned for recombinations. The colonies in which the deletion has taken place should present an amplification product with a reduced molecular size (it has lost the deleted fragment) than the amplification observed when the wild-type strain is used (*see* **Note 12**).

The procedure is schematized (*see* Fig. 8).

3.5 Identification of 4-Androstene-3,17-Dione and 1,4-Androstadiene-3,17-Dione as Metabolic Intermediates Using Specific Mutants of *P. putida* DOC21

Using the technique described in Subheading 3.4, the genes encoding the 3-ketosteroid-Δ^1-dehydrogenase (*stdH*) and the reductase component of the 3-ketosteroid-9α hydroxylase (*stdJ*) from *P. putida* DOC21 were deleted, collecting the mutant strains *P. putida* DOC21 Δ*stdH* and *P. putida* DOC21 Δ*stdJ*. As expected, mutant Δ*stdH* accumulated in the culture broths AD, whereas *P. putida* DOC21 Δ*stdJ* accumulated ADD when cultured in a medium containing testosterone or LCA (5 mM) as carbon sources

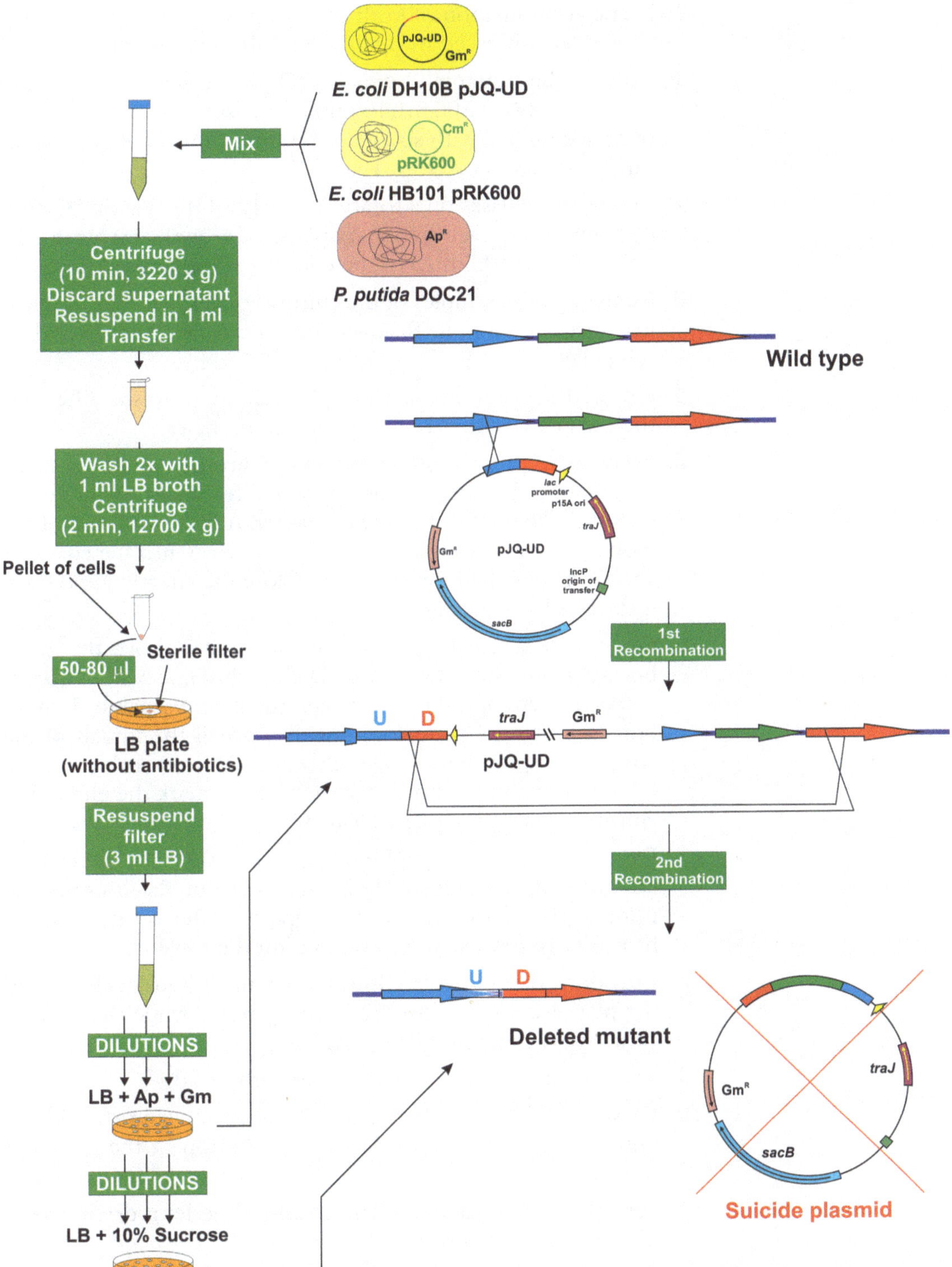

Fig. 8 Schematic representation of the procedure followed to delete a specific gene in *P. putida* DOC21. The two homologous recombinations that occur during the process are indicated. The suicide plasmid used in the representation (pJQ-UD) is the same whose construction is indicated in Fig. 7

[20]. The accumulation of both compounds in the culture broths could be easily observed by Thin-Layer Chromatography.

1. Culture the mutants *P. putida* DOC21 Δ*stdH* and *P. putida* DOC21 Δ*stdJ* in 500 mL Erlenmeyer flasks containing 100 mL of MM supplied with succinate (10 mM) and LCA (5 mM) as carbon sources (*see* **Note 13**).
2. Incubate the flasks on a rotary shaker (250 rpm) at 30 °C until 80 hours after reach stationary phase of growth (*see* **Note 14**). Take aliquots (0.5 mL) of the cultures at the desired times.
3. Centrifuge the samples of the cultures on a microfuge at maximal speed for 3 min. Recover the supernatant and discard de cell pellet.
4. Filtrate the supernatants through 0.22 μm pore size filters for the complete removal of cells and salt particles.
5. Previously, prepare a Thin-Layer Chromatography (TLC) developing chamber by disposing a filter paper lying against one of the major walls of the container. Add 80–100 mL of the mobile phase (benzene:acetone 5:3, by vol.) into the chamber and let it soak the blotting paper. Close the chamber and leave it for 1–2 h to saturate it.
6. Draw with a pencil a parallel line at 1.5 cm above the lower border from the TLC plate. Divide this line with separate marks of 1 cm of distance between them, and separate 1 cm of both sides of the plate. Write with a pencil the initials of the sample to apply below of each mark. Draw another line of 0.5 cm in the upper border of the plate to mark the end of the chromatographic step (*see* **Note 15**).
7. Drop 10 μL of each sample and patterns (AD, and ADD 5 mM each solved in sterile MM) in the marks of the bottom line using a glass Drummond micropipette. After spotting, use a hair drier to dry the spots evaporating the solvent.
8. Introduce the TLC plate into the chamber and let the top of the plate leaning against the blotting paper while the bottom contacts with the mobile phase.
9. Seal the chromatographic chamber with its lid.
10. Leave the TLC plate inside the chamber allowing the elution until the solvent reaches the line on the top of the plate (*see* **step 6**).
11. Take the TLC plate out of the chamber and let it dry in a ventilated place (exhaust hood).
12. Reveal the spots in the plate by spraying it with a solution of H_2SO_4 (30% by vol.) (*see* **Note 16**) and heating it on an oven at 100 °C for 10 min.
13. Compare the sample stains with the pattern ones.

See Fig. 9, it can be observed a TLC plate showing the accumulation of AD, and ADD in the culture broths of *P. putida* DOC21 Δ*stdH* and *P. putida* DOC21 Δ*stdJ* mutants, respectively. Commercial patterns of these compounds are also chromatographed.

3.6 Identification of StdA1 and StdA2 from P. putida DOC21 as ATP-Dependent Acyl-CoA Synthetases Involved in the Catabolism of the C-17 BA-Acyl Chain

Between the pool of *std.* genes (steroid degradation genes) found in the genome of *P. putida* DOC21, three of them (*stdA1*, *stdA2*, and *stdA3*) encoded acyl-CoA synthetases. These genes were deleted in *P. putida* DOC21 using the procedure described in Subheading 3.4, yielding the mutant strains Δ*stdA1*, Δ*stdA2*, and Δ*stdA3*. When cultured in MM containing BA, testosterone, or AD, their behavior was significantly different. Thus Δ*stdA1* and Δ*stdA2* strains were unable to grow with any BA, but they showed a similar growth to the wild type when cultured in MM containing testosterone or AD as carbon sources. However, Δ*stdA3* mutant grew poorly in media supplied with BA, and it does not grow when cultured in MM containing testosterone or AD as carbon sources. Moreover, when these mutants were cultured in MM containing succinate (10 mM) and CA, Δ*stdA1* accumulated Δ^1–3-ketocholate, and/or Δ^4–3-ketocholate, and $\Delta^{1,4}$–3-ketocholate, whereas Δ*stdA2* accumulated 7α,12α-dihydroxy-3-oxopregna-1,4-diene-20-carboxylate (DHOPDC). Moreover, when the mutant Δ*stdA3* was cultured in MM containing testosterone or bile acids, 3aα-*H*-4α(3'propanoate)7aβ-methylhexahydro-1,5-indanedione –HIP– (or the corresponding hydroxylated derivatives) were accumulated in the culture broths [17].

The assay of the enzymatic activities revealed that StdA1 was able to activate, in an ATP-dependent reaction, CA, 3-ketocholate, Δ^1–3-ketocholate, and/or Δ^4–3-ketocholate, and $\Delta^{1,4}$–3-ketocholate to their CoA thioester derivatives; StdA2 activated DHOPDC to DHOPDC-CoA; and StdA3 catalyzed the CoA thioesterification of HIP and its hydroxylated derivatives. Thus, it could be concluded that StdA1 and StdA2 are involved in the degradation of the C-17 acyl chain, whereas StdA3 initiates the degradation of the C and D rings of the steroid molecule (*see* Fig. 3). Furthermore, the results obtained with mutant Δ*stdA1* suggest that the introduction of insaturations in C-1 and C-4 in the A ring by StdH and by StdI could be accomplished without shortening of the C-17 acyl chain. These observations suggest that StdH and StdI are able to recognize as substrates testosterone, AD, 3-keto-derivatives of bile acids, and bile acids [17].

3.6.1 Preparation of the Substrates for StdA1 and StdA2 Assays

1. Culture the mutants Δ*stdA1* and Δ*stdA2* in 2000 mL Erlenmeyer flasks containing 500 mL of MM supplied with succinate (20 mM) as carbon and energy source and CA (5 mM) as source of intermediates. Incubate the flasks on a rotary shaker (250 rpm) at 30 °C until 80 h after reaching the stationary phase of growth.

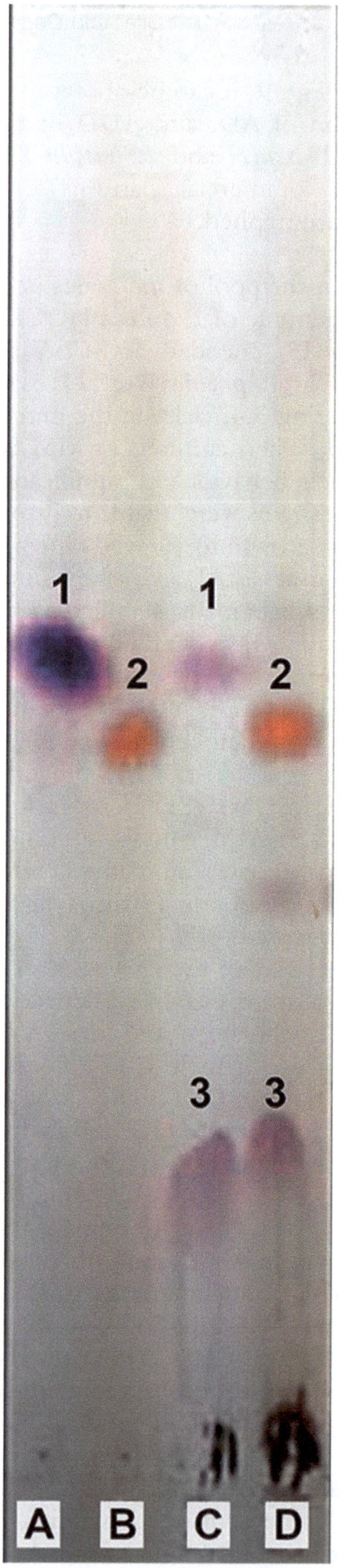

Fig. 9 Thin-Layer Chromatogram of commercial AD (**a**) and ADD (**b**), and the products accumulated in the culture broth of the mutants *P. putida* Δ*stdH* (**c**) and *P. putida* Δ*stdJ* (**d**) when these strains are cultured in MM

2. To obtain the culture broths containing the intermediates, centrifuge the cultures at 12,400 × *g* for 10 min at 4 °C. Recover the supernatant and discard de cell pellet.
3. Filtrate the culture broths through 0.22 μm pore size filters for the complete removal of cells and salt particles.
4. Extract the metabolites accumulated by mutants Δ*stdA1* (Δ^1–3-ketocholate, and/or Δ^4–3-ketocholate, and $\Delta^{1,4}$–3-ketocholate) and Δ*stdA2* (7α,12α-dihydroxy-3-oxopregna-1,4-diene-20-carboxylate, DHOPDC). The supernatant should be acidified to pH ~1 with HCl and extracted three times with equivalent volumes of ethyl acetate.
5. Combine organic fractions from each mutant and evaporate to dryness.
6. Suspend the residues in 2 mL of ethyl acetate and filter through a 0.22 μm filter.
7. Purify Δ^1–3-ketocholate, and/or Δ^4–3-ketocholate, and $\Delta^{1,4}$–3-ketocholate by HPLC using a semipreparative reversed-phase column (250 by 8 mm, Eurosphere II, 100–5 C18 H) with a total flow rate of 0.8 mL/min using as eluents K-Na-phosphate buffer (10 mM; pH 7.1) (eluent A) and acetonitrile (eluent B). A gradient method is used, starting with 20% eluent B for 2 min, increasing to 34% eluent B within 9 min, and returning to 20% eluent B within 1 min, followed by an equilibration of 6 min.
8. For the production of 3-ketocholate as a substrate for StdA1, perform a reaction containing 50 mM Tris–HCl (pH 8.0), 1 mM $NAD(P)^+$ or $K_3Fe[CN]_6$, and 0.3 U of α,β-hydroxysteroid dehydrogenase from *Pseudomonas testosteroni* (Sigma-Aldrich Co.). Start the reaction by adding 1 mM cholate. Cholate-dependent reduction of $NAD(P)^+$ and $K_3Fe[CN]_6$ is measured spectrophotometrically at 365 nm ($\varepsilon = 3.4\ mM^{-1}\ cm^{-1}$) and at 436 nm ($\varepsilon = 0.7\ mM^{-1}\ cm^{-1}$), respectively [16]. Extract the reaction products as indicated in **step 4**, and quantify residual cholate by HPLC, using an isocratic method with 70% eluent A and 30% eluent B.
9. DHOPDC extracted from Δ*stdA2* is purified by HPLC using a semipreparative reversed-phase column (250 by 8 mm, Eurosphere II, 100–5 C18 H) with a total flow rate of 0.8 mL/min. The gradient method used, starts with 20% eluent B for 2 min, increases to 70% eluent B in 1 min, followed by an equilibration of 6 min.

Fig. 9 (continued) containing succinate (10 mM) as carbon and energy source and LCA (5 mM) as precursor of intermediates. (*1*) AD spots; (*2*) ADD spots; (*3*) LCA remaining in the culture broths

10. The concentrations of Δ^1–3-ketocholate, and/or Δ^4–3-ketocholate, $\Delta^{1,4}$–3-ketocholate, and DHOPDC in the resulting solutions are estimated by measuring their absorbance at 245 nm, which is the characteristic maximum for steroid compounds containing a Δ^1-, Δ^4-, and $\Delta^{1,4}$–3-keto structure. Calculations are based on an averaged molar extinction coefficient of $\varepsilon_{245\ nm}$ of 14.7 cm^{-1} mM^{-1} [36].

3.6.2 StdA1 and StdA2 ATP-Dependent Acyl-CoA Synthetase Assays

1. Culture *P. putida* DOC21 and the Δ*stdA1* and Δ*stdA2* mutants in 2000 mL Erlenmeyer flasks containing 500 mL of MM containing succinate (10 mM) and CA (5 mM). Incubate on a rotary shaker (250 rpm, 30 °C) until the late exponential phase of growth.
2. Harvest the bacteria by centrifugation (9000 × *g*, 10 min at 4 °C).
3. Wash them with 50 mM K-Na-PO_4 buffer (pH 7.0) and centrifuge (9000 × *g*, 10 min at 4 °C).
4. Suspend the bacterial pellet in 3 mL of the same buffer.
5. Disrupt the bacteria by three passages through a French press at 138 MPa.
6. Remove the cell debris by centrifugation (17,900 × *g*, 30 min at 4 °C).
7. Desalt the homogenates through Sephadex G-25 PD-10 columns to eliminate low molecular weight compounds.
8. Froze at −20 °C until use.
9. Determine protein concentration by Bradford protein assay using bovine serum albumin as a standard.
10. Assay StdA1 and/or StdA2 activities by measuring the formation of different acyl-CoA thioesters in a mixture of reaction containing 50 mM MOPS buffer (pH 7.8), 1 mM CoA, 2 mM ATP, 3 mM $MgCl_2$, 0.6–1.0 mg/mL of cell-free extracts from the different strains, starting the reaction by adding 0.5 mM of the tested substrate (CA, 3-ketocholate, Δ^1–3-ketocholate, Δ^4–3-ketocholate, $\Delta^{1,4}$–3-ketocholate, or DHOPDC).
11. Withdraw samples immediately after the starting of the reaction and at defined time intervals (15 min–2 h).
12. Analyze the CoA thioesters produced by StdA1(cholyl-CoA, 3-ketocholyl-CoA, $\Delta^{1,4}$–3-ketocholyl-CoA) and StdA2 (DHOPDCCoA) by HPLC using a gradient method, starting with 10% acetonitrile +90% K-Na_phosphate buffer (10 mM, pH 7.1) for 2 min, increasing to 56% acetonitrile within 23 min, and returning to 10% acetonitrile within 1 min, followed by an equilibration of 6 min [18, 19] (*see* **Note 17**).

3.7 Identification of ATP-Dependent Acyl-CoA Synthetase StdA3 from *P. putida* DOC21 Involved in the Degradation of the C and D Rings of the Steroid Core

P. putida DOC21 Δ*stdA3* accumulated 3aα-*H*-4α(3′propanoate)7aβ-methylhexahydro-1,5-indanedione (HIP) when cultured in MM containing as source of precursors testosterone, AD or LCA. When this mutant is cultured in the presence of other bile acids, the respective hydroxylated derivatives of HIP are accumulated in the culture broth [3aα-*H*-4α(3′(*R*)-hydroxy-3′-propanoate)-7aβ-methylhexahydro-1,5-indanedione from CDCA or 3aα-*H*-4α(3′(*S*)-hydroxy-3′-propanoate)-7aβ-methylhexahydro-1,5-indanedione from UDCA -3′-OH-HIP-; 3aα-*H*-4α(3′-propanoate)-7α-hydroxy-7aβ-methylhexahydro-1,5-indanedione –7-OH-HIP- from DCA; and 3aα-*H*-4α(3′(*R*)-hydroxy-3′-propanoate)-7α-hydroxy-7aβ-methylhexahydro-1,5-indanedione –3′,7-diOH-HIP - formed from CA]. StdA3 has been characterized as the ATP-dependent acyl-CoA synthetase catalyzing the activation of the HIPs to their CoA derivatives for its further catabolism [17]. A similar activity has been identified in actinobacteria when the metabolism of cholesterol was studied [23].

3.7.1 Obtaining of the Substrates for StdA3 Assay

1. Culture the mutant Δ*stdA3* in 1 L of MM supplied with 20 mM of pyruvate (2 × 2000 mL Erlenmeyer flasks containing 500 mL MM each). Incubate the flasks on a rotary shaker (250 rpm) at 30 °C.
2. Harvest the cells by centrifugation (12,400 × *g* during 10 min at 4 °C) at the mid-exponential phase of growth (OD_{540} = 1.2–1.4).
3. Wash the cells with 400 mL of MM without carbon source; recover the cells by centrifugation at 12,400 × *g* during 10 min at 4 °C.
4. Suspend the cells in 400 mL MM containing 1–2 mM of the required steroid (BA or testosterone).
5. Incubate the bacterial cell suspensions with shaking (250 rpm) for 80–100 h at 30 °C.
6. Harvest the cells by centrifugation at 12,400 × *g* during 10 min at 4 °C and recover the supernatants.
7. Extract the metabolites accumulated by mutant Δ*stdA3* (HIP, or the corresponding hydroxylated derivatives) by acidification of the supernatants to pH ~ 3 with concentrated HCl and followed by three successive organic extractions with one equivalent of ethyl acetate.
8. Combine the organic fractions and evaporate the solvent to dryness.
9. Suspend the solid residue in 2 mL of ethyl acetate and pass through a 0.22 μm filter.
10. Purify by HPLC the extracted metabolites by loading the sample onto a 250 × 10 mm Luna 5 mm silica(2) column

(Phenomenex, Torrance, CA). Operate the HPLC system at 11.1 mL/min.

11. For HPLC separation of the HIPs, use a 60 min gradient of 0–100% ethyl acetate in hexane and 0.5% acetic acid.
12. Monitor the eluates at 290 nm.
13. Recover the fractions containing HIP (or the hydroxylated derivatives), evaporate them to dryness under nitrogen, and store until use at −80 °C.

3.7.2 StdA3 ATP-Dependent Acyl-CoA Synthetase Assay

1. Prepare cell-free extracts of wild-type strain and Δ*stdA3* mutant according to the **steps 1–9** from Subheading 3.6.2.
2. Assay StdA3 activity by measuring the formation of different acyl-CoA thioesters in 100 μL of mixture of reaction containing 100 mM HEPES (pH 7.3), 5 mM $MgCl_2$, 2 mM ATP, 1.2 mM CoA, 2 μM of StdA3, and 1 mM the tested substrate (3′,7-diOH-HIP, 3′-OH-HIP, 7-OH-HIP, or HIP).
3. Incubate the reactions for 2 h at 22 °C.
4. Analyze the thioester formation by HPLC by injecting 1 mL of samples on a Luna 3 μm PFP(2) 50 × 4.6 mm column equilibrated with 0.1 M ammonium acetate, pH 4.5. Elute the CoA thioesters using a 20 mL gradient from 0 to 90% methanol in 0.1 M ammonium acetate, pH 4.5, monitoring the eluate at 258 nm.

4 Notes

1. Incubate the filter containing the mixture of strains. The temperature selected is the optimal for the more restrictive strain (acceptor strain).
2. Avoid the formation of bubbles when disposing the mixture of cells over the filter; these bubbles could explode spreading the strains over the plate.
3. Taking into account that none of the parental strains carries both of the antibiotic resistances used for selection (Ap and Km), in these plates only could grow those *P. putida* DOC21 (Ap resistance) transconjugants, in which the transposon (giving the Km resistance) has been inserted in the genome.
4. If the acceptor strain would have a Km resistance, the selection should be made using these other antibiotic markers (bleomycin and streptomycin) also present in the transposon (*see* Fig. 4).
5. In *P. putida* DOC21 a deleterious effect inhibiting the bacterial growth has been observed when this strain is cultured in media containing Ap and BA. Although this effect has not

been investigated, it could be speculated that a mixture of bile acids (acting as detergents) and ampicillin could have a synergistic effect, sensitizing the bacteria against the antibiotic.

6. Remember that IS*50*L and IS*50*R are repeated and inverted sequences. Thus, the recombination through the sequence cloned in pJQ-Tn*5* could occur with the same probability over both IS*50* sequences but with an opposite direction, allowing the cloning of both, the sequences upstream and downstream from the Tn*5* insertion point (*see* Fig. 6).
7. Use various clones. Taking into account that the recombination event could happen with the same probability over the two IS*50* sequences, the resulting recombinants will correspond to both events.
8. The use of these restriction endonucleases (*Bam*HI, *Xba*I, *Sma*I, or *Sal*I) is due to the presence of target sequences in the multiple cloning site of the plasmid but the lack of these targets inside the IS*50* sequences.
9. After the restriction of the clones with the same endonuclease used to rescue the fragment adjacent to Tn*5*, only a band in an agarose gel should be observed. It corresponds to the pJQ200KS plus the insert (1269 bp from the IS*50* fragment plus the adjacent genomic DNA). If more restriction bands are observed, this could correspond to the illegitimate ligation of fragments from the genome. Discard these clones.
10. After the sequencing of the DNA placed at both sides of the Tn*5* insertion point, a directed repeat sequence of 9 bp will be observed. This repetition is due to the insertion of the transposon and should be corrected.
11. Never select a strain that showed Gm resistance even though grew in the sucrose selective plates. These strains could grow due to mechanisms different to the deletion of the desired gene, as, in example inactivation of the *sacB* gene or the promoter of this gene.
12. Other useful techniques for the determination of the deletion of the fragment could be Southern blot using the genomic DNA from the mutant, or Northern blot or real-time Q-PCR, using RNA extracted from induced bacteria.
13. Both mutants were unable to grow using BA, testosterone nor ADD as sole carbon sources. MM was supplied with 10 mM succinate (to support the bacterial growth) and 5 mM LCA or testosterone as a source of the intermediates (AD or ADD). Catabolism of succinate causes a strong repression over the use of steroids; thus, other non-steroid carbon sources to maintain the cellular growth could be profitable.
14. When succinate (10 mM) and a steroid compound (5 mM) are used as carbon sources, the bacteria do not start the biotrans-

formation of the steroid compounds into the precursors until succinate depletion occurs.

15. Handle the TLC plates with gloves on. Never touch the surface of the plate, even with gloves.
16. Take into account that commercial gas sprayers could be made with non-resistant to sulfuric acid plastic (mainly the tubes and the head of the sprayer). Prepare the H_2SO_4 30% (w/vol) solution for revealing, just in the moment of use and clean carefully the sprayer just after use. Alternatively, the H_2SO_4 solution could be disposed with a roller made with H_2SO_4 resistant materials but carefully avoiding the leakage of sulfuric acid.
17. Alternatively, analyze the products of the reaction assays by LC-tandem mass spectrometry (LC-MS/MS) as described [16].

Acknowledgments

This research was supported by the Ministerio de Economía y Competitividad (Madrid, España, Grant BIO2015-66960-C3-3R), by the CENIT Project RTC-2014-2249-1 (CDTI, Ministerio de Economía y Competitividad, Madrid, España), and by a Grant from the Junta de Castilla y León(Consejería de Educación, Valladolid, España; LE114U13).

References

1. Russell DW (2003) The enzymes, regulation, and genetics of bile acid synthesis. Annu Rev Biochem 72:137–174
2. Hofmann AF, Hagey LR (2008) Bile acids. Chemistry, pathochemistry, biology, pathobiology, and therapeutics. Cell Mol Life Sci 65:2461–2483
3. Björkhem I (1985) Mechanism of bile acid biosynthesis in mammalian liver. In: Danielsson H, Sjövall J (eds) Steroids and bile acids. Elsevier Science, New York, pp 231–260
4. Solaas K, Ulvestad A, Soreide O et al (2000) Subcellular organization of bile acid amidation in human liver: a key issue in regulating the biosynthesis of bile salts. J Lipid Res 41:1154–1162
5. Dawson PA, Lan T, Rao A (2009) Bile acid transporters. J Lipid Res 50:2340–2357
6. Hamilton JP, Xie G, Raufman JP et al (2007) Human cecal bile acids: concentration and spectrum. Am J Physiol Gastrointest Liver Physiol 293:G256–G263
7. Setchell KDR, Lawson AM, Tanida N et al (1983) General methods for the analysis of metabolic profiles of bile acids and related compounds in feces. J Lipid Res 24:1085–1100
8. Merino E, Barrientos A, Rodríguez J et al (2013) Isolation of cholesterol- and deoxycholate-degrading bacteria from soil samples: evidence of a common pathway. Appl Microbiol Biotechnol 97:891–904
9. Horinouchi M, Hayashi T, Kudo T (2012) Steroid degradation in *Comamonas testosteroni*. J Steroid Biochem Mol Biol 129:4–14
10. Van der Geize R, Yam K, Heuser T et al (2007) A gene cluster encoding cholesterol catabolism in a soil actinomycete provides insight into *Mycobacterium tuberculosis* survival in macrophages. Proc Natl Acad Sci U S A 104:1947–1952
11. Uhía I, Galán B, Morales V et al (2011) Initial step in the catabolism of cholesterol by *Mycobacterium smegmatis* mc^2 155. Environ Microbiol 13:943–959
12. Bragin EY, Shtatnikova VY, Dovbnya DV et al (2013) Comparative analysis of genes encoding key steroid core oxidation enzymes in fast-

growing *Mycobacterium* spp. strains. J Steroid Biochem Mol Biol 138:41–53

13. Van der Geize R, Grommen AW, Hessels GI et al (2011) The steroid catabolic pathway of the intracellular pathogen *Rhodococcus equi* is important for pathogenesis and a target for vaccine development. PLoS Pathog 7:e1002181
14. Drzyzga O, de las Fernández HL, Morales V et al (2011) Cholesterol degradation by *Gordonia cholesterolivorans*. Appl Environ Microbiol 77:4802–4810
15. Mohn WW, Wilbrink MH, Casabon I et al (2012) Gene cluster encoding cholate catabolism in *Rhodococcus* spp. J Bacteriol 194:6712–6719
16. Birkenmaier A, Holert J, Erdbrink H et al (2007) Biochemical and genetic investigation of initial reactions in aerobic degradation of the bile acid cholate in *Pseudomonas* sp. Strain chol1. J Bacteriol 189:7165–7173
17. Barrientos A, Merino E, Casabon I et al (2015) Functional analyses of three acyl-CoA synthetases involved in bile acid degradation in *Pseudomonas putida* DOC21. Environ Microbiol 17:47–63
18. Holert J, Jagmann N, Philipp B (2013) The essential function of genes for a hydratase and an aldehyde dehydrogenase for growth of *Pseudomonas* sp. Strain chol1 with the steroid compound cholate indicates an aldolytic reaction step for deacetylation of the side chain. J Bacteriol 195:3371–3380
19. Holert J, Kulić Ž, Yücel O et al (2013) Degradation of the acyl side chain of the steroid compound cholate in *Pseudomonas* sp. strain Chol1 proceeds via an aldehyde intermediate. J Bacteriol 195:585–595
20. Barrientos A, Merino E, Fernández de las Heras L, Perera J, Navarro Llorens JM, Olivera ER and Luengo JM (2017) Steroid catabolism in bacteria: Genetic and functional analyses of *stdI*, *stdH* and *stdJ* in *Pseudomonas putida* DOC21. Submitted for publication
21. Horinouchi M, Yamamoto T, Taguchi K et al (2001) Meta-cleavage enzyme gene *tesB* is necessary for testosterone degradation in *Comamonas testosteroni* TA441. Microbiology 147:3367–3375
22. Yam KC, D'Angelo I, Kalscheuer R et al (2009) Studies of a ring-cleaving dioxygenase illuminate the role of cholesterol metabolism in the pathogenesis of *Mycobacterium tuberculosis*. PLoS Pathog 5:e1000344
23. Casabon I, Crowe AM, Liu J et al (2013) FadD3 is an acyl-CoA synthetase that initiates catabolism of cholesterol rings C and D in actinobacteria. Mol Microbiol 87:269–283
24. Horinouchi M, Hayashi T, Koshino H et al (2006) ORF-18-disrupted mutant of *Comamonas testosteroni* TA441 accumulates significant amounts of 9,17-dioxo-1,2,3,4,10,19-hexanorandrostan-5-oic acid and its derivatives after incubation with steroids. J Steroid Biochem Mol Biol 101:78–84
25. Horinouchi M, Hayashi T, Koshino H et al (2014) Identification of 9α-hydroxy-17-oxo-1,2,3,4,10,19-hexanorandrost-6-en-5-oic acid and β-oxidation products of the C-17 side chain in cholic acid degradation by *Comamonas testosteroni* TA441. J Steroid Biochem Mol Biol 143:306–322
26. Horinouchi M, Hayashi T, Koshino H et al (2014) Identification of 9α-hydroxy-17-oxo-1,2,3,4,10,19-hexanorandrostan-5-oic acid in steroid degradation by *Comamonas testosteroni* TA441 and its conversion to the corresponding 6-en-5-oyl coenzyme A (CoA) involving open reading frame 28 (ORF28)- and ORF30-encoded acyl-CoA dehydrogenases. J Bacteriol 196:3598–3608
27. Martínez-Blanco H, Reglero A, Rodríguez-Aparicio LB et al (1990) Purification and biochemical characterization of phenylacetyl-CoA ligase from *Pseudomonas putida*. A specific enzyme for the catabolism of phenylacetic acid. J Biol Chem 265:7084–7090
28. Selvaraj G, Iyer VN (1983) Suicide plasmid vehicles for insertion mutagenesis in *Rhizobium meliloti* and related bacteria. J Bacteriol 156:1292–1300
29. Herrero M, de Lorenzo V, Timmis KN (1990) Transposon vector containing non-antibiotic selection markers for cloning and stable chromosomal insertion of foreign DNA in gram-negative bacteria. J Bacteriol 172:6557–6567
30. Quandt J, Hynes MF (1993) Versatile suicide vectors which allow direct selection for gene replacement in gram-negative bacteria. Gene 127:15–21
31. Bradford MM (1976) A rapid and sensitive method for the quantitation of microgram quantities of protein utilizing the principle of protein-dye binding. Anal Biochem 72:248–254
32. Ruvkun GB, Sundaresan V, Ausubel FM (1982) Directed transposon Tn*5* mutagenesis and complementation analysis of *Rhizobium meliloti* symbiotic nitrogen fixation genes. Cell 29:551–559
33. Olivera ER, Miñambres B, García B et al (1998) Molecular characterization of the phenylacetic acid catabolic pathway in *Pseudomonas*

putida U: the phenylacetyl-CoA catabolon. Proc Natl Acad Sci U S A 95:6419–6424

34. Arcos M, Olivera ER, Arias S et al (2010) The 3,4-dihydroxyphenylacetic acid catabolon, a catabolic unit for degradation of biogenic amines tyramine and dopamine in *Pseudomonas putida* U. Environ Microbiol 12:1684–1704
35. Berg DE, Weiss A, Crossland L (1980) Polarity of Tn*5* insertion mutations in *Escherichia coli.* J Bacteriol 142:439–446
36. Tenneson M, Baty J, Bilton R et al (1979) The degradation of cholic acid by *Pseudomonas* sp. NCIB 10590. Biochem J 184:613–618

Chapter 9

Selection of Biodegrading Phytosterol Strains

María-Angélica Mondaca, Maricel Vidal, Soledad Chamorro, and Gladys Vidal

Abstract

The phytosterol-biotransforming strains were selected from *Mycobacterium* sp., using a high concentration of β-sitosterol. The selection was made by culturing the strains in a medium enriched with 14 g β-sitosterol/L as the unique source of carbon. During 2 months, the bacterial cultures were transferred successively. The extraction of the biotransformation products was made with methanol and ethyl acetate. The qualitative and quantitative analysis was made by means of thin-layer chromatography, gas-liquid chromatography (GLC), and GLC-mass spectrometry. Under these conditions, it was observed that after seven transfers, the strains *Mycobacterium* sp. MB-3683 and the *Mycobacterium fortuitum* B-11045 increased their biotransformation capacity from 20% to 64% and from 34% to 55%, respectively. The products in the highest proportion identified for each trial were androstenedione and androstadienedione. The results suggest that the high substrate concentration could be a selective mechanism to obtain strains more efficient in the biotransformation of β-sitosterol into steroidal bases.

Key words Steroids, Biotransformation, Natural products, β-Sitosterol *Mycobacterium* sp.

1 Introduction

Growing industrial and human activity has led to an increase in the presence of steroids in the environment [1, 2]. The recalcitrant nature of these compounds, which is due to their structural stability, results in contamination of soil and water having a potential threat to biological processes like reproduction [1–3]. The use of high concentrations of phytosterol is an efficient way to select bacterial strains that can biotransform phytosterol. Several studies of steroid degradation pathways [4] suggest a likely convergence of steroids in common intermediates like 4-androsten-3,17-dione (AD) and 1,4-androstadien-3,17-dione (ADD) (Fig. 1) [5]. In this sense, steroid drugs are synthetized by chemical or microbial routes, both of which involve conversion of steroid precursors into drug intermediates and final conversion of intermediates. Microbial transformation cleaves complex side-chains of steroid

José-Luis Barredo, Ignacio Herráiz (eds.), *Microbial Steroids: Methods and Protocols*, Methods in Molecular Biology vol. 1645, DOI 10.1007/978-1-4939-7183-1_9, © Springer Science+Business Media LLC 2017

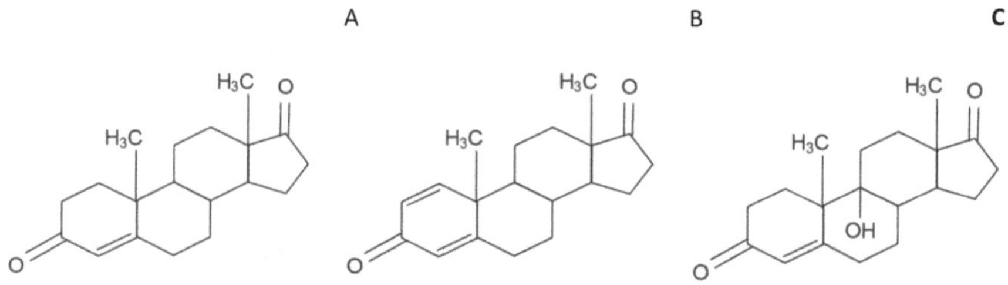

Fig. 1 Biotransformation products obtained during the selection of *Mycobacterium* sp. strains: (**a**) 4-Androstene-3,17-dione (AD). (**b**) 1,4-Androstadien-3,17-dione (ADD). (**c**) 9-Hydroxy-(4-androstene-3,17-dione) (9-OH-AD). The wild-type strain only produced AD and ADD. 9-OH-AD was also produced by the strains derived from *M. fortuitum* B-11045, beginning with the fourth assay

precursor steroids in a single step and incorporates desirable modifications in steroid nuclei [6, 7].

A better understanding of the genetic and biochemical catabolic routes of these compounds would enable us to use them as decontaminating biological agents and thereby mitigate the problems caused by steroid accumulation.

The goal of the present work was to select strains capable of transforming steroids using high concentrations of β-sitosterol.

2 Materials

2.1 Substrate

1. Fraction isolated from tall oil (Kraft mill effluent) rich in phytosterols. This fraction contains about 95% of sterols (85% correspond to β-sitosterol and 15% correspond to campesterol and stigmasterol).

2.2 Bacterial Strains

1. *Mycobacterium* sp. MB-3683 strains (Bombay) were obtained from *Mycobacterium* sp. NRRL B-3683 (supplied by Dr. Raltan Sood; University of Bombay) (*see* **Note 1**).
2. *Mycobacterium fortuitum* B-11045 strains according to Wovcha et al. [8].

2.3 Bacterial Growth Media

All solutions were prepared with purified and sterilized water.

1. Middlebrook agar 7H10: Formulation per liter of purified water: 0.025 g $MgSO_4 \cdot H_2O$, 5.0 g bovine albumin (fraction V), 0.04 g ferric ammonium citrate, 0.004 g catalase, 0.4 g sodium citrate, 0.001 g pyridoxine, 0.5 g $(NH_4)_2$ HSO_4, 0.001 g $ZnSO_4 \cdot 7H_2O$, 0.5 g monosodium glutamate, 0.001 g $CuSO_4 \cdot 7H_2O$, 0.0005 g biotin, 1.5 g $Na_2HPO_4 \cdot H_2O$, 1.5 g $KH_2PO_4 \cdot H_2O$, 0.0005 g $CaCl_2 \cdot H_2O$, 0.00025 g malachite green, 17.0 g agar, 5.0 g glycerol, 0.85 g NaCl, 2.0 g glucose, and 0.05 mL oleic acid.

2. Inoculum medium: 10 g yeast extract, 10 mL glycerol, 1.5 g $(NH_4)_2$ HPO_4, 1.3 g $K_2HPO_4 \cdot 3H_2O$, 0.2 g $MgSO_4$, 0.01 g $FeSO_4 \cdot 7H_2O$, and 0.02 g $ZnSO_4 \cdot 7H_2O$ per liter of distilled water.
3. Enrichment medium: 14 g β-sitosterol, 60 g Amberlite XAD-2 (Sigma-Chemical Company), 1.5 g $(NH_4)_2$ HPO_4, 1.3 g $K_2HPO_4 \cdot 3H_2O$, 0.2 g $MgSO_4$, 0.01 g $FeSO_4 \cdot 7H_2O$, and 0.02 g $ZnSO_4 \cdot 7H_2O$ per liter of distilled water.
4. β-sitosterol purity >98% (Sigma Chemical Co. MO, USA).
5. Amberlite XAD-2 resin surface area 330 m^2/g, pore diameter 9 nm and bead size 20–60 mesh (Sigma Chemical Co. MO, USA).

2.4 Equipment and Solvents

1. Thin-layer chromatography (TLC) performed on a sheet of glass (20 × 20 cm), which is coated with a thin layer of cellulose was conducted with methanol [8].
2. Gas-liquid chromatography (GLC) was conducted with a Varian Star 3400 Cx chromatograph and a Hewlett Packard (HP 5890 30 m length × 0.25 μm) GLC-mass spectrometer with a FID detector HP 5972 mass detector and HP-5MS column and helium as the carrier gas.
3. GLC-mass spectrometry was conducted with helium as the carrier gas.
4. Standard of AD purity >98% (Sigma Chemical Co. MO, USA).
5. Standard of ADD purity >97% (Sigma Chemical Co. MO, USA).
6. Incubator.
7. Orbital shaker.
8. 500 mL Erlenmeyer flasks.
9. Methanol purity ≥99.8%.
10. Ethyl acetate purity ≥99.8%.
11. Rotary evaporator.
12. Sephadex column.
13. 1H-nuclear magnetic resonance (NMR) Bruker AC 250-P spectrometer at 250 MHz.
14. CDCl3 solvent.
15. Nicolet 550 magno-IR spectrophotometer.

3 Methods

3.1 Selection of Strains Tolerant to High Concentrations of β-Sitosterol

1. Maintain *Mycobacterium* sp. strains in a freezer at −70 °C.
2. Seed 30 mL of Inoculum medium with 300 μL of the frozen vial and incubate at 30 °C for 7 days.

Table 1
***Mycobacterium* sp. strains selected with 14 g/L of β-sitosterol in the culture medium**

Strains	Biotransformation products (%)		
	AD	ADD	9-OH-AD
MB-3683(1)	51.2	48.0	0
MB-3683(2)	70.4	29.3	0
MB-3683(3)	68.5	30.1	0
MB-3683(4)	81.8	17.5	0
MB-3683(5)	61.2	38.5	0
MB-3683(6)	60.0	39.5	0
MB-3683(7)	72.3	26.7	0
MB-3683(27)	71.0	28.9	0
MB-3683(29)	95.8	4.1	0
MB-3683(30)	30.7	67.8	0
B-11045(31)	39.5	41.4	18.5
B-11045(32)	50.0	48.7	1.3
B-11045(33)	19.5	60.0	18.3
B-11045(34)	31.3	64.5	4.2
B-11045(35)	9.8	80.0	10.1
B-11045(39)	10.0	79.5	9.8
B-11045(40)	5.5	80.7	12.0
B-11045(41)	31.2	57.4	11.1
B-11045(42)	29.5	60.1	10.4

AD 4-androstene-3,17-dione, *ADD* 1,4-androstadien-3,17-dione, *9-OH-AD* 9-hydroxy-4-androstene-3,17-dione

3. Add 5 mL of inoculum culture (10%) to 50 mL of enrichment medium.
4. Incubate the bacterial cultures at 30 °C and 180 rpm for 7 days.
5. Transfer 5 mL of the bacterial culture to 50 mL of enrichment medium every 7 days over a period of 2 months.
6. Seed the tolerant (growing) strains in Middlebrook agar 7H10 with 14, 28, and 56 g/L of β-sitosterol (Table 1) (*see* **Note 2**).

Table 2
Biotransformation percentages for the selected strains of *Mycobacterium* sp. MB-3683 and *M. fortuitum* B-11045 cultured at a concentration of 14 g/L of β-sitosterol. Biotransformation percentages were determined with reference to the wild strain

Strains	Successive transfer	Biotransformation percentages (%)
Mycobacterium sp. MB-3683	1	22.39
	2	40.28
	3	39.40
	4	46.27
	5	59.54
	6	55.79
	7	64.28
M. fortuitum B-11045	1	34.06
	2	33.10
	3	40.77
	4	45.65
	5	43.35
	6	50.01
	7	55.35

3.2 β-Sitosterol Biotransformation

1. Inoculate 30 mL (10% of the final volume) of the strains previously selected in 500 mL Erlenmeyer flasks with a final volume of 300 mL of Enrichment medium (Table 2) (*see* **Notes 3** and **4**).
2. Incubate the flasks at 30 °C and 180 rpm for 7 days.

3.3 Extraction of Biotransformation Products

1. Sterilize the biotransformation cultures at 121 °C for 20 min.
2. Extract the biotransformation products after sterilizing the cultures from the amberlite with 150 mL methanol and 150 mL ethyl acetate.
3. Concentrate the total extracts concentrated in a rotary evaporator at reduced pressure and then take to a final volume of 10 mL (*see* **Notes 4** and **5**).
4. Concentrate 300 mL of total extracts in a rotary evaporator at reduced pressure at 25 °C for about 3–4 h to obtain a final volume of 10 mL (Table 2) (*see* **Notes 6** and **7**).

3.4 Analysis of Biotransformation Products

3.4.1 Analysis by TLC

1. Take each sample obtained from the biotransformation processes to be qualitatively analyzed initially by means of thin-layer chromatography (TLC) [9].
2. Place each sample on a silica gel 60 F 254 plate and leave dry for 15 min at room temperature.

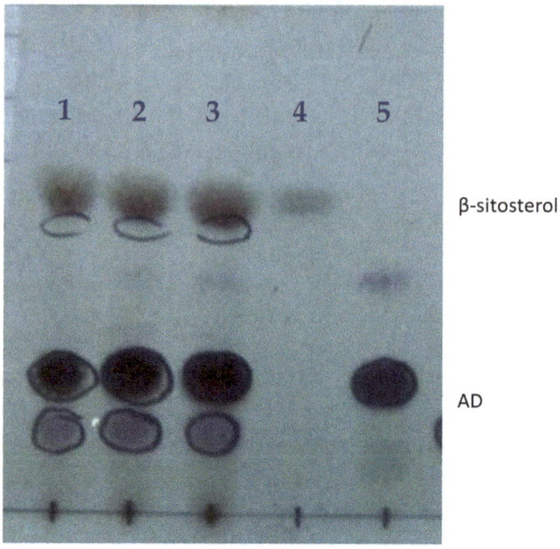

Fig. 2 Thin-layer chromatography of the *Mycobacterium* sp. strain MB-3683. Note the higher AD production by this strain. (*1*) Replica 1; (*2*) Replica 2; (*3*) Replica 3; (*4*) Standard β-sitosterol; (*5*) Standard AD

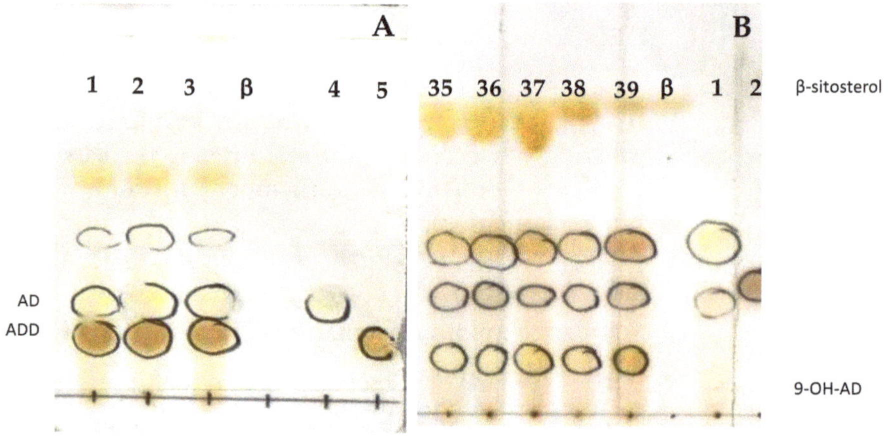

Fig. 3 Thin-layer chromatography of the *Mycobacterium fortuitum* strain B-11045. (**a**) Assay with the wild-type strain. (*1*) Replica 1; (*2*) Replica 2; (*3*) Replica 3; β. Standard β-sitosterol; (*4*) Standard AD; (*5*) Standard ADD. (**b**) *M. fortuitum* strains producing 9-OH-AD derived from B-11045 after the selection process. The band of 9-OH-AD is under the band of AD. (*35*) Strain 35; (*36*) Strain 36; (*37*) Strain 37; (*38*) Strain 38; (*39*) Strain 39; β. Standard β-sitosterol; (*1*) Standard AD; (*2*) Standard ADD

3. Place the silica gel plate inside the TLC box with mix of hexane/ethyl acetate (30/70) as a mobile phase.
4. Dry the silica plate and check under UV light. The selected strains of *Mycobacterium* sp. MB-3683 produce AD and ADD (Fig. 2). The selected strains derived from *M. fortuitum* B-11045 (from the fifth test) produce 9-hydroxy-androst-4-en-3,17-dione (9-OH-AD) (Fig. 3).

3.4.2 Steroid Quantification by GLC–Mass Spectrometry Analysis

1. Perform the gas-liquid chromatography (GLC) with helium as a carrier gas, flow rate 2 mL/min, injector temperature 250 °C, detector temperature 280 °C, and oven temperature 350 °C.
2. Prepare each sample as a mixture (1:1) with an internal standard, which in this case corresponds to cholesterol 1 mg/mL.
3. Inject 1 μL of sample.
4. For qualitative biotransformation take as a reference the area of standard solution of cholesterol, androstenedione, campesterol and β-sitosterol (concentration of 1 mg/mL of each of the components).
5. Determine the response factor of each of the components with reference to cholesterol.
6. Purify the compounds by chromatography on a Sephadex column and determine the structure of the pure compounds by 1H-nuclear magnetic resonance (NMR) and by infrared (IR) spectrophotometry.

4 Notes

1. The strains of *Mycobacterium* sp. were first trained to potentially transform steroid compounds. Mycobacteria have been used in this field, as well as successfully used to obtain hormones.
2. 14 g/L of β-sitosterol was used with the selected strain (30 strains from *Mycobacterium* sp. MB-3683 and 12 from *M. fortuitum* B-11045). Of these strains, 30 grew with concentrations of 28 g/L of β-sitosterol and 12 with 56 g/L of β-sitosterol (Table 1).
3. Our experiment in the use of high steroidal substrate concentrations as a selective mechanism for biotransforming strains was successful. In the first transfer, the total biotransformation percentages were less than 35% for the *Mycobacterium* sp. MB-3683 and *M. fortuitum* B-11045 strains. For the MB-3683 strains, the main transformation product was AD and, in the case of the B-11045 strains, the main product was ADD.
4. The 42 strains isolated in the previous stage were subjected to trials in biotransformation cultures, which were carried out in triplicate in order to determine an average value for their β-sitosterol biotransformation capacity. After seven successive transfers, the *Mycobacterium* sp. MB-3683 strains and the *M. fortuitum* B-11045 strains increased their biotransformation capacity from 22% to 64% and from 34% to 55%, respectively (Table 2).
5. From the fourth trial onward, the biotransformation percentages increased to 50% and the products for the strains derived from MB-3683 were AD and ADD, although some strains

began to produce only AD. In the case of the strains derived from B-11045, although the main product was also ADD, an additional compound, a hydroxylated AD derivative, was also identified (Fig. 1).

6. The production of 9-hydroxyandrostenedione by strains from *M. fortuitum* B-11045 could be the result of selective pressure, due to the high concentrations of β-sitosterol used. The high concentration of AD may have been responsible for the high level of enzymatic activity of 9-hydroxylase to convert AD into 9-hydroxyandrostenedione. Thus, these strains may use an alternative pathway in the degradation of phytosterols. In contrast, the strains from *Mycobacterium* sp. MB-3683 may use the traditional route, but the larger amount of substrate resulted in a higher yield in the end.
7. According to the results, the use of high substrate concentrations is an efficient methodology to select strains with phytosterol-biotransforming capacity. In addition, this approach has the advantage that it can be directed toward obtaining specific characteristics, such as a particular product or a high biotransformation capacity.

Acknowledgment

This work was supported by grant FONDEF D-96 1111, FONDECYT 1120664, and CONICYT/FONDAP/15130015.

References

1. Chamorro S, Hernández V, Matamoros V, Dominguez C, Becerra J, Vidal J, Piña B, Bayona J (2013) Chemical characterization of organic microcontaminant sources and biological effects in riverine sediments impacted by urban sewage and pulp mill discharges. Chemosphere 90:611–619
2. Pokhrel D, Viraraghavan T (2004) Treatment of pulp and paper mill wastewater—a review. Sci Total Environ 333(1–3):37–58
3. Vidal M, Becerra J, Mondaca MA, Silva M (2001) Selection of *Mycobacterium* sp. strain with capacity to biotransform high concentrations of β-sitosterol. Appl Microbiol Biotechnol 57:385–389
4. Kumar V, Dhall P, Kumar R, Singh YP, Kumar A (2012) Bioremediation of agro-based pulp mill effluent by microbial consortium comprising autochthonous bacteria. Sci World J 127014:1–7
5. Horinouchi N, Sakai E, Kawano T, Matsumoto S, Sasaki M, Hibi M, Shima J, Shimizu S, Ogawa J (2012) Construction of microbial platform for an energy-requiring bioprocess: practical 2′-deoxyribonucleoside production involving a C-C coupling reaction with high energy substrates. Microb Cell Fact 11(1):82–89
6. Rao S, Thakkar K, Pawar K (2013) Microbial transformation of steroids: current trends in cortical side chain cleavaje. Quest 1:16–20
7. Seidel L, Horhold C (1992) Selection and characterization of new microorganisms for the manufacture of 9-OH-AD from sterols. J Basic Microbiol 32:49–55
8. Wovcha MG, Antosz FJ, Knight JC, Kominek LA, Pyke TR (1978) Bioconversion of sitosterol to useful steroidal intermediates by mutants of *Mycobacterium fortuitum*. Biochim Biophys Acta 531:308–321
9. Shah K, Mehdi I, Khan A, Vora V (1980) Microbial transformation of sterols. I. Microbial transformation of phytosterols to androsta-1,4-diene-3,17-dione by *Arthrobacter simplex*. Eur J Appl Microbiol Biotechnol 10:167–169

Chapter 10

Stigmasterol Removal by an Aerobic Treatment System

Soledad Chamorro, Claudia Xavier, Víctor Hernández, José Becerra, and Gladys Vidal

Abstract

Stigmasterol is a phytosterol contained in Kraft mill effluent that is able to increase over 100% after aerobic biological treatment. This compound can act as an endocrine disrupter as its structure is similar to that of cholesterol. Furthermore, stigmasterol contained in Kraft mill effluent shows high toxicity (25-fold that of β-sitosterol) to aquatic organisms such as *Daphnia magna* (24–48 h). However, the operation of the aerobic treatment and biomass adaptation could be affecting their removal. The performances of activated sludge (AS), aerated lagoon (AL), and moving bed biofilm reactors (MBBR) are compared for removing stigmasterol contained in Kraft mill effluent. The AL operates at a hydraulic retention time of 6 h and removes up to 90% of phytosterols. So, a 96% of stigmasterol is removed by AL when the sterol retention load is 0.6 mg/L/day. However, stigmasterol concentrations increase from 29% to 37% at a low stigmasterol load rate (0.2 mg/L/day). On the other hand, the stigmasterol is removed between 65% and 87% by an AS under a hydraulic retention time of 3 h. Moreover, a 100% of stigmasterol can be removed by the MBBR when the hydraulic retention time is 2 days.

Key words Stigmasterol removal, Kraft mill effluent, Aerated lagoon, Activated sludge, Moving bed biofilm reactors

1 Introduction

Phytosterols and related compounds found in pulp and paper effluents have been identified as potential sources of endocrine disruption [1, 2]. The concentration of these compounds in effluents may vary depending on the raw material and the technology used in Kraft pulp production [3, 4]. Reported values range from 0.3 to 3.4 mg/L of phytosterol in effluents, depending on the kind of raw wood used [5]. Several authors [1–7] have noted that the phytosterols most commonly found in Kraft mill effluents are β-sitosterol, stigmasterol, stigmastanol, and campesterol. Studies have found that treating effluents with aerobic biological processes significantly reduces phytosterol concentrations [5]. Most bleaching Kraft mills in Chile have switched from using aerated lagoons

José-Luis Barredo, Ignacio Herráiz (eds.), *Microbial Steroids: Methods and Protocols*, Methods in Molecular Biology vol. 1645, DOI 10.1007/978-1-4939-7183-1_10, © Springer Science+Business Media LLC 2017

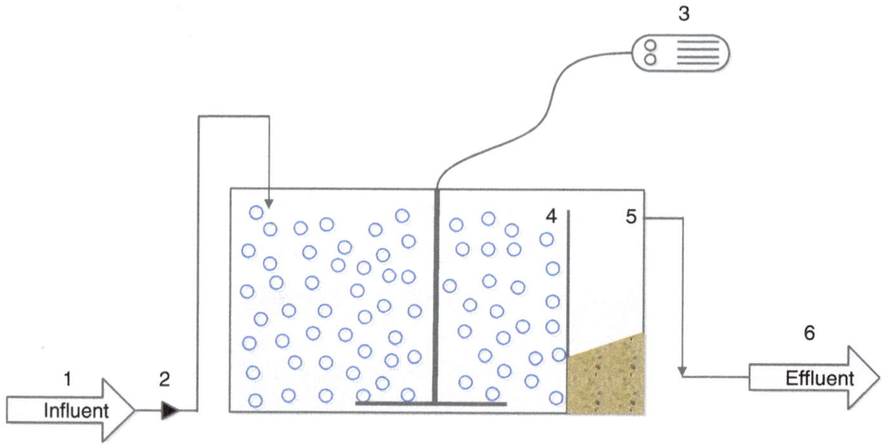

Fig. 1 Aerated lagoon system. Influent (*1*); feeding pump (*2*); air compressor (*3*); aerobic reaction zone (*4*); settler (*5*); effluent (*6*)

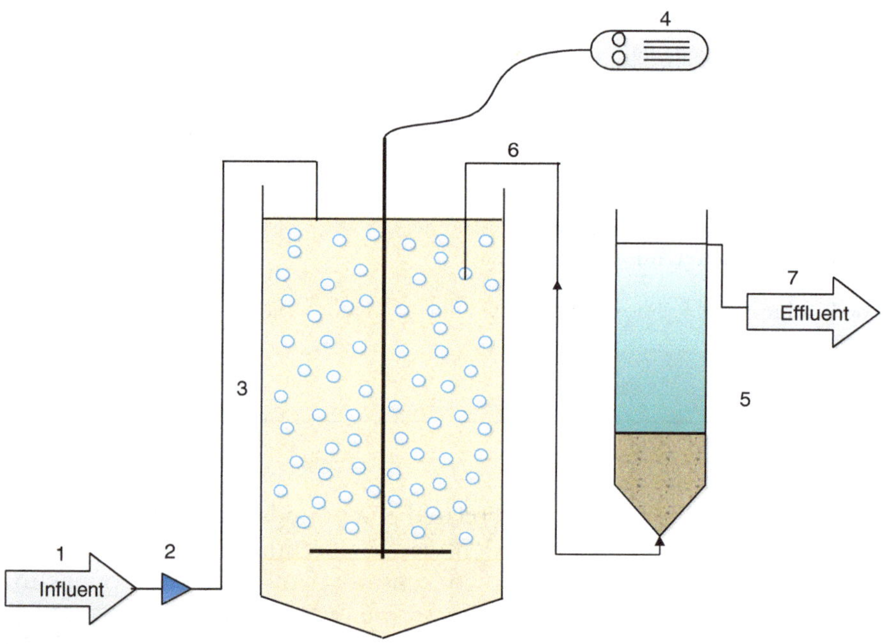

Fig. 2 Activated sludge system. Influent (*1*); feeding pump (*2*); aerobic reaction zone (*3*); air compressor (*4*); settler (*5*); biomass recycling (*6*); effluent (*7*)

(AL) (Fig. 1) for treating effluent to activated sludge (AS) (Fig. 2) and moving bed biofilm reactors (MBBR) (Fig. 3) [8]. AL and AS can remove 41% of sterols by degrading or transforming them and can eliminate 100% of acute toxicity [9, 10]. However, increases of 192% to 367% of some phytosterols (stigmasterol) have been observed with AS [11]. Meanwhile, MBBR has been used successfully in the last decade to treat cellulose mill effluent. This approach

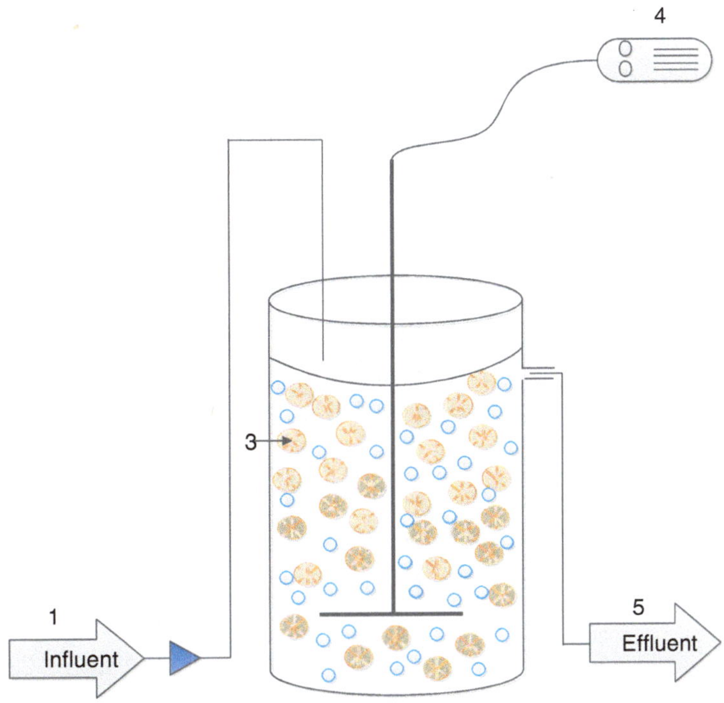

Fig. 3 MBBR system. Influent (*1*); feeding pump (*2*); carrier Kaldness (*3*); air compressor (*4*); effluent (*5*)

reduces the probability of bulking during the operation through the use of a carrier for biofilm formation [8]. Nevertheless, little is known about the capacity of MBBR to remove or degrade phytosterol.

2 Materials

2.1 Equipment

1. Aerated lagoon system (AL). Area (0.44 L) and a settling (nonaerated) area (0.22 L), Fig. 1 (*see* **Note 1**).
2. Activated sludge system (AS) composed of an aerobic area (0.90 L) and a settling area (0.45 L) (Fig. 2).
3. Moving bed biofilm reactor (MBBR). Volume 0.509 L (Fig. 3).
4. Polyethylene carrier (Kaldnes K3).
5. Peristaltic pump.
6. Oxygen diffuser system.
7. Membrane filters (0.45 μm).
8. Binder-free glass microfiber filters (4.7 cm in diameter and 0.4 μm particle retention). Grade GF/C (Whatman).
9. 1 × 1 cm quartz cell.

10. UV-VIS spectrophotometer.
11. Membrane filter (0.2 μm).
12. HPLC.
13. Reverse phase C-18 column.
14. Gas chromatography CG-MS with mass detector.
15. Column 5% phenyl-methylsiloxane, 30 m long, with an internal diameter of 0.5 μm.

2.2 Raw Materials and Solvents

1. Influent from a local Kraft mill.
2. Inoculum of aerobic microorganisms.
3. Urea 99.8% (used as nitrogen supplement).
4. Di-potassium hydrogen phosphate trihydrate (used as phosphorous supplement).
5. NaOH 0.1 M.
6. HCl.
7. Stigmasterol purity >95% (Sigma Chemical Co. MO, USA).
8. β-sitosterol purity >98% (Sigma Chemical Co. MO, USA).
9. Buffer 0.2 M KH_2PO_4 prepared as follows: dissolve 27.22 g of KH_2PO_4 in milli-Q water, adjust the pH to 6.0 with NaOH 30%, and dilute to 1 L.
10. Phenol purity ≥99% (Merck, Darmstadt, Germany).
11. Milli-Q water.
12. Methanol MS SupraSolv® purity ≥99.8% (Merck, Darmstadt, Germany).
13. Ethyl acetate MS SupraSolv® purity ≥99.8% (Merck, Darmstadt, Germany).
14. n-Hexane SupraSolv® purity ≥98% (Merck, Darmstadt, Germany).
15. Dichloromethane MS SupraSolv® purity ≥99.8% (Merck, Darmstadt, Germany).

3 Methods

The method to remove stigmasterol by aerobic treatment has the following steps: (1) Seed an inoculum of aerobic microorganisms to treatment. (2) Feed the treatment (AL or MBBR) with Kraft mill influent. (3) Maintain the reactor in optimum conditions, temperature, hydraulic retention time (HRT) between 0.3 and 2 days and supplement the organic matter, nitrogen, and phosphorous. (4) Collect the influent, effluent, and biomass. (5) Extract with solid-phase cartridges previously conditioned. (6) Assess the phytosterol by gas chromatography with mass detector.

3.1 Inoculum

1. Collect 5 L of aerobic microbial consortium from a Kraft mill plant.
2. Determine the volatile suspended solid (VSS) according to APHA [12] as follows. Condition a membrane filter disc (0.45 μm) at constant weight, drying it in an oven at 105 °C overnight. The weight of the membrane disc will be the total suspended solids determination (TSS-1). Take 5 mL of aerobic microbial consortium and filter through the membrane filter (0.45 μm). Place the residual sample in the membrane filter disk in a muffle at 550 °C for 1 h (TSS-2). Record the weights of (TSS-1) and (TSS-2) in a spreadsheet and report the concentration as g/L of total suspended solids.
3. Inoculate the reactor with 5 g of volatile suspended solid per liter of aerobic microbial consortium.
4. Keep 2 weeks at room temperature (20 °C) for conditioning in the reactor.

3.2 Primary Treatment System of Kraft Mill Influent

1. Remove the solids from the influent by sedimentation during 30 min at room temperature.
2. Neutralize the influent by adjusting to pH 7.0 with NaOH 0.1 M or HCl, as appropriate. Now the influent is ready to be used in further treatments.

3.3 Influent

1. Collect 30 L of influent after primary treatment.
2. Store the influent in a 30 L PVC tank, in darkness and refrigerated at 3 °C.
3. Supplement the influent with urea and di-potassium hydrogen phosphate trihydrate to obtain a BOD_5:N:P ratio of 100:5:1.
4. Keep the pH in the range from 6.0 to 7.5.

3.4 Aerated Lagoon (AL) Treatment

1. Feed the system with a peristaltic pump.
2. Keep the dissolved oxygen concentration above 6 mg O_2/L.
3. Conduct the operation at room temperature (10–22 °C).
4. Supplement the influent with stigmasterol (95%) (concentration ranging from 0.2 to 1 mg/L) (*see* **Notes 2** and **3**).

3.5 Activated Sludge (AS) Treatment

1. Conduct the operation at room temperature (10–22 °C).
2. Keep the dissolved oxygen concentration above 6 mg O_2/L.
3. Supplement the influent with stigmasterol and β-sitosterol to reach a concentration of 2 mg/L, respectively (*see* **Note 4**).

3.6 Moving Bed Biofilm Reactors (MBBR Treatment): Start-Up and Operation

1. Incorporate the polyethylene carrier (Kaldnes K3) with a density and specific surface area of 100 kg/m^3 in the MBBR reactor (*see* **Note 5**).
2. Maintain the dissolved oxygen concentration above 8 mg O_2/L (*see* **Note 6**).

Table 1
Conditions of treatment system for stigmasterol removal

	AL			AS		MBBR	
Aerobic treatment system	**Step I**	**Step II**		**Step I**	**Step II**	**Step I**	**Step II**
Stigmasterol influent (mg/L)	–	0.2	1.1	0.07	2.0	0.02	0.05
SLR (mg/L/day)	–	0.2	1.1	–	–	–	–
OLR (gCOD/L/day)	0.81	0.96	0.96	0.26–0.66	0.45–9.0	0.155	0.343
HRT (h)	24	12.4	12.4	1.87–1.57	1.49–0.13	2.0	2.0
Removal (%)	–	−37	96	91.1	100	100	100

AL aerated lagoon, *AS* activated sludge, *MBBR* moving bed biofilm reactor, *SLR* stigmasterol load rate, *OLR* organic load rate, *HRT* hydraulic retention time

3. Conduct the operation at room temperature (10–22 °C).
4. Characterize the phytosterol of influent and effluent (*see* **Note 7**).

3.7 Performance of All Reactors

1. The operation for AL, AS, and MBBR is divided into two phases (Table 1) (*see* **Notes 1** and **8**).
2. Filter the samples through a 0.45 μm membrane.
3. Determine the spectrophotometric measurements of filtered samples performed at wavelengths, 280 nm for lignin, 346 nm for the lignosulfonic acid, 440 nm for the colors.
4. Determine the total phenolic compound concentrations by UV absorbance in a 1 × 1 cm quartz cell at 215 nm, and transform into concentrations using a calibration curve with phenol as a standard solution.

3.8 Extraction Techniques and CG-MS Sample Determinations

1. Filter 500 mL samples of each effluent through binder-free glass microfiber filters (4.7 cm in diameter and 0.4 μm particle retention).
2. Extract the samples by solid-phase extraction (SPE) using C-18 reverse phase cartridges previously conditioned with two volumes (ca 12 mL) of milli-Q water and one volume each of methanol, ethyl acetate and n-hexane.
3. Perform the runs at 4 mL min/L (prepare two cartridges for each sample).
4. Analyze the influent and effluent composition from one of the cartridges following elution with two volumes each of methanol, ethyl acetate, and n-hexane [1].

3.9 Determination of Phytosterols

1. Filter the Kraft mill influent and effluent through a 0.2 μm membrane filter.

2. Extract the compounds from 100 mL of samples with 20 mL dichloromethane at pH 7 [2].
3. Analyze by gas-chromatography CG-MS with mass detector using a column 5% phenyl-methylsiloxane, 30 m long, with an internal diameter of 0.5 μm.
4. The detection limit is 1 μg/L (*see* **Note 9**).

4 Notes

1. The AL operation is divided into two phases. In phase I, the system is fed with only Kraft mill influent to establish steady-state performance conditions. The HRT of the system is maintained for 24 h, with organic load rates (OLR) of approximately 0.81 g COD/L/day. In phase II, the lagoon is fed with Kraft mill effluent supplemented with progressively higher concentrations of stigmasterol. Concentrations ranged from 0.2 to 1 mg/L. In the second phase, the OLR increases to 0.96 g COD/L/day and the HRT is reduced stepwise from 24 to 12 h.
2. AL performance indicates removal efficiencies for BOD_5 and COD ranging from 48% to 96% and from 25.0% to 65.2%, respectively. Simultaneously, the average removal efficiency of total phenolic compounds was 20.7% and 28.5% for Phases I and II, respectively. Color and aromatic compounds were polymerized in the aerobic system.
3. The AL removed 96% of β-sitosterol and stigmasterol when the sterol retention load was 0.6 mg/L/day (*see* Table 1), while stigmasterol concentrations increased by 29% to 37% at a low stigmasterol load rate (0.2 mg/L/day). This phenomenon was also described, showing that stigmasterol concentrations can increase by more than 300% in the AL system [11]. Khan and Hall [13] also found that aerated systems are not efficient at removing stigmasterol.
4. The removal efficiency of the AS system was higher than the AL system. The AS system removed over 95% of BOD_5 and from 57% to 67% of COD. However, phenolic compounds increased by 15% and coloring compounds by 10–40%.
5. The activated sludge system removed from 65% to 87% of phytosterols at an HRT of 3 h.
6. Attached biomass systems like MBBR efficiently reduce organic matter, with average values of 61.5% and 87.4% for COD and BOD_5, respectively. However, the removal efficiency of specific compounds like color and total phenolic compounds is less than 15% [14, 15].
7. The stigmasterol removal by MBBR system is 100% (Table 1) with operational parameters of a 2-day hydraulic retention

time and an organic matter load of 0.343 ± 0.001 kg COD/ m^3/day.

8. The system operated at biomass concentrations higher than 10 g VSS/L. Excess sludge was withdrawn to keep a 30 days SRT (Sludge Retention Time).
9. These results show that stigmasterol removal is a consequence of the reactor performance, mainly the biomass adaption.

Acknowledgment

This work was supported by FONDECYT 3120216, FONDECYT 1120664, and CONICYT/FONDAP/15130015.

References

1. Chamorro S, Pozo G, Jarpa M, Hernández V, Becerra J, Vida G (2010) Monitoring endocrine activity in kraft mill effluent treated by aerobic moving bed bioreactor system. Water Sci Technol 62(1):154–161
2. Chamorro S, Hernández V, Matamoros V, Dominguez C, Becerra J, Vidal J, Piña B, Bayona J (2013) Chemical characterization of organic microcontaminant sources and biological effects in riverine sediments impacted by urban sewage and pulp mill discharges. Chemosphere 90:611–619
3. Chamorro S, Hernández V, Monsalvez E, Becerra J, Mondaca MA, Piña B, Vidal G (2010) Detection of estrogenic activity from kraft mill effluents by the yeast estrogen screen. Bull Environ Contam Toxicol 84:165–169
4. Vidal G, Becerra J, Hernández V, Decap J, Xavier C (2007) Anaerobic biodegradation of sterol contained in Kraft mil effluents. J Biosci Bioeng 104(6):476–480
5. Kostamo A, Holmbom B, Kukkonen JVK (2004) Fate of wood extractives in wastewater treatment plants at kraft pulp mill and mechanical pulp mills. Water Res 37:972–982
6. Dykstra CM, Giles HD, Banerjee S, Pavlostathis SG (2014) Biotransformation of phytosterols under aerobic conditions. Water Res 58:71–81
7. Fernandes P, Cruz A, Angelova B, Pinheiro HM, Cabral JM (2003) Microbial conversión of steroid compounds: recent developments. Enzyme Microb Technol 32:688–705
8. Diez MC, Castillo G, Aguilar L, Vidal G, Mora ML (2002) Operational factors and nutrients effect on activated sludge treatment for phenolic compounds degradation from *Pinus radiata* kraft mill effluent. Bioresour Technol 83:131–138
9. Odegaard H (2006) Innovations in wastewater treatment: the moving bed biofilm process. Water Sci Technol 53(9):17–33
10. Chamorro S, Xavier C, Hernández V, Becerra J, Vidal G (2009) Aerobic removal of stigmaterol contained in kraft mill effluents. Electron J Biotechnol 12(2):1–7
11. Cook DL, LaFleur L, Parrish A, Jones J, Hoy D (1997) Characterization of plant sterols from 22 US pulp and paper mills. Water Sci Technol 35:297–303
12. APHA-AWWA-WPCF (1985) Standard methods for examination of water and wastewater, 16th edn. American Public Health Association, Washington, DC
13. Khan Z, Hall ER (2003) Occurrence and removal of plant sterol in pulp and paper mill effluents. J Environ Eng Sci 2(1):17–26
14. Navia R, Levet L, Mora ML, Vidal G, Diez MC (2003) Allophanic soil adsorption system as a bleached kraft mill aerobic effluent posttreatment. Water Air Soil Pollut 148:323–333
15. Villamar C, Jarpa M, Decap J, Vidal G (2009) Aerobic moving bed bioreactor performance: a comparative study of removal efficiencies of kraft mill effluents from *Pinus radiata* and *Eucalyptus globulus* as raw material. Water Sci Technol 59(3):507–514

Chapter 11

Production and Biotransformation of Phytosterol Microdispersions to Produce 4-Androstene-3,17-Dione

Rodrigo A. Mancilla, Roberto Pavez-Díaz, and Alejandro Amoroso

Abstract

The current state of knowledge regarding phytosterols biotransformation to produce the steroid intermediate 4-androstene-3,17-dione (AD) shows different technologies. However, the initial concentration of phytosterols in batch cultures is limited due to its low solubility in aqueous media, causing serious difficulties for scaling up because of the low mass transfer. In this chapter, we describe a fermentation method of a phytosterol microdispersion with *Mycobacterium* sp. B3805 in the context of an integral technology for AD production. The microdispersion generation is based on a patent application that claims the production of an aqueous phytosterol microdispersion with an average size particle of 370 nm, and high stability and solubility in water at high phytosterols concentrations (Harting et al., 2012/US0046254). Our results indicate that up to 20 g/L phytosterols can be biotransformed with this technology allowing a good dispersion and stability of reactants in the fermentation broth.

Key words Biotransformation, Phytosterols, Dispersions, Solubility, Tall oil, 4-Androstene-3,17-dione, AD, *Mycobacterium* sp.

1 Introduction

Pulp industry has great relevance for global economy. The Kraft process for pulp production from wood produces some wastes, within these, a black liquor that must be treated to mitigate its environmental impact. This black liquor contains several sub-products rich in different molecules as phytosterols, rosin acid, and others. In this context, a fraction of black liquor called tall-oil has been utilized to produce industrially phytosterols with yields as good as the pure sitosterol [1–4].

Research and development trends in the pharmaceutical industry are being driven by market forces that demand more economical and efficient processes. Among the steroids, 4-androstene-3,17-dione (AD), and 1,4-androstadiene-3,17-dione (ADD) have the special status of being the starting compounds for the synthesis of main steroidal drugs [5, 6]. At present, raw materials as phytosterols are

José-Luis Barredo, Ignacio Herráiz (eds.), *Microbial Steroids: Methods and Protocols*, Methods in Molecular Biology vol. 1645, DOI 10.1007/978-1-4939-7183-1_11, © Springer Science+Business Media LLC 2017

preferred for steroid synthesis because of their ease of transformation by bacteria and low cost. However, the industrial bottleneck of AD(D) production by fermentation is the low phytosterol solubility in water [6–10]. Moreover, it has been reported in related works that the conversion efficiency is also affected by the size, stability, and the shape of the phytosterol molecules available for fermentation [11–14].

In this work, we demonstrate that an aqueous dispersion of phytosterols (370 nm particle size) currently used in the food industry, called A10, can be efficiently used to produce AD improving the solubility of phytosterols in flasks. We show that A10 biotransformation in a ratio (biomass/phytosterol) of 0.1, using resting cells and a complex biotransformation media with yeast extract and glucose proportional to A10 concentration (1:1) plus the surfactant DK-Ester P-160 (HLB 16), yields the best process to produce AD from A10. This technology depicts significant physicochemical stability, as no flocculation, sedimentation, and creaming that normally occurs during the process. Thus, AD production is not limited by the restricted contact between reactants and microorganism.

2 Materials

2.1 Aqueous Phytosterol Dispersion (A10)

1. Phytosterols (Phy) from tall-oil effluent of a paper pulp industry (purity over 97%, with a β-sitosterol content of 78%) from Arboris Inc.
2. One liter laboratory Parr Reactor provided with a turbine stirrer, nitrogen, and vacuum line (Parr Instrument Company, Moline, IL, USA).
3. Homogenizer APV Gaulin, model MR-15.
4. Sodium salts of sunflower fatty acids (SSSFA).
5. Tween 80.
6. Distilled water.
7. Mastersizer to measure the particle size of the dispersions obtained.

2.2 Growth and A10 Biotransformation with *Mycobacterium* sp. B3805

1. Growth medium 1: 5 g/L peptone and 3 g/L beef extract.
2. Growth medium 2: 20 g/L glucose, 6 g/L yeast extract, 0.8 g/L K_2HPO_4, 0.6 g/L $MgSO_4{\cdot}7H_2O$, 0.14 g/L $FeSO_4{\cdot}7H_2O$, and 5 g/L NH_4NO_3.
3. Fermentation medium for A10 bioconversion: 20 g/L glucose, 6 g/L yeast extract, 1.24 g/L $MgSO_4{\cdot}7H_2O$, 0.24 g/L $FeSO_4{\cdot}7H_2O$, 5 g/L NH_4NO_3, 0.04 g/L $Cl_2Co{\cdot}6H_2O$,

0.36 g/L MoO_3, 10.28 g/L DK-Ester P-160, and 100 mM K_2HPO_4/KH_2PO_4 buffer pH 7.

4. Phytosterols A10 (370 nm particle size).
5. Tween 80.
6. Temperature Controlled Orbital Shaker.
7. Laboratory centrifuge.
8. Autoclave.
9. Balance.

2.3 Quantification of Steroids

1. Gas Chromatography with Flame Ionization Detector (GC-FID) (Agilent Technologies, model 7693A or equivalent).
2. Column HP-5 (5%-Phenyl)-methylpolysiloxane (or equivalent).
3. Phytosterols >99% pure (Sigma Aldrich).
4. Ethanol.
5. Toluene.

3 Methods

3.1 10% Aqueous Phytosterol Dispersion (A10)

1. Charge a 1-L Parr reactor, provided with a turbine stirrer, nitrogen and vacuum lines and controlled electric heating, with 500 g of soft water, 50 g of phytosterols, 0.5 g of Tween 80, and 0.5 g of SSSFA.
2. Close the reactor and keep stirring for 2 min at 30 rpm.
3. Increase the stirrer speed to 700 rpm and heat the reactor content to 160 °C for 10 min. The result is a liquid mixture formed by the phytosterols, emulsifiers, and the aqueous medium.
4. Feed the former dispersion to an APV Gaulin MR-15 homogenizer and homogenize in one stage at 300 bar. Let the homogenized mixture cool down to room temperature to form the final phytosterol microdispersion [15]. This microdispersion is stable and has an average particle size of less than 370 nm (*see* **Note 1**).
5. Additionally, it is recommended to prepare a process control called Non-A10 which has not been passed through the homogenizer.

3.2 Mycobacterium sp. B3805 Biomass Growth

All *Mycobacterium* sp. B3805 growth processes must be performed under the same operating conditions for 3 days at 30 °C, 200 rpm, and pH 7 until it reaches the pre-stationary phase of growth. Additionally, presence of contamination by Ziehl-Neelsen stain must be verified in each transfer of the microorganism (*see* **Note 2**).

1. Seed (in sterility) 100 mL flasks supplemented with 20 mL of growth medium 1 with *Mycobacterium* sp. B3805 from a cryovial (1 mL) or a colony from Petri plates (*see* **Note 3**).
2. Incubate for 3 days at 30 °C and 200 rpm.
3. Gently transfer 1/10 volume of growth medium 1 to a new 500 mL flask with 100 mL of growth medium 2, and incubate for 3 days at 30 °C and 200 rpm.
4. After this, transfer 25% volume of growth media 2 to a new flask of 1 L containing 250 mL of growth medium 2 supplemented with 0.5 g/L A10 and 0.8% Tween 80 (*see* **Note 4**).
5. Centrifuge the fermented broth at 3000 × *g* for 15 min.
6. Discard the supernatant and rehydrate the pellet with at least 3 volumes of H_2O (*see* **Note 5**).

3.3 Biotransformation of 20 g/L of A10 into AD

The required materials to perform A10 bioconversion (biomass, sterile water, A10, and fermentation media) are prepared under sterile conditions, either under a laminar flow cabinet or a Bunsen burner.

1. To make the biotransformation of 20 g/L phytosterols, mix 2 g of A10 (previously pasteurized for 15 min in boiling water bath), 2.85 g of *Mycobacterium* sp. B3805, 1.53 g of fermentation medium, and 2.62 g of sterile water in a 100 mL flask (*see* **Note 6**). Note that for other A10 concentrations the biotransformation conditions will be different (*see* **Note 7**).
2. Carry out the biotransformation reactions in 100 mL flasks with 10 mL of fermentation media at 30 °C and 200 rpm. Reaction time is determined by the desired AD yield percentage as shown in Fig. 1 [16].

3.4 Quantification of Sterols and AD

1. In each flask add 10 mg of cholesterol as internal standard (is) once finalized the biotransformation.
2. Add gradually while stirring 60 mL of ethanol/toluene (3:1) until substrates and products in the bioconversion flask are completely dissolved.
3. Take 1 mL of the mixture and transfer to a chromatography vial (*see* **Note 8**).
4. Prior to the injection of samples in GC-FID and when conditions vary (maintenances, column shift, cleaning, etc.), it will be necessary to measure the response factors (rf) of sterols. This is determined using the pure compounds as standard (std) (>98%) and comparing them with cholesterol 99% purity using the following equation:

$$rf = \frac{\text{is Area}}{std\text{ Area}} \times \frac{std\text{ Weight}\,(mg)}{\text{is Weight}\,(mg)} \times \frac{\text{is Purity\%}}{std\text{ Purity\%}}$$

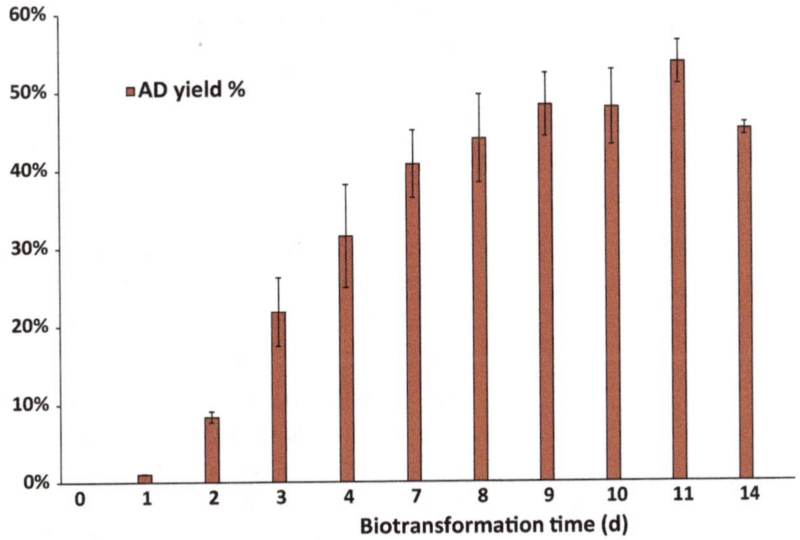

Fig. 1 4-Androstene-3,17-dione production by biotransformation of A10 microdispersion at 20 g/L phytosterols by *Mycobacterium* sp. B3805 as a function of the reaction time (days)

5. The chromatography conditions are as follows: Carrier: Helium 68.4 (64 cm/s) constant flow. Injection Split (30:1), 0.5 μL. Detector: FID 300 °C Temperature: 200 °C (10 min), 5 °C/min (20 min), 300 °C (10 min).
6. The absolute mass (AM) of the analyzed compounds is determined using the following equation:

$$AM = \frac{\text{Molecule Area}}{\text{is Area}} \times \frac{\text{is Weight}\,(mg)}{\text{Sample Weight}\,(mg)} \times rf$$

7. The bioconversion parameters of phytosterols into AD are defined by the following equations:

$$AD\ \text{yield}\,(\%) = \frac{AD\ \text{weight}}{\text{initial substrate weight}} \times \frac{\text{substrate}\ MW}{AD\ MW} \cdot 100$$

4 Notes

1. We find it is recommendable to analyze A10 size every 6 months. MasterSizer X is suitable equipment to measure A10 size. Moreover, storing A10 at environmental temperature reduces its flocculation.
2. The primary protocol used in the acid-fast Ziehl Neelsen stain reaction is as follows: Fix the smear of the 100 μL sample over

the glass slide by heating. Pour carbol fuschin over the smear and heat gently until fumes appear. Do not overheat nor allow it to dry. Then wash off with water until the effluent is clear. The slide is then flooded with a mild solution of 1% sulfuric acid to destain the carbol fuchsin and wash off with water until the effluent is clear. Pour methylene blue, wait for 1 min, wash again with water, and allow it to air dry. Examine under oil immersion lens in the microscope. *Mycobacterium* sp. appear fuchsia; the negative response to this test is blue.

3. Using *Mycobacterium* sp. B3805 from agar plates helps the culture grow faster.
4. The approximated concentration of biomass is 7 g/L of *Mycobacterium* sp. B3805. Once the biomass production phase is finished, the concentration of biomass in the culture is obtained interpolating the absorbance at 650 nm on the corresponding biomass calibration curve.
5. In this way biomass or resting cells are obtained to inoculate flasks. Also growing cells can be used, but poorer results have been observed.
6. The fermentation medium is sterilized by autoclaving at 121 °C for 15 min. Moreover, this is prepared concentrated 13 times (130 g/L glucose) to reach with low volume the desired concentration in flasks. Keep a stock of sterile water and pasteurized sterols (A10) 100 g/L.
7. The biotransformation stoichiometry is dependent on A10 concentration in the reactions and it is always based on the parameter RB (reason biomass/sterols) $RB = \frac{[\text{Biomass}] \times Vol.\text{Biomass}}{[\text{Sterols}] \times Vol.\text{Sterols}}$ which is set at RB = 0.1 [17]. An RB different to 0.1, particularly higher to 0.1, will inhibit the A10 biotransformation at concentrations above 20 g/L. Additionally, always keep the fermentation medium used in a ratio 1:1 (A10/glucose) and the remaining component in proportion to glucose. When the fermentation medium or ratio (A10/glucose) is changed, the A10 dispersion tends to suffer physicochemical instability phenomena. Finally, A10/glucose ratio 1:1 means for example that for each 5 g/L of sterols in the fermentation media, 5 g/L of glucose should be used.
8. When there are problems with peak resolution in the chromatograms, it is advisable to derivatize the sample by taking 1 mL through evaporating under nitrogen gas and mantle thermostated in chromatographic vial for subsequent derivatization with silanizing N, O Bis (trimethylsilyl) acetamide in the presence of pyridine as a catalyst. Close the chromatographic vial, vortex and heat to 85 °C for 15 min, stir again and inject into GC-FID.

Acknowledgment

This work was supported by Grant N° 12CTI-16838 from CORFO and by Grant operational expenses N°21110224 from Conicyt. R.A. Mancilla was a doctoral fellow supported by CONICYT-Chile. The authors would like to thank Msc Raúl Aravena from Nutrartis S.A. for his support formulating dispersions, Dr. Cecilia Brañez from Naturalis S.A. for conceptual contribution to the project, and Karen Marquez and Camila Ahumada from PUCV research staff for technical assistance.

References

1. Fernandes P, Cabral JM (2007) Phytosterols: applications and recovery methods. Bioresour Technol 98(12):2335–2350
2. Markovitz E, Harting T (2001) High efficiency process for the recovery of the high pure sterols. US Patent 6,465,665
3. Fuenzalida MA, Markovitz A (2002) Fractionation process for the unsaponifiable material derived from black-liquor soaps. US Patent 6,462,210
4. Fuenzalida MA, Markovitz A, Leiva R, Markovitz E (2001) Process for obtaining unsaponifiable compounds from black-liquor soaps, tall oil and their by-products. US Patent 6,297,353
5. Wang F, Yao K, Wei D (2011) From soybean phytosterols to steroid hormones. In: Soybean and Health. 2011;11:232–252
6. Donova MV, Egorova OV (2012) Microbial steroid transformations: current state and prospects. Appl Microbiol Biotechnol 94(6):1423–1447
7. Stefanov S, Yankov D, Beschkov V (2006) Biotransformation of phytosterols to androstenedione in two phase water-oil systems. Chem Biochem Eng Q 20(4):421–427
8. Zhang XY, Peng Y, Su ZR, Chen QH, Ruan H, He GQ (2013) Optimization of biotransformation from phytosterol to androstenedione by a mutant *Mycobacterium neoaurum* ZJUVN-08. Biomed Biotechnol 14(2):132–143
9. Xu YG, Guan YX, Wang HQ, Yao SJ (2014) Microbial side-chain cleavage of phytosterols by mycobacteria in vegetable oil/aqueous two-phase system. Appl Biochem Biotechnol 174(2):522–533
10. Kutney J, Herrington E, Spassov G (2003) Process for fermentation of phytosterols to androstadienedione. EP 1,507,867
11. Izadi Z, Nasirpour A, Garousi G (2012) Optimization of phytosterols dispersion in an oil/water emulsion using mixture design approach optimization of phytosterols dispersion in an oil/water emulsion using mixture design approach. J Dispers Sci Technol 33(12):37–41
12. Malaviya A, Gomes J (2008) Nutrientbroth/PEG200/TritonX114/Tween80/chloroform microemulsion as a reservoir of solubilized sitosterol for biotransformation to androstenedione. J Ind Microbiol Biotechnol 35(11):1435–1440
13. Mcclements DJ (2012) Advances in fabrication of emulsions with enhanced functionality using structural design principles. Curr Opin Colloid Interface Sci 17(5):235–245
14. Noh SK, Kim MK, Yoon WT, Park KM, Park SO (2004) Method for preparation of Androst-4-ene-3,17-dione and Androsta-1,4-ene-3,17-dione. US Patent 20040152153
15. Harting T, Markovits A, Fuenzalida MA (2012) Dispersion of phytosterols. US Patent 20120046254
16. Chamy R, Mancilla R, Marquez K, Vivanco E (2014) Proceso para producir 4-Androstene-3-,17-Diona a partir de esteroles en un medio con un microorganismo. CL Patent 201402152
17. Olivares A, Acevedo F (2011) Effect of inoculation strategies, substrate to biomass ratio and nitrogen sources on the bioconversion of wood sterols by *Mycobacterium* sp. World J Microbiol Biotechnol 27(11):2513–2520

Chapter 12

β-Sitosterol Bioconversion to Androstenedione in Microtiter Plates

Marco P.C. Marques and Pedro Fernandes

Abstract

Microtiter plates are routinely used as low-cost miniaturized bioreactors due to the large number of experiments that can be conducted simultaneously under similar conditions and replicate all functions of bench-scale reactors at dramatically smaller volumes. These plates, due to the standard footprint, can be integrated with liquid-handling systems and associated equipment expanding considerably their application and use. However, care has to be taken to operate the microtiter plates in optimized mixing and oxygen transfer conditions, preventing medium evaporation in prolonged experiment runs. Here, we describe the production of 4-androstene-3,17-dione (androstenedione; AD), a key pharmaceutical steroid intermediate, by *Mycobacterium* sp. NRRL B-3805 via the selective cleavage of the side-chain of β-sitosterol using 24-well microtiter plates.

Key words Microtiter plate, *Mycobacterium* sp. NRRL B-3805, 4-Androstene-3,17-dione, AD, β-Sitosterol, Steroid, Evaporation, High-throughput screening

1 Introduction

Miniaturizing bioprocess unit operation devices is a well-established approach to find novel routes for process intensification and improved process economics. Devices, which allow a high level of parallelization while operating at reduced volumes, are especially attractive for fast and cost-effective process development.

When dealing with miniaturizing upstream unit operations, one can encounter shaken systems (Erlenmeyer flaks or microtiter plates), miniature, and microbioreactors [1–3]. Microtiter plates are widely used since they offer in a reduced footprint, typically 8.56 × 12.76 cm ($w \times l$), the possibility to carry out 6 to 1536 experiments in working volumes ranging from 3 mL to 3 μL, respectively [4]. These devices are inexpensive when compared to other miniaturized devices, relatively simple to handle, and due to the standard footprint enable automation and integration with liquid-handling systems and associated equipment (e.g., HPLC

José-Luis Barredo, Ignacio Herráiz (eds.), *Microbial Steroids: Methods and Protocols*, Methods in Molecular Biology vol. 1645, DOI 10.1007/978-1-4939-7183-1_12, © Springer Science+Business Media LLC 2017

and spectrometric plate readers). Furthermore, microtiter plates replicate all functions of bench-scale reactors, having monitoring capabilities of key process variables (e.g., pH and oxygen) [5], control of pH and fed-batch operations [6], under similar operational conditions [7].

Microtiter plates have been extensively used for evaluating large biocatalyst libraries, generation of design data (e.g., for medium screening, assessing pH and temperature influence in reaction yield and evaluating culturing and induction time) to be used in process and economic models and for fermentation and biocatalysis process development [4, 8].

In this chapter, we show the use of microtiter plates to perform multi-step biocatalytic reactions using whole cells. The model system is the selective cleavage of the side-chain of β-sitosterol by *Mycobacterium* sp. NRRL B-3805, leading to the synthesis of 4-androstene-3,17-dione (androstenedione; AD) as major product and 1,4-androstadiene-3,17-dione (androstadienedione; ADD) as minor product [9].

AD and ADD are key pharmaceutical steroid intermediates and their yearly reported market is 1000 tons [10]. This multi-step reaction is especially challenging at the small scale due to the low aqueous solubility of β-sitosterol (need for in situ substrate supply strategies) and low reaction rates (prevention of significant medium evaporation is necessary).

2 Materials

Prepare all solutions with deionized water and analytical grade reagents. All reagents should be stored at room temperature (~22 °C) unless stated otherwise. No special equipment is required to perform these experiments, besides an autoclave, a shaking incubator, a vortex, and a HPLC for sample quantification. Use of personal protection equipment (PPE) is recommended throughout the entire work.

2.1 Culture Media and Equipment for Mycobacterium sp. NRRL B-3805 Growth

1. PDA culture plates: Add 1 L of water to a glass beaker. Weigh 42 g of potato dextrose agar (PDA) and transfer to the beaker. Mix with the aid of a magnetic bar. Transfer this solution to a 2 L screw capped bottle (*see* **Note 1**). Autoclave for 15 min at 121 °C (total cycle approx. 2 h). Allow the bottle to cool down until it is bearable to manipulate it without any heatproof gloves. Transfer approx. 20 mL of the autoclaved agar medium into a plastic Petri dish. Repeat this process until the entire autoclaved agar medium is used (*see* **Note 2**). Leave the plates without the cover in a safety cabinet until the agar start to solidify (approx. 1 h). After this period turn the Petri dishes upside down to prevent condensation and leave for a couple of

hours to further solidify. After complete solidification place the lids on the Petri dishes and store them at 4 °C until use.

2. Sodium-potassium Phosphate Buffer (100 mM, pH 7): Add 1 L of water to a glass beaker. Weigh 14.20 g of Na_2HPO_4 (molecular weight 141.96 g/mol) and transfer to the beaker. Mix with the aid of a magnetic bar. Transfer this solution to a 2 L screw capped bottle. Repeat this process for KH_2PO_4 by adding 1 L of water to a glass beaker. Weigh 13.61 g of KH_2PO_4 (molecular weight 136.09 g/mol) and transfer to the beaker where it is mixed with a magnetic bar. Transfer this solution to a 2 L screw capped bottle. Transfer the Na_2HPO_4 solution to a large container. Mix with the aid of a magnetic bar and use a pH probe to measure continuously the pH. Add slowly the KH_2PO_4 solution until pH 7 is reached. Transfer the final buffer (100 mM, pH 7) to a 2 L screw capped bottle and store at room temperature until use.
3. Complex Growth Medium: Add 1 L of 100 mM sodium-potassium phosphate buffer pH 7 to a glass beaker. Weigh 10 g of yeast extract, 10 g of glycerol, 0.2 g of $MgSO_4{\cdot}7H_2O$, 1 g of micronized β-sitosterol, and 0.8 g of Tween® 20, and transfer to the beaker. Mix the solution with the aid of a magnetic bar. Transfer 20 mL of this solution to 100 mL shaken flasks (denominated inoculum flasks) and 400 mL to 2 L shaken flasks (denominated growth flasks) (*see* **Notes 1** and **3**). Use cotton wool plugs to close the shaken flasks and cover with aluminum foil. Autoclave for 15 min at 121 °C (total cycle approx. 2 h). Store at 4 °C until use.
4. Defined Growth Medium: Add 1 L of 0.1 M sodium-potassium phosphate buffer pH 7 to a glass beaker. Weigh 20 g of glycerol, 4 g of NH_4Cl, 0.2 g of $MgSO_4{\cdot}7H_2O$, and 0.8 g of Tween® 20, and transfer to the beaker. Mix with the aid of a magnetic bar. Transfer this solution to a 2 L screw capped bottle (*see* **Note 1**). Cap the bottles loosely and autoclave for 15 min at 121 °C (total cycle approx. 2 h). Tighten the bottle cap and store at 4 °C until use.
5. Ethanol β-sitosterol Stock Solution: Add 10 mL of absolute ethanol to a glass beaker. Weigh 99.53 mg of β-sitosterol and transfer to the beaker. Mix gently with the aid of a magnetic bar. Transfer this solution to a 10 mL screw capped bottle and store at 4 °C until use.
6. Poly(dimethylsiloxane) β-sitosterol Stock Solution: Add 10 mL of poly(dimethylsiloxane) (*see* **Note 4**) to a screw capped bottle. Weigh 145.15 mg of β-sitosterol and transfer to the bottle. Mix gently with the aid of a magnetic bar. Place the bottle in an oven at 100 °C. After 10 min, take the bottle out of the oven, using PPE. Mix with the aid of a magnetic bar. Repeat this process until β-sitosterol is completely dissolved.

7. Shaking incubator with orbital motion. The incubator needs to have controlled temperature and needs to accommodate 2 L shaken flasks.

2.2 AD Extraction and HPLC Analysis

1. Extraction solvent with internal standard: Add 1 L of n-heptane to a screw capped bottle. Weigh 0.2 g of progesterone and transfer to the bottle. Mix gently with the aid of a magnetic bar. When completely dissolved store at 4 °C (*see* **Note 5**).
2. HPLC equipment with UV detector.
3. LiChrospher® Si 60 column (particle size 5 μm, Merck, Germany).
4. Mobile phase, n-heptane and ethanol (9/1, v/v): Add 1.8 L of heptane and 0.2 L of absolute ethanol to a 2 L screw capped bottle. Mix gently with the aid of a magnetic bar. Flush the bottle for 20 min with helium to remove all air bubbles.
5. AD stock solutions (*see* **Note 6**): Add 8.6 mg of 4-androstene-3,17-dione and 15 mL of n-heptane to a 25 mL screw capped bottle. Mix gently with the aid of a magnetic bar. Prepare ten Eppendorf tubes®, marked 1–10, containing each 500 μL of n-heptane. Transfer 500 μL of the initial solution present in the screw capped bottle to the first Eppendorf tube®. Transfer 500 μL from Eppendorf tube® 1 to Eppendorf tube® 2. Repeat this process until all the Eppendorf tubes® are filled. Transfer all the solutions to HPLC vials.
6. ADD stock solution (*see* **Note 6**): Add 8.5 mg of 1,4-androstadiene-3,17-dione and 15 mL of n-heptane to a 25 mL screw capped bottle. Follow the same procedure so as to prepare AD stock solutions.
7. Bench centrifuge.
8. Vortex.

2.3 Sterilized Materials

1. Box of blue (1 mL) and yellow (200 μL) pipet tips autoclaved (15 min at 121 °C, total cycle approx. 2 h).
2. Sterile microtiter plates.
3. Sterile Petri dishes.

3 Methods

Carry out all procedures at room temperature (~22 °C) and in aseptic conditions unless otherwise specified.

3.1 Preparation of Resting Cells of Mycobacterium sp. NRRL B-3805

1. Defrost stock cultures (cell bank) of *Mycobacterium* sp. NRRL B-3805 and add 100 μL of the stock culture to a PDA culture plate.
2. Spread the culture using a loop according to the streaking method (*see* **Note 7**).

3. Place the culture plate upside down in a 30 °C incubator for at least 24 h.
4. Take an isolated colony of *Mycobacterium* sp. NRRL B-3805 (*see* **Note 8**) and add to the inoculum flasks containing complex growth medium.
5. Place the inoculum flasks (100 mL flask) in a shaken incubator at 200 rpm and 30 °C.
6. After 24 h (late exponential growth phase; *see* **Note 9**) transfer the entire content of the inoculum flask to growth flasks (2 L flasks) containing with complex growth medium.
7. Place the growth flasks (2 L flask) in a shaken incubator at 200 rpm and 30 °C (*see* **Note 10**).
8. After about 72 h (late exponential growth phase; *see* **Note 9**) harvest the cells by vacuum filtration (*see* **Note 11**).
9. Thoroughly wash the cells on the filter with the prepared phosphate buffer (100 mM, pH 7).
10. Store the wet cell paste at −20 °C until use (*see* **Note 12**).

3.2 Bioconversion Trials Using Resting Cells of Mycobacterium sp. NRRL B-3805

1. Add 15 mL of phosphate buffer (100 mM, pH 7) to a glass beaker (*see* **Note 13**).
2. Weigh 0.7 g of the wet cell paste and transfer to the bottle.
3. Mix gently with the aid of a magnetic bar until a yellowish solution is obtained.
4. Add 550 μL of the *Mycobacterium* sp. NRRL B-3805 solution to each well of a 24-well microtiter plate (*see* **Notes 14–16**).
5. Add 55 or 41 μL of the β-sitosterol stock solutions in ethanol or poly(dimethylsiloxane), respectively (*see* **Note 17**).
6. Cover the microtiter plate with a breathable sealing tape and place lid (*see* **Note 16**).
7. Place the microtiter plate in a shaking incubator at 250 rpm and 30 °C.
8. Use a sacrificial well approach by sampling the entire content of the microtiter plate well.
9. Transfer the sample to a 1.5 mL Eppendorf tube® and add 550 μL of the extraction solvent with internal standard.
10. Mix the content in a vortex for 30 s at maximum speed.
11. Place the Eppendorf tubes® in a centrifuge for 2 min, 4 °C and 10,000 × *g*.
12. Collect the organic phase (top phase) and place in a HPLC vial.
13. Analyze samples in a HPLC (*see* **Note 18**).

3.3 Bioconversion Trials Using Growing Cells of Mycobacterium sp. NRRL B-3805

1. Prepare inoculum as in Subheading 3.1 (**steps 1–5**).
2. After 24 h (late exponential growth phase; *see* **Note 9**) transfer the entire content of the inoculum flask to the defined medium bottle (500 mL bottle).
3. Gently mix the bottle.
4. Add 550 μL of the inoculated defined medium to each well of a 24-well microtiter plate (*see* **Notes 14–16** and **19**).
5. Add 55 μL or 41 μL of the β-sitosterol stock solutions in ethanol or poly(dimethylsiloxane), respectively (*see* **Note 17**).
6. Cover the microtiter plate with a breathable sealing tape and place lid (*see* **Note 16**).
7. Place the microtiter plate in a shaking incubator at 250 rpm and 30 °C.
8. Use a sacrificial well approach by sampling the entire content of the microtiter plate well.
9. Conduct similar sample processing and analysis as in Subheading 3.2 (**steps 9–13**).

4 Notes

1. PYREX® or other borosilicate glassware should be used for autoclaving since this type of glassware can withstand the stress of autoclave pressures and temperatures.
2. As a rule of thumb, a volume of 500 mL of potato dextrose agar can be used to fill 20 Petri dishes.
3. The medium can be stored at 4 °C for at least 1 week. However, as a good practice, fresh medium should be prepared every time a new experiment is conducted.
4. Poly(dimethylsiloxane) (PDMS) suitable for high temperatures should be used to prepare the stock solutions [11]. The use of a "low temperature" PDMS is not recommended since the silicone can potentially solidify at room temperature.
5. The extraction solvent with internal standard should be allowed to come to room temperature (22 °C) and be mixed every time it is used since crystallization of progesterone can occur at low temperatures.
6. For the HPLC calibration curve, stock solution concentration of both AD and ADD can be used up to 1.5 mM. Higher concentration can present deviation from linearity.
7. The streaking method allows obtaining isolated colonies from an inoculum by streaking a plate several times in succession. This will create on a single plate several areas with increased dilution. For this method, a sterilized loop is necessary. The

loop, containing the inoculum, is streaked across the surface of the sterile PDA culture plate. Following the initial streak, the subsequent streaks are made at 90° angles being the loop sterilized in between. If a disposable loop is used then the sterilization is unnecessary between streaks. A total of four streaks are advised.

8. *Mycobacterium* sp. NRRL B-3805 should present in PDA agar plates colonies yellowish in color, irregular form, raised elevation, rugose surface, and earthy smell. *Mycobacterium* sp. NRRL B-3805 can however change their phenotype according to specific growth conditions. It is therefore recommended that identification protocols should be carried out regularly [12].

   Obtaining an isolated colony can be difficult with *Mycobacterium* sp. NRRL B-3805. For inoculation, collecting a loopful of several colonies is recommended.

9. Measuring the optical density (OD) in *Mycobacterium* sp. NRRL B-3805 cultures can be difficult since this bacteria forms aggregates. As an approximation, late exponential growth phase corresponds to an OD in the range of 0.6–0.9 (640 nm). A more reliable method of determining *Mycobacterium* sp. NRRL B-3805 is based on total protein or on dry cell weight. However, these methods are time consuming and data are not obtained in real time.

10. Flasks containing large volumes (e.g., the growth flasks) should be placed in the incubator platform such that it is balanced and all weight is evenly distributed.

11. Whatman® qualitative filter paper, Grade 1 can be used to filter *Mycobacterium* sp. NRRL B-3805 cultures. The cake resistance will increase and filtration efficiency will decrease over time. To aid the filtration regularly and smoothly remove the cake from the filter with a spatula.

12. The wet cell paste contains about 190 mg dry cell weight per g of wet paste.

13. Different bioconversion conditions can be tested. For example, different salts can be added to the phosphate buffer to increase bioconversion yield [9].

14. Different filling volumes can be used [9]. However, $k_{L}a$ (volumetric oxygen mass transfer coefficient) will depend on filling volume [7]. As a rule of thumb, the lower the volume the higher the oxygen transfer. Details are given in Fig. 1.

15. If lower volumes are used, special care should be given to the higher evaporation rate of the culture medium. The medium should be replenished every day with autoclaved deionized water. The amount to be replenished in each well can be determined by weighting the microtiter plate and dividing the value by the number of wells used in the experiment.

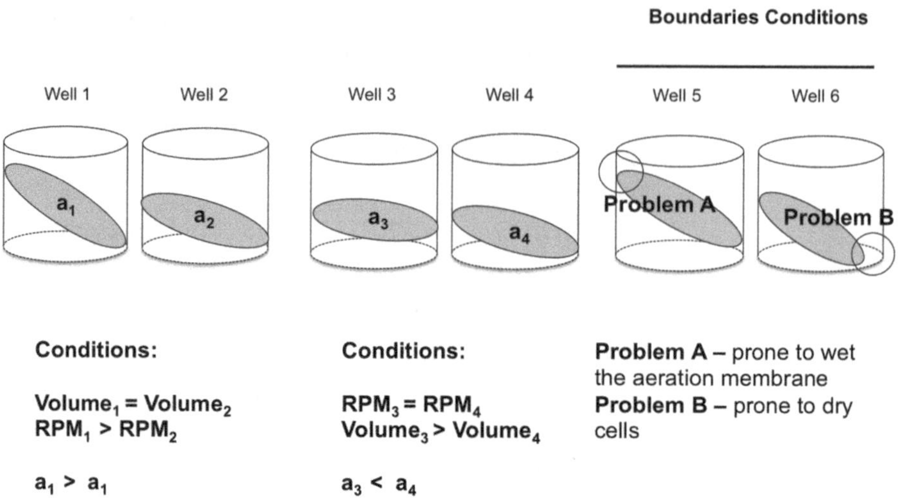

Fig. 1 Correlation of gas-liquid interface surface area (a_i) with the filling volume (Volume) and the shaking frequency (RPM) of a microtiter plate well. Excessive or reduced filling volume results in either wetting of the aeration membrane or uncontrolled concentration of the media, respectively

A control well using standard bioconversion conditions should be used in each microtiter plate. We recommend 550 μL of the phosphate buffer (100 mM, pH 7) containing *Mycobacterium* sp. NRRL B-3805 cells supplemented with 55 μL of the ethanol β-sitosterol stock solution.

16. The breathable sealing tape will seal the microtiter plate, preventing medium evaporation and cross-contamination of the wells due to spilling. Detailed steps are given in Fig. 2.

17. Different organic solvent can be used or aqueous:organic ratio [13, 14].

18. Quantification of steroids should be performed by HPLC. A possible method consists of using a LiChrospher® Si 60 column (particle size 5 μm, Merck, Germany) with mobile phase composed of n-heptane and ethanol (9:1, v/v). An isocratic elution of 1 mL/min should be used with detection at 215 nm (β-sitosterol) and 254 nm (4-androstene-3,17-dione, AD, and 1,4-androstadiene-3,17-dione, ADD).

19. Microtiter plates with integrated oxygen and pH sensors can be used to monitor bacterial growth [7, 15]. If the readout of the sensors is performed using fluorescence microplate readers, the medium evaporation rate can increase. Lowering the readout frequency can circumvent high evaporation rate of the culture medium. Nonetheless, it is advised to weigh the microtiter plate on an hourly basis, and to replenish the medium with autoclaved deionised water, if necessary.

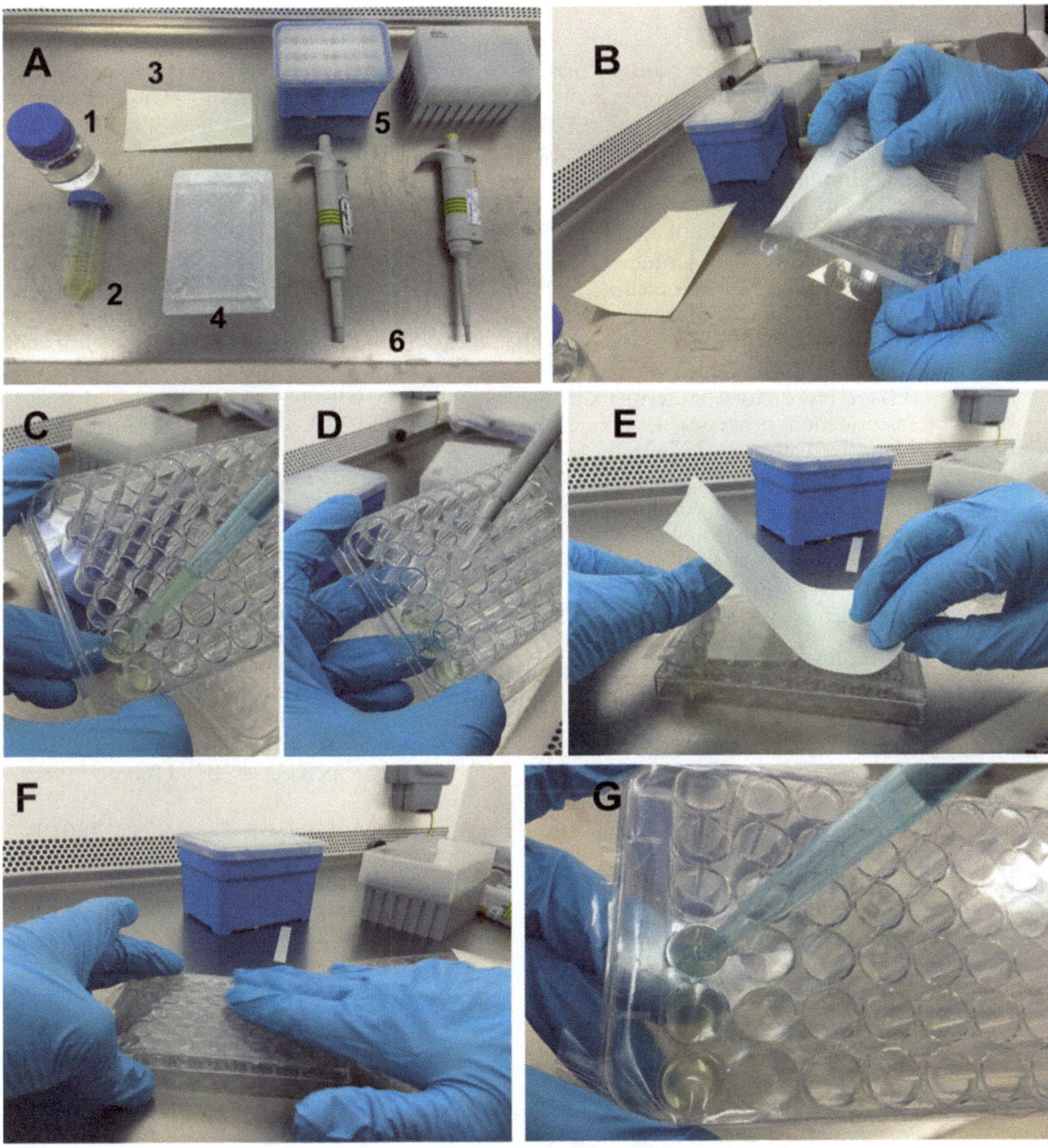

Fig. 2 Preparation of the microtiter plate. (**a**) Material needed: *1.* Steroid solution in ethanol; *2. Mycobacterium* sp. NRRL-3805 cell suspension; *3.* Aeration membrane; *4.* Microtiter plate; *5.* Pipette tips box; *6.* Micropipettes. (**b**) Open the microtiter plate in the safety cabinet. Use gloves during handling. (**c**) Fill the wells with the mycobacterial cell suspension. Tilt the microtiter plate in such a manner that tip of the pipette touches the bottom of the well. (**d**) Add the steroid solution the wells. Immerge the tip in the mycobacterial cell suspension. (**e**) Place the sealing tape on the top of the microtiter plate according to the manufacturer's instructions. (**f**) Certify that the sealing tape adequately covers all the wells. (**g**) When sampling, puncture the sealing tape with the tip and collect the entire volume of the well

References

1. Kirk TV, Szita N (2013) Oxygen transfer characteristics of miniaturized bioreactor systems. Biotechnol Bioeng 110:1005–1019
2. Marques MPC, Cabral JMS, Fernandes P (2009) High throughput in biotechnology: from shake-flasks to fully instrumented microfermentors. Recent Pat Biotechnol 3:124–140
3. Fernandes P, Carvalho F, Marques MPC (2011) Miniaturization in biotechnology: speeding up the development of bioprocesses. Recent Pat Biotechnol 5:160–173
4. Fernandes P, Cabral JMS (2006) Microlitre/millilitre shaken bioreactors in fermentative and biotransformation processes – a review. Biocatal Biotransformation 24:237–252
5. Kensy F, John GT, Hofmann B, Buchs J (2005) Characterisation of operation conditions and online monitoring of physiological culture parameters in shaken 24-well microtiter plates. Bioprocess Biosyst Eng 28:75–81
6. Funke M, Buchenauer A, Mokwa W, Kluge S, Hein L, Mueller C, Kensy F, Buechs J (2010) Bioprocess control in microscale: scalable fermentations in disposable and user-friendly microfluidic systems. Microb Cell Fact 9:86
7. Marques MPC, Magalhães S, Cabral JMS, Fernandes P (2009) Characterization of 24-well microtiter plate reactors for a complex multistep bioconversion: from sitosterol to androstenedione. J Biotechnol 141:174–180
8. Lye GJ, Ayazi-Shamlou P, Baganz F, Dalby PA, Woodley JM (2003) Accelerated design of bioconversion processes using automated microscale processing techniques. Trends Biotechnol 21:29–37
9. Marques M, de Carvalho C, Claudino M, Cabral J, Fernandes P (2007) On the feasibility of the microscale approach for a multistep biotransformation: sitosterol side chain cleavage. J Chem Technol Biotechnol 82:856–863
10. Fernandes P, Cabral JMS (2010) Steroid Bioconversion. In: Flickinger MC (ed) Encyclopedia of industrial biotechnology-bioprocess, bioseparation, and cell technology, vol 7. John Wiley & Sons, New York
11. Carvalho F, Marques MPC, de Carvalho CCCR, Cabral JMS, Fernandes P (2009) Sitosterol bioconversion with resting cells in liquid polymer based systems. Bioresour Technol 100:4050–4053
12. Ribn W (2012) Biochemical isolation and identification of mycobacteria. In: Jimenez-Lopez JC (ed) Biochemical testing. InTech. doi:10.5772/34309
13. Marques M, Carvalho F, Magalhães S, Cabral J, Fernandes P (2009) Screening for suitable solvents as substrate carriers for the microbial side-chain cleavage of sitosterol using microtitre plates. Process Biochem 44:556–561
14. Marques M, Carvalho F, de Carvalho C, Cabral J, Fernandes P (2010) Steroid bioconversion: towards green processes. Food Bioprod Process 88:12–20
15. Marques M, Cabral J, Fernandes P (2010) A microwell platform for the scale-up of a multistep bioconversion to bench-scale reactors: sitosterol side-chain cleavage. Biotechnol J 5:402–412

Chapter 13

Bioconversion of Phytosterols into Androstenedione by *Mycobacterium*

Kjell D. Josefsen, Anna Nordborg, and Håvard Sletta

Abstract

The chapter describes the bioconversion of phytosterols to androstenedione (AD) with *Mycobacterium* spp. in shake flasks and fermenters, as well as LC-MS based methods for analysis of phytosterols and steroid products.

Phytosterols are derived as a by-product of vegetable oil refining and of manufacture of wood pulp. Phytosterols contain the same four-ring nucleus as steroids, and may be converted to high-value steroids by removing the side chain at C17 and minor changes at other sites in the ring structure.

Many bacteria, including *Mycobacterium* spp., are able to degrade phytosterols. Mutants of *Mycobacterium* spp. unable of ring cleavage can, when growing on phytosterols, accumulate the steroid intermediates androstenedione (AD) and/or androstadienedione (ADD).

The practical challenge with microbial conversion of phytosterols to steroids is that both the substrate and the product are virtually insoluble in water. In addition, some steroids, notably ADD, may be toxic to cells.

Two main strategies have been employed to overcome this challenge: the use of two-phase systems, and the addition of chemically modified cyclodextrins. The latter method is used here.

Defined cultivation and bioconversion media for both shake flask and fermenter are given, as well as suggestions to minimize the practical problems caused by the water-insoluble phytosterol. Sampling, sample extraction, and quantification of substrates and products using LC-MS analysis are described.

Key words *Mycobacterium*, Phytosterols, Steroids, Transformation, Analysis, Fermentation, LC-MS

1 Introduction

Phytosterol is a collective term for plant-derived sterols and stanols. Stanols are saturated sterols without double bonds in the sterol ring structure. They are triterpenes and important structural components of plant membranes. Phytosterols are structurally related to cholesterol but have different side chain configurations (Fig. 1). More than 100 types of phytosterols have been reported in plant species, but the most abundant are β-sitosterol, stigmasterol, and campesterol [1]. Commercially, phytosterols are isolated from vegetable oils (soybean oil, rapeseed oil, sunflower oil, etc.) as a

José-Luis Barredo, Ignacio Herráiz (eds.), *Microbial Steroids: Methods and Protocols*, Methods in Molecular Biology vol. 1645, DOI 10.1007/978-1-4939-7183-1_13, © Springer Science+Business Media LLC 2017

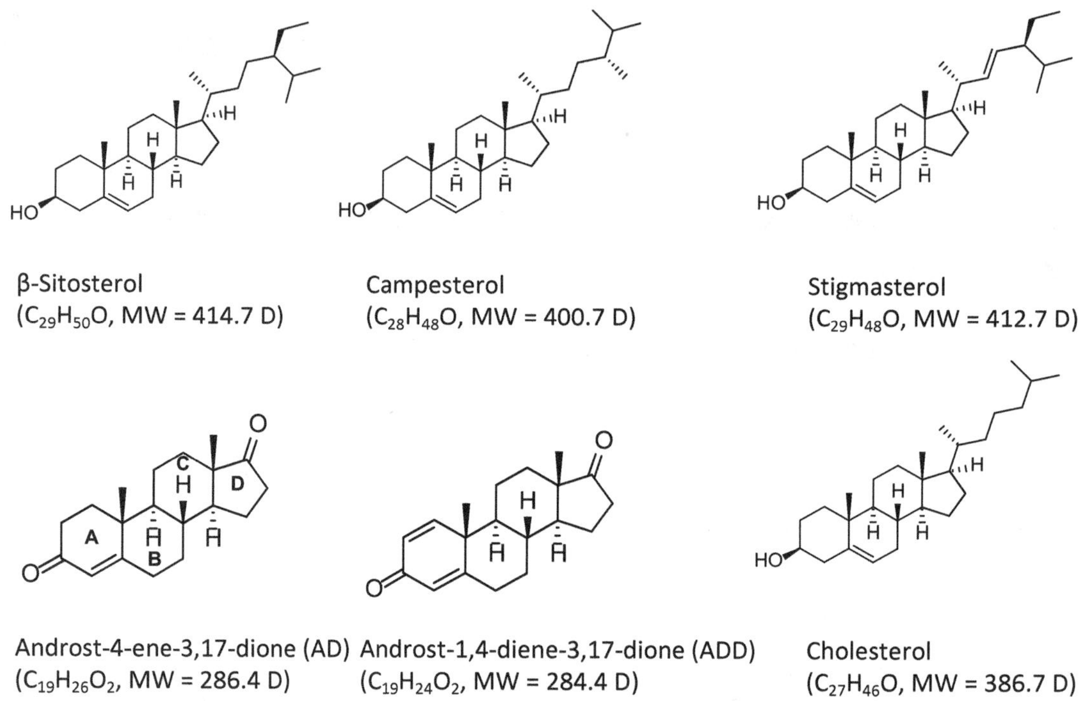

Fig. 1 Chemical structure of β-sitosterol, campesterol, stigmasterol, cholesterol, androstenedione (AD), and androstadienedione (ADD). The conventional designation of the rings (A–D) is indicated in the AD structure

by-product of traditional vegetable oil refining, and from the so-called "tall oil," a by-product of the manufacture of wood pulp [2]. These commercial products contain a mixture of different sterols and stanols, and the composition reflects the source and the production process (Table 1). Phytosterols contain the same four-ring nucleus as steroids, and may be converted to androst-4-ene-3,17-dione (androstenedione, AD) or androst-1,4-diene-3,17-dione (androstadienedione, ADD) by removing the side chain at C17 and minor modifications at other sites in the structure.

AD and ADD are steroid intermediates used in the pharmaceutical industry for production of anabolic drugs. The conversion of phytosterols into AD or ADD may be achieved chemically, enzymatically, and with microorganisms, including *Mycobacterium* spp. Here we focus on microbial conversion. For a recent review of chemical and enzymatic processes, *see* Al Jasem et al. [3]. Theoretically, and according to their molecular mass, 1.0 kg β-sitosterol can yield 0.69 kg AD.

1.1 Microbial Transformation of Phytosterols to Steroids

The initial steps in the microbial degradation of phytosterols are outlined in Fig. 2. The key to microbial conversion of phytosterols to steroids is to prevent ring opening. The two key reactions in the ring opening process are the addition of an OH-group in the 9α-position by a 9α-hydroxylase and the introduction of a double

Table 1
Composition (%) of different commercial phytosterols [2]

Phytosterol/stanol	Tall oil	Vegetable oil
β-Sitosterol	36–80	43–52
β-Sitostanol	6–34	0–3
Campesterol	4–25	24–28
Campestanol	0–14	0–2
Stigmasterol	<1	14–24
Brassicasterol	<1	0–9
Minor sterols	0–2	<3

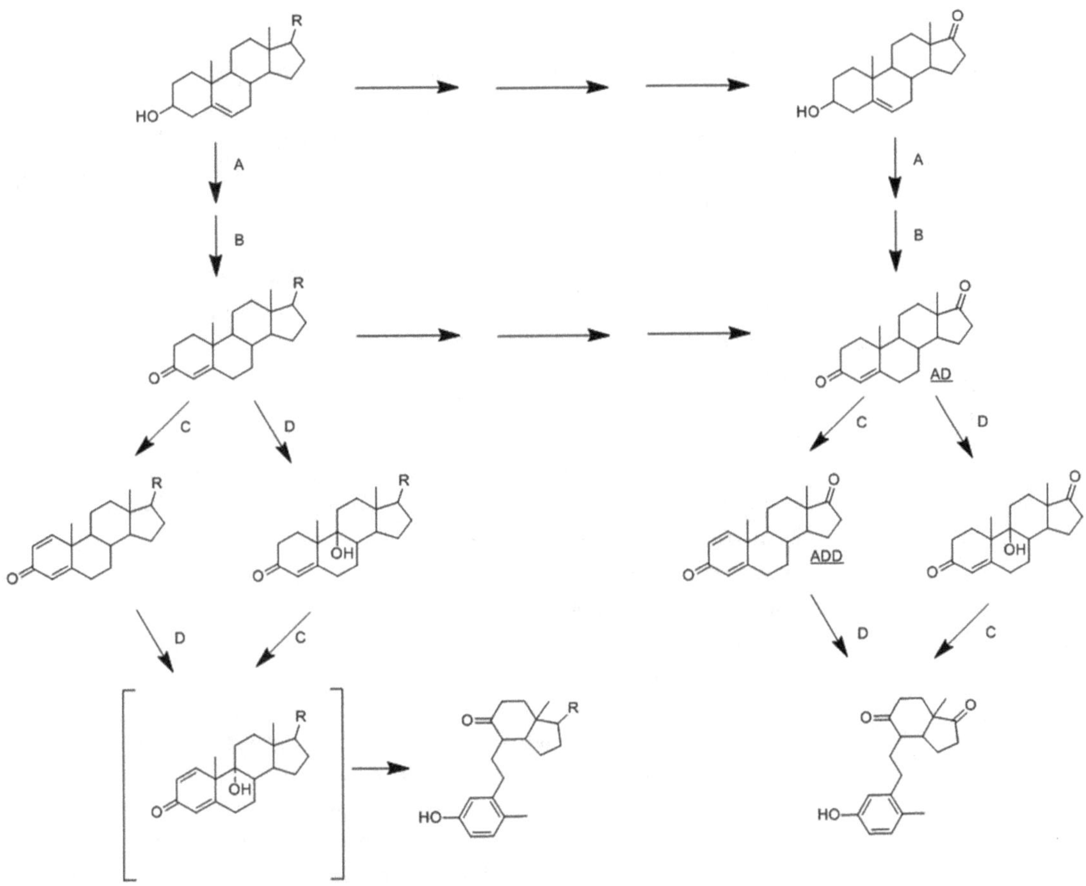

Fig. 2 Outline of the initial steps in bacterial degradation of phytosterols including ring opening. Based on [4] and references therein. A = β-hydroxysteroid dehydrogenase, B = 3-keto steroid isomerase, C = 9α-hydroxylase, D = 1,2-dehydrogenase

bond between C1 and C2 in ring A by a 1,2-dehydrogenase. When these two reactions are combined, they cause aromatization and cleavage of ring B [4]. The activity of these two enzymes is relatively independent of the side chain at C17. Similarly, the multistep enzymatic degradation of the side chain at C17 is relatively independent of the chemical structure at ring A [4]. Thus, if the activity of the 9α-hydroxylase and/or the 1,2-dehydrogenase can be inhibited, the degradation of the four-ring sterol nucleus can be prevented, and AD and/or ADD will accumulate. Initially, in the 1960s, research focused on the use of inhibitory agents such as 8-hydroxyquinoline, Ni^{2+}, and α,α-dipyridyl, or on chemical modification of the sterol structure itself to prevent attack; for a review *see* [5]. However, focus soon shifted to mutants devoid of 9α-hydroxylase and/or 1,2-dehydrogenase, and a number of mutants unable or only poorly able of ring cleavage were isolated and patented during the 1970s. Mutants of *Mycobacterium* spp. were prominent among these.

The bioconversion of phytosterols to AD with mutants of *Mycobacterium* spp. may be achieved either by (1) first producing the cell mass on another carbon substrate, e.g., glycerol or glucose, followed by a bioconversion phase with phytosterol as C-source, or by (2) cultivating the mutants on phytosterol as C-source from start. Both strategies have their advantages and disadvantages. The cultivation of the mutants can be performed both in shake flasks and fermenters, but pH control can be a challenge in shake flasks, particularly when *Mycobacterium* is cultivated on defined media with C-sources such as glycerol or glucose. Here pH may become very low (<4) due to acid production. In cultivation media containing complex ingredients such as yeast extract or peptone, pH changes are more moderate. The bioconversion rate is not rapid and will, depending on the concentration of phytosterol and the cell concentration from start, usually take days. Thus, producing the cell mass on another C-source may speed up the bioconversion step, but the process also becomes more complicated than the one-pot cultivation process with phytosterol as C-source from start. It should also be taken into account that the cells need a transition period before they can attack phytosterols if cultivated on another C-source.

1.2 Phytosterols and Steroids are Almost Insoluble in Water

The main challenge with bioconversion of phytosterols to AD(D) is that both the substrate and the products are poorly water soluble. The estimated solubility of β-sitosterol at 25 °C is 20 μg/L [6], while the solubility of the related molecule cholesterol is about 26 μg/L at 30 °C [7]. However, the measured "solubility" of mixed phytosterols is considerably higher, possibly due to micelle formation. Yuan et al. [8] found a "solubility" of approx. 11 mg/L for mixed phytosterols at 30 °C. For a bioconverting bacterium, it is access to the substrate that is critical, not if the compound is truly dissolved or not. However, 11 mg/L is also a very low concentration and in systems where phytosterol is dispersed in water-based media, direct contact between the cells and the phytosterol is probably required for efficient substrate uptake. Small, finely dispersed phytosterol particles and thus a large surface to volume ratio is an advantage. Fine phytosterol particles can easily be made with a mortar or otherwise. Unfortunately, during heat sterilization, the finely dispersed phytosterol particles often aggregate into larger particles that are difficult to disperse again. An alternative is to dissolve the phytosterols in a water-miscible solvent, e.g., ethanol [9], sterilize the solution, and slowly drip it into the sterile medium. The phytosterol will then precipitate as small particles. The disadvantage is that if the solvent concentration becomes too high, this will affect the cells negatively.

The solubility of steroids in water is typically 20–50 mg/L [10]. The solubility of testosterone and AD at 25 °C is 23 and 58 mg/L, respectively [11]. Again less rigorous solubility studies indicate a much higher "solubility", possibly due to micelle formation or other phenomena. Khomutov et al. [12] found a solubility of 9α-hydroxy-4-androstene-3,17-dione and ADD at 30 °C of 632 and 464 mg/L, respectively.

ADD has been shown to have toxic effects on mycobacteria and other bacteria [13–15] and AD is suspected of having toxic effects [16]. ADD is often produced as a by-product also when AD is the main product. When cells convert phytosterols to steroids in a water based system, it may be difficult for the cells to "get rid of" the formed water-insoluble steroids. The cells may export them from the interior, but the hydrophobic outer layers on mycobacteria and the cell membrane will be a more hydrophobic environment than the water phase, and steroids may aggregate around the cells rather than dissolve or disperse in the water phase. This is exactly the place you do not want a cell-toxic compound.

To overcome the challenge of a water insoluble substrate and an almost water insoluble and possibly toxic product, several strategies have been employed. The two main approaches are two-phase systems and the use of cyclodextrins.

1.3 Two-Phase Systems

A number of different aqueous–organic two-phase systems have been studied. The principle is that the phytosterols are dissolved in a water-immiscible organic substance such as vegetable oils [17, 18], ionic liquids [8], polypropylene glycol [18, 19], silicone oils [18, 19], or phthalates [20–22]. The phytosterol–oil mixture is added to the medium and stirring or shaking disperses the oil as small droplets that the cells can adhere to and thus get in contact with the phytosterol. The produced steroids may dissolve in the same oil phase and thus be removed from the cell surface. Important aspects to consider is the solubility of both phytosterols and steroids in the oil, the compatibility of the oils with the cells, the viscosity of the medium with oil, and effects on downstream processing and purification of the product. Notably, emulsion formation due to the presence of surface-active biological material and intense mixing, may complicate downstream separation. In large scale, the cost of the oil and effects on the working as well as the surrounding environment must also be taken into account.

1.4 Cyclodextrins

Cyclodextrins (CDs) are cyclic oligosaccharides consisting of six, α-CD, seven, β-CD, eight, γ-CD, or more glucopyranose units linked by α-(1,4) bonds [23]. They form a torus-like three-dimensional structure with a hydrophobic inner cavity and a hydrophilic outer surface. The hydrophobic cavity provides a microenvironment into which appropriately sized nonpolar moieties can enter to form inclusion complexes. One or two guest molecules can be entrapped by one, two, or three CDs [23 and references therein]. Complex formation is a dimensional fit between host cavity and guest molecule. No covalent bonds are broken or formed during the formation, and the binding of the guest molecule within the host CD is not fixed or permanent but a dynamic equilibrium. Binding strength depends on how well the "host–guest" complex fits together and on local interactions between surface atoms.

Native CDs have limited water solubility. The solubility of β-CD is 18.5 g/L (16 mM) [24]. By chemical modification, the CDs can become much more water-soluble. The water solubility of randomly methylated β-cyclodextrin (Me-β-CD) and 2-hydroxypropyl-β-cyclodextrin (HP-β-CD) is >500 and >600 g/L, respectively [24]. Both these modified CDs have been used to solubilize steroids, and possibly also to some extent phytosterols in microbial phytosterol bioconversion processes. Unfortunately, the cost of these modified β-CDs is high, in the order of 300–350 USD/kg (bulk prices) versus 5 USD/kg for native β-CD [24]. This makes them too costly for industrial use although attempts have been made to recirculate HP-β-CD [25]. However, in laboratory scale where the aim is to study and understand processes and phenomena, they are very useful.

The modified β-CDs may in high concentrations have a negative effect on the growth of *Mycobacterium* [26, 27]. Thus, the concentration of modified β-CDs in the medium from start should not be too high (>50 g/L Me-β-CD, >70 g/L HP-β-CD). More can be added during the bioconversion process if needed. A general guideline is that at least one Me-β-CD/HP-β-CD molecule should be available for each steroid molecule produced. As one phytosterol molecule theoretically can generate one steroid molecule, this means that on weight basis at least 3.3 g Me-β-CD or 3.9 g HP-β-CD should be added for each gram of phytosterol.

1.5 Cultivation of Mycobacterium

The cultivation of *Mycobacterium* spp. and bioconversion of phytosterols have been performed in a number of different media. Generally, we may separate cultivation media into three groups, rich media where all or most components are complex (yeast extract, peptone, tryptone, etc.), semidefined media where one or a few complex medium components are added to an otherwise defined medium, and defined media where all constituents are pure chemicals. Defined media are an advantage for basic studies as all carbon sources and their quantities are exactly known, but cell growth is usually more rapid on rich or semidefined media. Preparation of and cultivation on defined media is more challenging than preparation and use of more complex media. When using defined media, precipitates may be formed during preparation and the pH changes during cultivation are usually larger.

1.6 Sampling

The poor solubility of both the substrate and the product(s) makes it difficult to obtain representative samples from the cultures, in particular from fermenters as clumps of phytosterols and cells aggregate at various locations in the fermenter and are almost impossible to disperse in the water phase before sampling. In shake flask studies, an alternative is to harvest a whole shake flask at every sampling point, in particular if the aim is to obtain a good mass balance. However, the latter strategy implies relatively large samples that may complicate the extraction process.

1.7 Analysis

Phytosterols and steroids can be extracted from the samples with ethyl acetate, diethyl ether, hexane, or chloroform, and analyzed by gas chromatography (GC), TLC [28], or HPLC [29], with or without coupling to mass spectrometry [30, 31]. When analyzing by HPLC using UV, identification of the analytes is based on comparison with retention times for standards [32] and combination with mass spectrometry has therefore been used to aid analyte identification [33]. Mass spectrometric detection gives more reliable identification than UV-based detection, and use of multiple reaction monitoring (MRM) mode ensures specific detection of the selected analytes and greatly improved limit of detection. In HPLC coupled to mass spectrometry, electrospray ionization has been used for both steroids and phytosterols, but ionization efficiency might be a problem for some steroids and derivatization protocols have been established to help overcome these issues [34]. For phytosterols, atmospheric pressure chemical ionization (APCI) has also been used [35, 36]. Due to the desire to analyze steroids and phytosterols simultaneously, APCI was chosen as the ionization method in this work.

2 Materials

2.1 Culture Media

1. Semidefined growth medium (per liter in distilled water or similar quality): Glycerol (99.5%) 10 g, yeast extract 10 g, $(NH_4)_2HPO_4$ 1.5 g, K_2HPO_4 0.5 g, KH_2PO_4 0.5 g, $MgSO_4{\bullet}7H_2O$ 0.2 g, $FeSO_4{\bullet}7H_2O$ 5 mg, and $ZnSO_4{\bullet}7H_2O$ 2 mg. All components are mixed, pH adjusted to a range of 7.0–7.2 and heat-sterilized (121 °C, 20 min; based on the recipe from Prof. Marina V. Donova, GK Skryabin Inst. of Biochemistry and Physiology of Microorganisms, Russian Acad. Sci., Pushchino, Moscow Region, Russia).
2. Solid semidefined growth medium: Add 15 g/L agar to the semidefined growth medium.
3. SINTEF's defined cultivation medium for shake flasks (per liter in distilled water or similar quality): glycerol (99.5%) 10.0 g, PIPES [piperazine-*N,N'*-bis(2-ethanesulfonic acid)] 4.0 g, NH_4Cl 2.5 g, $MgSO_4{\bullet}7H_2O$ 0.50 g, K_2HPO_4 0.50 g, NaH_2PO_4 0.30 g, aspartic acid 0.10 g, SINTEF's trace mineral solution 4.0 mL, pH 7.0–7.2 (*see* **Note 1**). The buffer PIPES ($pK_a = 6.76$ at 25 °C) dissolves slowly and medium preparation should start by mixing water (983 mL) and PIPES and adding 4.5 mL 3 M NaOH. Leave with stirring (0.5–2 h) until all PIPES has dissolved and pH decreased to between 7 and 8 (*see* **Note 2**). Add glycerol, ammonium chloride, phosphate salts and aspartic acid and adjust pH to a range of 7.0–7.2. Heat (121 °C, 20 min) or filter (0.2 μm) sterilize the bulk medium. After sterilization, add the following filter-sterilized (0.2 μm) solutions (per liter final medium): 4.0 mL of SINTEF's trace mineral solution, 4.0 mL of 0.3 M NaOH (*see* **Note 3**), and 4.0 mL of $MgSO_4$ solution (6.25 g $MgSO_4{\bullet}7H_2O$ in distilled water to 50 mL; *see* **Note 4**).
4. SINTEF's defined bioconversion medium for shake flasks (per liter in distilled water or similar quality): Phytosterol 2.0 g, methyl-β-cyclodextrin 8.0 g, glycerol (99.5%) 0.5 g, PIPES 2.0 g, NH_4Cl 1.5 g, KH_2PO_4 2.0 g, $Na_2HPO_4{\bullet}2H_2O$ 3.5 g, $MgSO_4{\bullet}7H_2O$ 0.25 g, aspartic acid 0.1 g, SINTEF's trace mineral solution 2.0 mL, pH 7.0–7.2 (*see* **Note 5**). The buffer PIPES ($pK_a = 6.76$ at 25 °C) dissolves slowly and medium preparation should start by mixing water (991.7 mL) and PIPES and adding 2.3 mL 3 M NaOH. Leave with stirring for 0.5–1 h until PIPES has dissolved and pH decreased to between 7 and 8 (*see* **Note 2**). Add ammonium chloride, phosphate salts, glycerol, aspartic acid and methyl-β-cyclodextrin and adjust pH at a range of 7.0–7.2. Distribute the medium in shake flasks, e.g., 25–100 mL per shake flask, and add phytosterol corresponding to the volume in the shake flask, e.g., 50–200 mg/shake flask. Heat-sterilize (121 °C, 20 min) the

shake flasks with medium. After sterilization, add the following filter-sterilized (0.2 μm) solutions (per liter final medium): 2.0 mL of SINTEF's trace mineral solution, 2.0 mL of 0.3 M NaOH (*see* **Note 3**), and 2.0 mL of $MgSO_4$ solution (6.25 g $MgSO_4$•$7H_2O$ in distilled water to 50 mL; *see* **Note 4**).

5. SINTEF's defined cultivation medium for fermenter (per liter in distilled water or similar quality): Glycerol (99.5%) 40 g, NH_4Cl 9.0 g, $MgSO_4$•$7H_2O$ 1.5 g, NaH_2PO_4 1.2 g, K_2HPO_4 1.5 g, aspartic acid 0.20 g, and SINTEF's trace mineral solution 15.0 mL, pH 7.0–7.2 (*see* **Note 6**). The medium is prepared by dissolving glycerol, ammonium chloride, phosphate salts, and aspartic acid in 973 mL distilled water, and heat-sterilize in the fermenter. After sterilization, add the following filter-sterilized (0.2 μm) solutions (per liter final medium): 15.0 mL of SINTEF's trace mineral solution, and 12.0 mL of $MgSO_4$ solution (6.25 g $MgSO_4$•$7H_2O$ in distilled water to 50 mL; *see* **Note 4**). pH is adjusted and controlled at pH 7.0–7.2 with 3 M HCl and 3 M NaOH. pH will mostly decrease during the fermentation. Antifoam (Clerol FBA 622) 10 mL is mixed with 90 mL distilled water and heat-sterilized. From start 5 mL of this mixture is added per liter medium. More is added as required during the fermentation (*see* **Note 7**).
6. SINTEF's defined bioconversion medium for fermenter (per liter in distilled water or similar quality): Phytosterol 10 g, methyl-β-cyclodextrin 40 g, NH_4Cl 9.0 g, $MgSO_4$•$7H_2O$ 1.5 g, NaH_2PO_4 1.2 g, K_2HPO_4 1.5 g, aspartic acid 0.20 g, and SINTEF's trace mineral solution: 15.0 mL, pH 7.0–7.2 (*see* **Note 8**). Dissolve methyl-β-cyclodextrin, ammonium chloride, phosphate salts, and aspartic acid in 823 mL distilled water. This solution is heat-sterilized in the fermenter. After sterilization, add the following solutions (per liter final medium): (1) 15.0 mL of filter-sterilized (0.2 μm) SINTEF's trace mineral solution, (2) 12.0 mL of filter-sterilized (0.2 μm) $MgSO_4$ solution (6.25 g $MgSO_4$•$7H_2O$ in distilled water to 50 mL; *see* **Note 4**), (3) 100 mL of heat-sterilized phytosterol dispersion, and (4) 50 mL heat-sterilized washing water. pH is adjusted and controlled at pH 7.0–7.2 with 3 M HCl and 3 M NaOH. pH will mostly decrease during the fermentation. 10 mL of antifoam (Clerol FBA 622) is mixed with 90 mL distilled water and heat-sterilized. From start 5 mL of this mixture is added per liter medium. More is added as required during the fermentation (*see* **Note 7**).
7. Heat-sterilized phytosterol dispersion. Phytosterol (10 g) is dispersed as small particles (ULTRA-TURRAX) in 100 mL hot distilled water in a small bottle or an Erlenmeyer flask and heat-sterilized (121 °C, 20 min). It is an advantage to transfer the phytosterol dispersion to the fermenter while it is still hot after sterilization.

8. Heat-sterilized washing water. Distilled water (50 mL) is heat-sterilized in a small bottle or an Erlenmeyer flask and used to wash out remaining particles of phytosterol in the flask above before it is transferred to the fermenter. It is an advantage if the water is still hot after heat sterilization.
9. SINTEF's trace mineral solution (per liter): $FeCl_3$ 3.0 g, $Na_2EDTA{\bullet}2H_2O$ 500 mg, $MnCl_2{\bullet}4H_2O$ 400 mg, $ZnCl_2$ 300 mg, $CuCl_2{\bullet}2H_2O$ 250 mg, $CoCl_2{\bullet}6H_2O$ 20 mg, $Na_2MoO_4{\bullet}2H_2O$ 10 mg, conc. HCl (37%) 25 mL, and distilled water (or similar quality) 975 mL (*see* **Note 9**).

2.2 Sterol and Steroid Analysis

1. Eluent A for LC-MS analysis (per liter in distilled water): 25 mM formic acid. 0.94 mL of formic acid (analytical grade >99%) is added to distilled water in a 1 L volumetric flask. The mixture is thoroughly mixed and diluted to volume with distilled water. The resulting 25 mM formic acid solution is transferred to an eluent bottle (*see* **Note 10**).
2. Eluent B for LC-MS analysis: Methanol (LC-MS grade; *see* **Note 11**).
3. Extraction solvent: Ethyl acetate (analytical or LC-MS grade).
4. Androst-4-ene-3,17-dione (AD).
5. Androst-1,4-diene-3,17-dione (ADD).
6. 3β-Hydroxyandrost-5-ene-17-one (DHEA).
7. 11α-Hydroxyandrost-4-ene-3,17-dione (11-α-OH-AD).
8. Testosterone.
9. Campesterol.
10. Stigmasterol.
11. β-Sitosterol.
12. Brassicasterol.
13. Stock standards—Phytosterols: 1–2 mg/mL in chloroform. Weigh out accurate amounts of the individual phytosterols and add solvent (chloroform) to a phytosterol concentration of 1–2 mg/mL. Solutions are stored at −20 °C (*see* **Note 12**).
14. Stock standards—Steroids: 1–2 mg/mL in ethyl acetate. Weigh out accurate amount of the individual steroids and add solvent (ethyl acetate) to a steroid concentration of 1–2 mg/mL. Solutions are stored at −20 °C (*see* **Note 12**).
15. Working standards: 1000 μM in ethyl acetate. Dilute the individual stock standards to 1000 μM in ethyl acetate. Prepare a mixed working standard of all analytes at individual concentrations of 1000 μM in ethyl acetate. Solutions are stored at −20 °C.
16. Calibration standards: 0, 0.1, 0.5, 1, 5, 10, 25, 50, and 100 μM in methanol. Prepare calibration standards by diluting the

mixed 1000 μM standard in methanol. Also prepare a blank of only methanol. Solutions are stored at −20 °C.

17. LC-MS Instrumentation: High-pressure gradient pumps equipped with degasser, autosampler, and column thermostat, connected to a triple quadrupole MS. For example; Agilent 1200 system connected to an Agilent triple quadrupole equipped with an atmospheric pressure chemical ionization (APCI) ion source.
18. LC-column Zorbax SB-C8 (4.6 × 100, 3.5 μm) or equivalent.

3 Methods

3.1 Preparation of Standard Inoculum

1. Obtain the bacterium from a culture collection, usually as a freeze-dried sample (*see* **Note 13**).
2. After opening the ampoule, add a few drops of sterile semidefined growth medium to disperse the freeze-dried cells.
3. Seed a shake flask of 250 mL with 50 mL of semidefined growth medium.
4. As a backup, transfer some of the dispersed cells to semidefined agar medium, and incubate the agar plates at 30 °C.
5. Incubate the shake flask at 30 °C at a moderate shaking rate (e.g. 150–200 rpm with 2.5 cm orbit) for 48–96 h, until late logarithmic growth phase (*see* **Note 14**).
6. When the culture is in late logarithmic growth phase, transfer a small volume (1–3 mL) to a new 500 mL shake flask with 100 mL defined cultivation medium (100 mL; *see* **Note 15**).
7. Incubate the shake flask at 30 °C at a moderate shaking rate (e.g. 150–200 rpm with 2.5 cm orbit) until a suitable culture density is obtained (*see* **Note 16**).
8. Transfer a suitable volume of culture, e.g., 60 mL, to a sterile flask with screw cap and add 40 mL sterile 50% (v/v) glycerol solution. Mix well.
9. Distribute the mixture of culture and glycerol in sterile cryotubes (1–3 mL per tube, at least 20–25 cryotubes) and freeze at −80 °C (*see* **Notes 17** and **18**).

3.2 Purity Check of Standard Inoculum

1. Select at random 2–3 cryotubes with frozen standard inoculum. Thaw them rapidly (*see* **Note 19**) and transfer 1 mL thawed culture from each cryotube to a 500 mL shake flask with 100 mL semidefined medium.
2. Incubate the shake flasks at 30 °C and a moderate shaking rate for 1–3 days until the culture is in late logarithmic growth phase.
3. Transfer a small volume (0.1–0.5 mL culture) to a new shake flask (500 mL) with 100 mL semidefined medium and again incubate until late logarithmic growth phase.

4. Repeat the transfer procedure one or two times to give any contaminant present in the original inoculum a chance to increase to a significant fraction of the culture.
5. From the late logarithmic growth phase culture, make a ten-fold dilution series in defined cultivation medium without C-source, and plate aliquots (0.1 mL) from suitable dilutions on semidefined agar medium.
6. Incubate the agar-plates at 30 °C until colonies appear (2–4 days) and inspect the plates for suspicious or atypical colonies that may belong to a contaminating organism in the culture (*see* **Note 20**).

3.3 Shake Flask Cultivation

1. Seed a shake flask of 500 mL with 100 mL of defined cultivation medium for shake flasks with 1–3 mL standard inoculum (*see* **Note 19**).
2. Incubate the shake flask at 30 °C and a moderate shaking rate for 48–96 h, until late logarithmic growth phase (*see* **Note 21**).
3. Use the shake flask culture to inoculate shake flasks for bioconversion studies or fermenters.

3.4 Shake Flask Bioconversion of Phytosterol to AD

1. Seed a shake flask of 500 mL with 100 mL of defined bioconversion medium for shake flasks with 2–30 mL of the shake flask culture above.
2. If a large inoculum is used (>5 mL), centrifuge down the cells (3200 × *g*, 10 min, 15 °C) and disperse them in a small volume (2–5 mL) of defined bioconversion medium without phytosterol before transferring them to the shake flask (*see* **Note 22**).
3. Incubate the shake flask at 30 °C, 100 rpm, for 48–120 h (*see* **Note 21**).

3.5 Fermenter Cultivation

1. Prepare the defined cultivation medium in a 1–10 L fermentation vessel and sterilize at 121 °C for 20 min.
2. Inoculate from a shake flask with defined cultivation medium. An inoculum volume of 10–30 mL/L fermenter medium is often suitable.
3. Incubate at 30 °C with pH controlled to 7.0–7.2 (*see* **Note 23**).
4. Control the dissolved oxygen level at 20–30% of saturation by stirring rate.
5. If foaming becomes a serious problem, add some antifoam from a sterile antifoam solution.

3.6 Fermenter Bioconversion of Phytosterol to AD

1. Prepare the defined bioconversion medium in a 1–10 L fermentation vessel and sterilize at 121 °C for 20 min.
2. Inoculate from a shake flask or an inoculum fermenter. If the inoculum volume is small (≤30 mL/L fermenter medium)

transfer the culture directly to the fermenter. If the inoculum volume is large, centrifuge down the cells (3200 × *g*, 10 min, 15 °C) and resuspend them in a small volume (10–30 mL/L fermenter medium) of defined bioconversion medium before inoculation (*see* **Note 22**).

3. Incubate at 30 °C with pH at a range of 7.0–7.2 (*see* **Note 23**).
4. Control dissolved oxygen level at 20–30% of saturation by stirring rate.
5. If foaming becomes a serious problem, add some antifoam from a sterile antifoam solution.

3.7 Sample Collection

1. Collect samples of 5 mL from the cultures at suitable intervals (12–24 h) to monitor the conversion of phytosterols to steroids (*see* **Note 24**).
2. If sampling a shake flask, try to disperse as much as possible of the cells and phytosterols on the glass wall in the culture by manual shaking of the flask, before sampling. Sterile inoculation loops may be used to manually detach lumps of cells and/or phytosterol if necessary.
3. If sampling a laboratory fermenter, first shake the fermenter manually if possible (*see* **Note 23**). Thereafter, increase the stirring to high speed (e.g. 1000–1500 rpm for a 1–3 L laboratory fermenter) for a short period (15–30 s, watch out for foaming) to disperse as much as possible of the phytosterol in the water phase before sampling.
4. Transfer the sample to a centrifuge tube
5. Centrifuge the sample immediately (18,000 × *g*, 10 min, 10 °C) or place on ice until centrifugation in order to minimize further bioconversion.
6. Transfer the supernatant to a new tube.
7. Extract the cell–phytosterol pellet and the supernatant as described below, either immediately or freeze (−20 °C) both fractions for later extraction.

3.8 Sample Extraction and Preparation

1. Resuspend the pellet in the centrifugation tube in 0.5 mL distilled water and transfer the suspension to a glass extraction tube (*see* **Note 25**). Wash the centrifugation tube with 0.5 mL distilled water to ensure complete transfer of the pellet, and transfer the washing water to the extraction tube. If the pellet has been frozen, thaw it first, but avoid heating it to room temperature to minimize bacterial activity in the pellet.
2. Add 10 mL ethyl acetate to the extraction tube. Close the tube with the cap and mix vigorously (whirl mixer).
3. From the supernatant, transfer a 1.0 mL aliquot to the same type of glass extraction tube as above. If the supernatant has

been frozen, it should be thawed, heated to room temperature and mixed well before the aliquot is transferred.

4. Add 10 mL ethyl acetate to the extraction tube. Close the tube with the cap and mix vigorously (whirl mixer).
5. Place the extraction tubes in a rotary shaker for 20 min (room temperature).
6. After extraction, centrifuge the tubes (3300 × *g*, 10 min, room temperature) to aid layer separation.
7. The ethyl acetate layer (top) is collected and reduced in volume under a stream of air (at elevated temperature, 40 °C).
8. The dried samples are reconstituted in ethyl acetate and diluted in methanol. Typically, 1 + 9, 1 + 99, and 1 + 999 dilutions are prepared for each sample.

3.9 LC-MS Analysis

1. Establish the initial LC operating conditions: Flow rate 0.5 mL/min, eluent A (20%), eluent B (80%), column temperature, 45 °C.
2. Establish the mass spectrometer operating conditions: APCI ionization; positive mode; multiple reaction monitoring; gas temperature 250 °C, APC1 heater 300 °C; gas flow rate 4 L/min; nebulizer 40 psi; and capillary 3500 V (+). All parameters are optimized for the selected analytes. MRM settings are given in Table 2.

Table 2
MRM settings for steroids and phytosterols

Analyte		*M*	Precursor	Product	Fragmentor (V)	CE (V)	RT (min)
DHEA	$C_{19}H_{28}O_2$	288.4244	289.4	271.4	100	4	3.95
			289.4	253.3	100	10	
AD	$C_{19}H_{26}O_2$	286.4085	287.4	97.3	100	15	3.75
11-α-OH-AD	$C_{19}H_{26}O_3$	302.41	303.4	267.2	100	15	2.71
Testosterone	$C_{19}H_{28}O_2$	288.4244	289.4	97.3	100	15	3.95
ADD	$C_{19}H_{24}O_2$	284.39	285.4	121.3	100	15	3.43
Campesterol	$C_{28}H_{48}O$	400.68	383.4	161.3	140	15	7.66
Stigmasterol	$C_{29}H_{48}O$	412.69	395.4	297.3	140	15	7.61
β-sitosterol	$C_{29}H_{50}O$	414.71	397.4	161.3	140	15	7.95
Brassicasterol	$C_{28}H_{46}O$	398.66	381.4	297.3	140	15	7.29

DHEA 3β-hydroxyandrost-5-ene-17-one, *AD* androst-4-ene-3,17-dione, *ADD* androst-1,4-diene-3,17-dione, *11-α-OH-AD* 11α-hydroxyandrost-4-ene-3,17-dione

3. Allow the LC-system and the mass spectrometer to equilibrate. This typically takes up to 2 h.
4. Inject a blank sample (methanol). The injection volume, 2 μL, is used also for samples and standards. Gradients for elution of standards and samples are:

Time (min)	A (%)	B (%)
0	20	80
1	20	80
4	1	99
7	1	99
8	20	80
10	20	80

Post run before next run 1 min.

5. Inject a series of standards (0–100 μM).
6. Inject samples.
7. For every 20 samples, inject a 10 μM standard and a blank sample.
8. Finally inject a series of standards.

3.10 Calculation of Results

1. Use the standards to calculate the peak area per concentration of the analyte. This is usually performed by the LC-MS software. Plot the calibration graph and verify linearity in sample concentration range.
2. Divide the peak area of the analyte in the sample with the peak area per concentration determined from the standards to obtain the concentration of the analyte in the analyzed sample.
3. The concentration of the analyte in the original culture is calculated by taking into account preconcentrations and dilutions during sample workup.

4 Notes

1. The shake flask medium is designed to yield approx. 4 g cells (d.w.) per liter. pH may decrease to a final pH < 4 due to acid production.
2. Using a mortar to generate a fine PIPES powder will speed up dissolution.
3. The trace mineral solution contains HCl (0.3 M) to ensure that the trace minerals remain dissolved during storage. NaOH is added to neutralize the acid in the trace mineral solution so that pH in the medium does not change significantly.

4. Precipitation of $MgNH_4PO_4$ ($k_{sp} = 3 \times 10^{-13}$ at 25 °C) is always a challenge when preparing defined cultivation media. By adding $MgSO_4$ after heat sterilization of the medium, precipitation is minimized. The effect of this salt on pH is limited. If $MgSO_4$ is added to the medium before filter sterilization, precipitated components may be removed during filtration. The $MgSO_4$ solution may be stored in a closed bottle for several weeks, but should be filter-sterilized immediately before use. Even with these precautions, a slight precipitation may be observed in the final medium, but this is not critical for the use of the medium. In fermentation media, precipitation will be more extensive due to higher concentrations of phosphate, ammonium, and magnesium. An alternative is to add one-third of the ammonium chloride and trace mineral solution from start. The rest is added as one or two batches during the fermentation. The need for ammonium and trace minerals is closely connected to cell growth. Thus, when the cell concentration in the fermenter reaches 5–7 g d.w./L, more ammonium chloride and trace minerals should be added. If phytosterol is present in the medium, cell concentration can be difficult to determine as optical density or dry weight, but the cell concentration can be estimated from the accumulated CO_2 expired in the off gas by assuming that 50% of the metabolized carbon goes to CO_2, the rest to cell mass, and that carbon constitute 50% of the cell mass. Thus, an accumulated expiration of 210 mmol CO_2 from a fermenter with 1 L medium corresponds to roughly 5 g dry weight cells/L. The estimate may be refined by correcting for CO_2 in the air (approx. 0.04%), but unless the aeration rate is very high or the fermentation time very long (several days), the influence of CO_2 in the air on this crude estimate is negligible. It is preferable to add more nutrients before they limit cell growth.
5. The shake flask medium is designed to be C-source (phytosterol) limited and more phytosterol (1–2 g/L) and corresponding amounts of Me-β-CD may be added if desired. The molar ratio between phytosterol and Me-β-CD should be at least 1:1. A too high concentration (>50 g/L) Me-β-CD may have negative effects on cell growth. A ratio between phytosterol and Me-β-CD of 1:4 by weight corresponds to a molar ratio of approx. 1:1.3. More phytosterol and/or Me-β-CD may be added during the fermentation. Based on observations it appears that when Me-β-CD is complexed with steroids it affects the cells less.
6. The fermenter medium is designed to yield approx. 15 g/L cells (d.w.). If a higher final cell density is required the concentration of the components should be increased correspondingly, but oxygen transfer may become a challenge if the final

cell density becomes too high. The problems with precipitation of medium components (*see* **Note 4**) increase with increasing concentration of medium components.

7. The selected antifoam has been chosen based on previous experience. Other antifoams may work equally well.
8. The bioconversion fermenter medium is designed to be C-source (phytosterol) limited and more phytosterol (20–50 g/L) and corresponding amounts of Me-β-CD may be added if desired. However, because a too high concentration of Me-β-CD (>50 g/L from start) may inhibit cell growth, more phytosterol and Me-β-CD should be added at intervals during the bioconversion process.
9. The trace mineral solution can be stored for at least a year at room temperature in a closed bottle. It is composed so that 1 mL solution provides roughly the amount of trace minerals required by 1 g (d.w.) bacterial cells.
10. Prepare Eluent A (25 mM formic acid in water) fresh prior to analysis.
11. All reagents should be of analytical or HPLC grade unless otherwise stated. All solvents used for the HPLC eluents should be of mass spectrometric grade. Water used should be deionized, e.g., on a Milli-Q system.
12. Solubility of especially the phytosterols is limited in the eluents used in the LC-MS analysis. Therefore, always prepare the stock solutions at high concentrations in a good solvent (such as chloroform for phytosterols and ethyl acetate for steroids), and prepare only the diluted standards in ethyl acetate or methanol.
13. Strains from culture collections are usually received as freeze-dried samples, and the culture collection will usually provide recommended resurrection procedures and initial cultivation medium. In other cases, the strain may be received as colonies on agar medium or dispersed in buffer or medium. It is advisable to first cultivate the strain in a rich or semidefined medium.
14. Freeze-dried bacteria often require some time in liquid medium before they start to grow. If, in addition, the survival rate in the freeze-dried sample is low, it may take some time before visible growth is observed. The necessary incubation time is therefore difficult to predict. Observe the culture daily. Visible growth should normally be obtained within a week. If not, the backup suspension of freeze-dried cells plated on agar medium may have yielded colonies.
15. With some mycobacteria we have experienced that it is difficult to obtain growth if the bacterium is transferred directly from a rich or semidefined medium to a defined medium. In such

cases an adaption process may be employed. The culture is transferred from the rich or semidefined medium to a defined medium containing a small amount of for example 33%, of the complex components from the former medium, and then in a second step to a medium with 5–10% of the complex medium components before final transfers to the completely defined medium.

16. Some mycobacteria produce considerable amounts of acid when growing on defined medium and the pH in shake flask cultures can become very low (pH < 4). It may then be preferable to harvest the culture at an early stage, i.e., in mid-logarithmic growth phase rather than late logarithmic growth phase to avoid the most extreme pH reduction. The disadvantage is that the standard inoculum will be thinner and it will take longer before a dense culture is obtained from the inoculum.

17. The idea behind standard inoculum is to minimize inoculum effects in further studies. The standard inoculum should preferably be prepared on a medium resembling the medium to be used in later studies. Different "schools" recommend from 15–25% final concentration of glycerol, but in our experience, this is not very critical when freezing bacteria. It is advisable to incubate the cryotubes in a refrigerator for an hour or two before transferring them to the −80 °C freezer to give the glycerol time to penetrate into the cells. By this method, 30–100 cryotubes with the same frozen culture is easily prepared. An alternative to freezing at −80 °C is to maintain the organisms on agar slants. The culture is inoculated onto agar slants in a test tube and cultivated at 30 °C until small colonies are visible. Then the test tube is put in a refrigerator. The disadvantage with this method is that new slant cultures have to be prepared at intervals (weeks to months) and over time genetic drift in the organisms may make it slightly different from the organisms you started with. The risk of contamination is also higher.

18. Glycerol-preserved bacteria frozen at −80 °C remain viable for several years, often tens of years. However, freezers do not last that long. To avoid losing all the cryotubes with standard inoculum if the freezer breaks down, it is recommended, if possible, to divide the cryotubes between two freezers.

19. Every time a new experiment is started, a cryotube with frozen culture is taken from the freezer and inoculated into a shake flask with defined medium. The frozen culture should be thawed rapidly, either in lukewarm water or by clutching it in your hand. However, remember that the culture temperature should not exceed 30 °C.

20. Microscopy of the cultures along the purity test may also be used to ensure that the culture is pure, and has the added advantage of making the experimenter familiar with the appearance of the organism in culture. This is a great advantage if, at a later stage in the studies contamination is suspected. If contamination of the standard inoculum is found or suspected, pick three to four separate colonies from the agar plates that appear to be of the right type and use these to prepare a new standard inoculum. The old standard inoculum is discarded. Alternatively, if it is suspected that the contamination may have occurred during the purity test, repeat the purity test with three new cryotubes.
21. A challenge when cultivating mycobacteria in shake flasks is wall growth. Many mycobacteria like to cling to the glass wall at or above the water line. This tendency is amplified when the cells are cultivated on phytosterols as the insoluble phytosterols often adhere to the glass wall above the water line. Commercially available shake flasks with baffles vary in their baffle design. We have found that shake flasks with the baffles relatively high upon the wall are better at keeping the cells in the water phase (less splashing), especially if combined with a low to moderate shaking rate (e.g. 100–200 rpm with 2.5 cm orbit). As the bacterium grows slowly, in particular when growing on phytosterols, the oxygen transfer rate does not need to be very high and a low shaking rate can be employed.
22. The inoculum size in bioconversion studies depend on whether you want the cells to grow and multiply on the phytosterol substrate or use a system where the bulk of the cell mass has been produced on another substrate, here glycerol. To avoid changing the concentration of phytosterol significantly and transferring significant amounts of glycerol to the conversion medium with the inoculum, the cells from the inoculum culture are centrifuged down and redispersed in a small volume of defined bioconversion medium without phytosterol before transfer to the bioconversion shake flask or fermenter.
23. Many *Mycobacterium* spp. have a tendency to adhere to the wall above the water line so "splashing" inside the fermenter should be minimized. This can be achieved by keeping the aeration rate low and the volume in the fermenter well above or well below the uppermost stirrer in the fermenter. Because the cell growth is slow, oxygen transfer is seldom a challenge. In fermentations with water-soluble carbon sources such as glycerol, the problem with wall growth can be kept reasonably under control. However, when water-insoluble phytosterols are added, the problem increases considerably. Often, large lumps of cells and phytosterol accumulate above the water line in the fermenter. Small laboratory fermenters (1–3 L) can

often, after the stirring is turned off, be loosened and manually shaken, and this has often been an efficient method to bring most of the phytosterol back into the water phase. As the bioconversion process proceeds and more and more phytosterol is converted into AD(D) that is solubilized by Me-β-CD, the problem with phytosterol–cell aggregates diminish.

24. We have found 5 mL to be a suitable sample volume, but this will depend on the system and the number of samples planned from the culture. Representative sampling is a challenge. In cultures containing enough Me-β-CD [≥1 mol Me-β-CD/mol AD(D)], almost all steroids are solubilized, but the phytosterols are mainly present as insoluble particles and clumps, usually intermixed with bacterial cells.

25. We use glass centrifuge tubes (approx. 13 mL) with short conical bottom (Assistant, Germany) closed with plastic caps (art. no. 726267, Ulstrup Plast, Denmark), but other types of tubes and caps may also be suitable. The volumes in the extraction procedure are based on a sample volume of 5 mL. If a much larger sample volume, i.e., 20 mL or more, is used, the amount of cell pellet will increase and the volumes may need to be adjusted.

References

1. Moreau RA, Whitaker BD, Hicks KB (2002) Phytosterols, phytostanols, and their conjugates in foods: structural diversity, quantitative analysis, and health-promoting uses. Prog Lipid Res 41:457–500
2. Cantrill R, Kawamura Y (2008) Phytosterols, phytostanols and their esters, chemical and technical assessments. In: Proceedings of 69th JEFCA, pp 1–13. Available at http://www.fao.org/fileadmin/templates/agns/pdf/jecfa/cta/69/Phytosterols.pdf
3. Al Jasem Y, Khan M, Taha A, Thiemann T (2014) Preparation of steroidal hormones with an emphasis on transformations of phytosterols and cholesterol – a review. Med J Chem 3:796–830
4. Szentirmai A (1990) Microbial physiology of sidechain degradation of sterols. J Ind Microbiol 6:101–116
5. Malaviya A, Gomes J (2008) Androstenedione production by biotransformation of phytosterols. Bioresour Technol 99:6725–6737
6. The Human metabolome database. http://www.hmdb.ca/metabolites/hmdb00852
7. Saad HY, Higuchi WI (1965) Water solubility of cholesterol. J Pharm Sci 54:1205–1206
8. Yuan J-J, Guan Y-X, Wang YT, Wang HQ, Yao S-J (2016) Side-chain cleavage of phytosterols by *Mycobacterium* sp. MB 3683 in a biphasic ionic liquid/aqueous system. J Chem Technol Biotechnol. Published online in Wiley online library: wileyonlinelibrary.Com. doi:10.1002/jctb.4865
9. Pendharkar GB, Anjum SD, Patil S (2014) Enhanced biotransformation of phytosterols, a byproduct of soybean refineries, to a key intermediate used for synthesis of steroidal drugs. Asian J Pharm Clin Res 7:178–180
10. Yalkowski SH, He Y (2003) Handbook of aqueous solubility data. CRC Press, Boca Raton, FL
11. NIH PubChem open chemistry database. https://pubchem.ncbi.nlm.nih.gov/
12. Khomutov SM, Dovbnya DV, Donova MV (2001) Dissolution of a mixture of steroids in cyclodextrin solutions: a model description. Pharm Chem J 55:627–629
13. Lee C-Y, Liu W-H (1992) Production of androsta-l,4-diene-3,17-dione from cholesterol using immobilized growing cells of *Mycobacterium* sp. NRRL B-3683 adsorbed on solid carriers. Appl Microbiol Biotechnol 36:598–603

14. Smith M, Zahnley J, Pfeifer D, Goff D (1993) Growth and cholesterol oxidation by *Mycobacterium* species in Tween 80 medium. Appl Environ Microbiol 59:1425–1429
15. Pérez C, Falero A, Llanes N, Hung BR, Herve ME, Palmero A, Marti E (2003) Resistance to androstanes as an approach for androstandienedione yield enhancement in industrial mycobacteria. J Ind Microbiol Biotechnol 30:623–626
16. Donova MV (2007) Transformation of steroids by actinobacteria: a review. Appl Biochem Microbiol 43:1–14
17. Xu Y-G, Guan Y-X, Wang H-Q, Yao S-J (2014) Microbial side-chain cleavage of phytosterols by mycobacteria in vegetable oil/aqueous two-phase system. Appl Biochem Biotechnol 174:522–533
18. Stefanov S, Yankov D, Beschkov V (2006) Biotransformation of phytosterols to androstenedione in two phase water-oil system. Chem Biochem Eng Q 20:421–427
19. Carvalho F, Marques MPC, de Carvalho CCCR, Cabral JMS, Fernandes P (2009) Sitosterol bioconversion with resting cells in liquid polymer based systems. Bioresour Technol 100:4050–4053
20. Marques MPC, Carvalho F, de Carvalho CCCR, Cabral JMS, Fernandes P (2010) Steroid bioconversion: towards green processes. Food Bioprod Process 88:12–20
21. Cruz A, Fernandes P, Cabral JMS, Pinheiro HM (2002) Effect of phase composition on the whole-cell bioconversion of β-sitosterol in biphasic media. J Mol Catal B Enzym 19–20:371–375
22. Cruz A, Angelova B, Fernandes P, Cabral JMS, Pinheiro HM (2004) Study of key operational parameters for the side-chain cleavage of sitosterol by free mycobacterial cells in bis-(2-ethylhexyl) phthalate. Biocatal Biotransformation 22:189–194
23. Del Valle EM (2004) Cyclodextrins and their uses: a review. Process Biochem 39:1033–1046
24. Brewster ME, Loftsson T (2007) Cyclodextrins as pharmaceutical solubilizers. Adv Drug Deliv Rev 59:645–666
25. Gao X-Q, Feng J-X, Wang XD, Hua Q, Wei DZ (2015) Enhanced steroid metabolites production by resting cell phytosterol bioconversion. Chem Biochem Eng Q 29:567–573
26. Donova MV, Nikolayeva VM, Dovbnya DV, Gulevskaya SA, Suzina NE (2007) Methyl-β-cyclodextrin alters growth, activity and cell envelope features of sterol-transforming mycobacteria. Microbiology 153:1981–1992
27. Shen Y, Wang M, Zhang L, Ma Y, Ma B, Zheng Y, Liu H, Luo J (2011) Effects of hydroxypropyl-β-cyclodextrin on cell growth, activity, and integrity of steroid-transforming *Arthrobacter simplex* and *Mycobacterium* sp. Appl Microbiol Biotechnol 90:1995–2003
28. Gulla V, Banerjee T, Patil S (2008) Quantitative TLC analysis of steroid drug intermediates formed during bioconversion of soysterols. Chromatographia 68:663–667
29. Kollerov VV, Shutov AA, Fokina VV, Sukhodolskaya GV, Donova MV (2008) Biotransformation of 3-keto-androstanes by *Gongronella butleri* VKM F-1033. J Mol Catal B Enzym 55:61–68
30. Gomez C, Fabregat A, Pozo OJ, Marcos J, Segura J, Ventura R (2014) Analytical strategies based on mass spectrometric techniques for the study of steroid metabolism. Trends Anal Chem 53:106–116
31. Marwah A, Marwah P, Lardy H (2001) High-performance liquid chromatographic analysis of dehydroepiandrosterone. J Chromatogr A 935:279–296
32. Sánchez-Machado DI, López-Hernández J, Paseiro-Losada P, López-Cervantes J (2004) An HPLC method for the quantification of sterols in edible seaweeds. Biomed Chromatogr 18:183–190
33. Rocco A, Fanali S (2009) Analysis of phytosterols in extra-virgin olive oil by nano-liquid chromatography. J Chromatogr A 1216:7173–7178
34. Ahonen L, Keski-Rahkonen P, Saarelainen T, Paviala J, Ketola RA, Auriola S, Poutanen M, Kostianen R (2012) Comparison of liquid chromatography-microchip/mass spectrometry to conventional liquid chromatography–mass spectrometry for the analysis of steroids. Anal Chim Acta 721:115–121
35. Mo S, Dong L, Hurst WJ, van Breemen RB (2013) Quantitative analysis of phytosterols in edible oils using APCI liquid chromatography–tandem mass spectrometry. Lipids 48:949–956
36. Palmgren JJ, Töyräs A, Mauriala T, Mönkkönen J, Auriola S (2005) Quantitative determination of cholesterol, sitosterol, and sitostanol in cultured Caco-2 cells by liquid chromatography–atmospheric pressure chemical ionization mass spectrometry. J Chromatogr B 821:144–152

Chapter 14

Scale-Up of Phytosterols Bioconversion into Androstenedione

Sonia Martínez-Cámara, Esther Bahíllo, José-Luis Barredo, and Marta Rodríguez-Sáiz

Abstract

Phytosterols, generated as a by-product of vegetable oils or wood pulp, contain the cyclopentane-perhydrophenanthrene nucleus, and can be converted into steroid intermediates by removing the C17 side chain. This chapter shows the scale-up, from flask to fermentor, of the phytosterols bioconversion into 4-androstene-3,17-dione (androstenedione; AD) with *Mycobacterium neoaurum* B-3805. Due to the fact that phytosterols and AD are nearly insoluble in water, two-phase systems and the use of chemically modified cyclodextrins have been described as methods to solve it. Here we use a water–oil two-phase system that allows for the bioconversion of up to 20 g/L of phytosterols into AD in 20 L fermentor.

Key words Bioconversion, Phytosterols, Fermentation, 4-Androstene-3,17-dione, Androstenedione, AD, *Mycobacterium neoaurum*

1 Introduction

Androst-4-ene-3,17-dione (androstenedione; AD) is a key intermediate in the synthesis of steroids such as testosterone, estradiol, progesterone, cortisone, prednisone, or prednisolone, which have great interest for the pharmaceutical companies [1–4]. Indeed, the market for steroid drugs is second only after antibiotics market in the pharmaceutical business; the global market for steroids is over US $10 billion and more than 1,000,000 tons annually. Moreover, the combined worldwide market for AD and androsta-1,4-diene-3,17-dione (androstadienedione; ADD), according to Forbes Medi-Tech Inc. (FMT; Vancouver, BC, Canada), is approximately US $750 million annually, but that is just for raw materials. Products derived from these starting materials are valued at nearly $4 billion per year. The growing demand of the world market for these and other steroidal products have focused pharmaceutical research on producing high value steroid precursors through the

José-Luis Barredo, Ignacio Herráiz (eds.), *Microbial Steroids: Methods and Protocols*, Methods in Molecular Biology vol. 1645, DOI 10.1007/978-1-4939-7183-1_14, © Springer Science+Business Media LLC 2017

use of efficient eco-friendly technologies with low production costs, as an alternative to the more environmentally unfriendly chemical processes [5].

The success of a fermentation-based process depends on the process cost (raw materials, utilities, manpower, and indirect costs), the fermentation yield and the market price. In this way, the use of phytosterols, a by-product of oil or wood pulp industry with similar characteristics to cholesterol such as structural behavior and functionality, as a starting material to produce steroid intermediates with *Mycobacterium* sp. in a single fermentation step, allows for the development of highly competitive processes in the global market [6–8].

Mycobacterium sp. possesses the ability to catabolize naturally occurring 3β-hydroxy-Δ5-sterols (cholesterol, β-sitosterol, etc.) or a mixture of sterols (phytosterols) to carbon dioxide and water, producing AD and ADD as the main intermediates during the degradation [9–11]. New mutants have been obtained, by classical mutagenesis or synthetic biology, to block the catabolic pathway of phytosterol favoring the accumulation of steroid intermediates of pharmaceutical value (Fig. 1) [12, 13]. Among them, *Mycobacterium neoaurum* NRRL-3805, an AD-producing strain obtained by mutagenesis from a soil isolate producing ADD is found [14, 15]. This mutant transforms phytosterol into AD with a low content of ADD because of the partial failure of the dehydrogenation at C1. The genome has been recently sequenced showing a circular chromosome of 5,421,338 bp containing 4844 protein-coding genes [16].

However, the industrial biotechnological processes for AD and other steroids are severely limited by the poor water solubility of phytosterols and the inhibition or toxicity that both substrate and products can have upon the microorganism [17]. Steroid-producing companies are focused on dealing with these shortcomings, including the use of surfactants, liquid polymers, cyclodextrins, oils, etc. to increase the phytosterol loading and achieve a more competitive process [18–21].

Carrying out a large-scale experimental fermentation is therefore costly, so pilot scale-up (from 5 to 40 L) is a powerful tool before industrializing the laboratory results. A pilot plant mimics the industrial scale fermentation process on a smaller scale, being cost-effective, environmentally aware, and the best option to explore potential changes in a production operation. Potential critical parameters of the AD bioconversion process, such as fermentor design, raw materials, development of the vegetative phases, phytosterol loading, type of fermentation (batch, semi-batch, continuous, etc.), pH, oxygen, temperature, and viscosity can be studied deeply at pilot scale before scaling up to industrial.

Finally, the pilot scale research allows for the evaluation of the costs of different AD upstream and downstream technologies, and

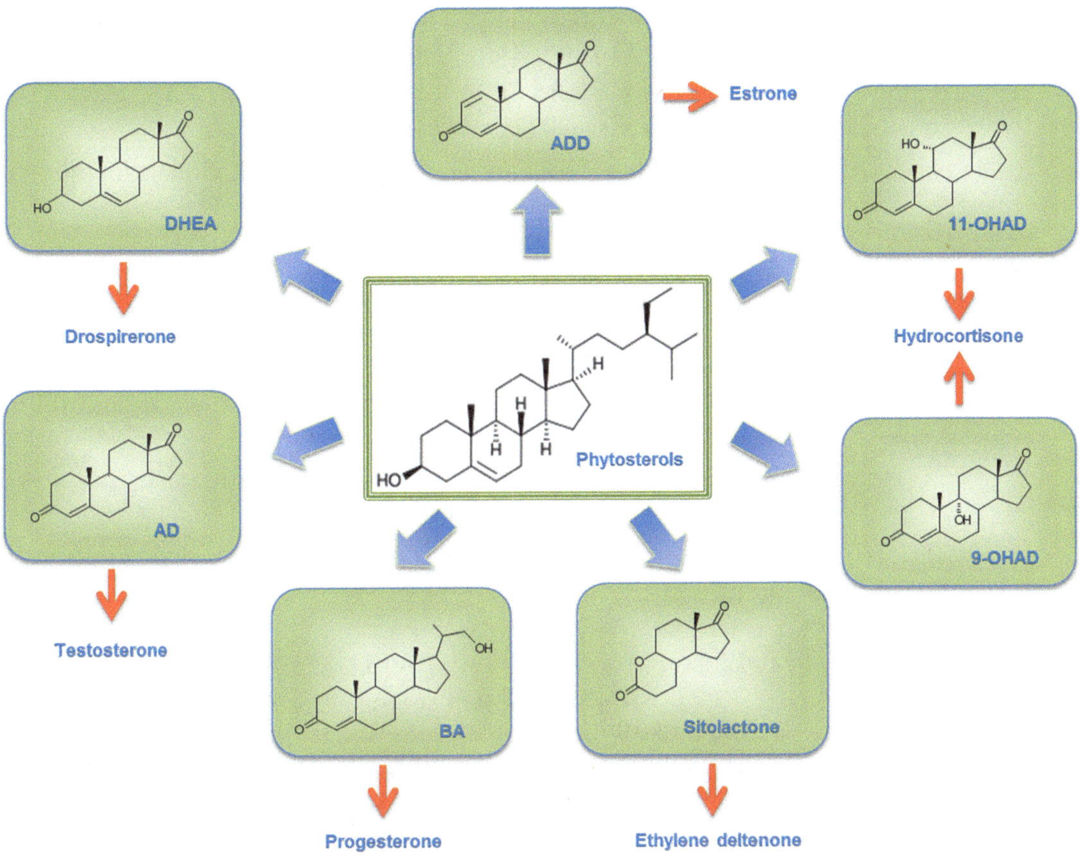

Fig. 1 Steroid production processes using phytosterols as a starting material. *Blue* and *red arrows* show bioconversion and chemical processes, respectively

a decision to be taken on their application in order to obtain a product which complies with the quality commercial requirements.

In this chapter, we show a protocol for the bioconversion of phytosterols into AD at 5 and 20 L scale, emphasizing those steps considered as critical points of the process.

2 Materials

2.1 Vegetative Stages

2.1.1 First Vegetative and Second Vegetative Stages (Flask Scale)

1. *Mycobacterium neoaurum* B-3805 (CECT 3018; DSM 2967; NCIMB 11678).
2. Erlenmeyer Flasks: 500 mL for first vegetative stage and 2 L for second vegetative stage.
3. Growth medium: 54 g/L molasses, 5.4 g/L $NaNO_3$, 0.6 g/L $NH_4H_2PO_4$, 20 g/L soya oil. Adjust to pH 7.0.
4. Flow laminar cabinet.
5. Orbital Shaker.

2.1.2 Third Vegetative Stage (5 L Biorreactor)

1. Growth medium: 54 g/L molasses, 5.4 g/L $NaNO_3$, 0.6 g/L $NH_4H_2PO_4$, and 20 g/L soya oil. Adjust to pH 7.0.
2. Fermentor Bioflo 3000, or equivalent, with a 5 L glass vessel (New Brunswick Scientific).
3. Inoculation flask: Sterile Erlenmeyer 2 L flask with silicone tube and stainless steel needle.

2.2 AD Bioconversion at 5 L Scale

1. Bioconversion medium: 2 g/L NH_4NO_3, 1 g/L KH_2PO_4, 2 g/L K_2HPO_4, 0.2 g/L KCl, 0.3 g/L $CaCl_2$, 150 g/L soya oil, and 20 g/L phytosterol. Add 1 mL/L from a micronutrient solution (11 g/L $ZnSO_4$, 1 g/L $FeSO_4$, 6 g/L $MnSO_4$, 0.04 g/L $CuSO_4$, 0.3 g/L $CoSO_4$, 0.03 g/L H_3PO_4, and 0.001 g/L KI). Adjust to pH 7.0.
2. Fermentor Bioflo 3000, or equivalent, with a 5 L glass vessel (New Brunswick Scientific).
3. Phytosterol from pine tree (Arboris, USA).
4. Blender.

2.3 AD Bioconversion at 20 L Scale

1. Bioconversion medium: 2 g/L NH_4NO_3, 1 g/L KH_2PO_4, 2 g/L K_2HPO_4, 0.2 g/L KCl, 0.3 g/L $CaCl_2$, 150 g/L soya oil, and 20 g/L phytosterol. Add 1 mL/L from a micronutrient solution (11 g/L $ZnSO_4$, 1 g/L $FeSO_4$, 6 g/L $MnSO_4$, 0.04 g/L $CuSO_4$, 0.3 g/L $CoSO_4$, 0.03 g/L H_3PO_4, and 0.001 g/L KI). Adjust to pH 7.0.
2. Fermentor Biostat C, or equivalent, with a stainless steel vessel of 20 L (about 15 L working volume) (Sartorius).
3. Phytosterol from pine tree (Arboris, USA).
4. Sterile tubing system for vegetative culture transfer.

2.4 AD Analysis

2.4.1 AD Analysis by HPLC

1. Androstenedione (4-androstene-3,17-dione, AD) (Sigma-Aldrich) 99.5% purity.
2. Butyl acetate.
3. Acetonitrile HPLC grade.
4. Analytical balance.
5. HPLC equipment. Waters Alliance e2695 HPLC system with Waters 2998 Photodiode Array (PDA) Detector, or equivalent (Waters).
6. Column RP C18 Nucleosil 250 × 4.6 mm, 5 μm (with guard column), or equivalent.
7. Mobile phase: Acetonitrile–water (52:48) with acetic acid 0.01%.

2.4.2 AD Analysis by TLC

1. Ethyl acetate.
2. TLC chamber for 20 × 20 cm aluminum sheets.

3. TLC Silica gel 60 F354 aluminum sheets 20 × 20 cm (Merck).
4. Mobile phase: Heptane–ethyl acetate (1:1).
5. Development solution: 20% H_2SO_4.
6. Heater plate.

2.5 General Equipment

1. Desktop centrifuge.
2. Swinging head centrifuge.
3. Microcentrifuge.
4. pH meter.
5. Autoclave.
6. Microscope.
7. Graduate tubes.
8. Magnetic stirrer.

3 Methods

3.1 Vegetative Stages

The vegetative stages are necessary in any fermentation process to obtain enough biomass for the successful productive stage. The number of seed stages and incubation time depend on the growth rate of the microorganism and on the size of the fermentor. In the case of AD bioconversion process, about 10% (v/v) seeding is required to inoculate the productive stage. Therefore, a bioconversion process in a 5 L fermentor with a working volume of 3 L will need 300 mL of vegetative stage.

From a 1 mL frozen suspension as initial seeding, two growth stages (first vegetative and second vegetative) are necessary to get enough biomass to inoculate the fermentative stage. The process in a 20 L fermentor with a working volume of 15 L will require an additional seed stage (third vegetative) to get 1.5 L.

3.1.1 Mycobacterium neoaurum B-3805 First Vegetative Stage (100 mL)

1. Take a colony of *M. neoaurum* B-3805 from a Petri plate or 1 mL of frozen suspension kept at −80 °C, and seed a 500 mL flask with 100 mL of growth medium (*see* **Notes 1** and **2**).
2. Incubate in an orbital shaker with 5 cm stroke at 31 °C and 250 rpm for 68–74 h (*see* **Note 3**).
3. Take a sample of 10 mL in a graduated tube in sterile conditions and measure the pH. The expected value after 72 h incubation is about 6.8 ± 0.2.
4. Centrifuge the sample in a swinging head centrifuge at 4000 × *g* for 10 min.
5. Measure the pellet volume (packed cell volume; PCV) and express the biomass content as a percentage of the centrifuged volume. The optimal PCV at this stage is about 5–6%.

3.1.2 *Mycobacterium neoaurum B-3805 Second Vegetative Stage (300 mL)*

1. Take 30 mL of first vegetative culture and seed a 2 L flask with 300 mL of growth medium (*see* **Note 2**).
2. Incubate in an orbital shaker of 5 cm stroke at 31 °C and 200 rpm for 24–30 h (*see* **Note 4**).
3. Take a sample of 10 mL in a graduated tube in sterile conditions and measure pH and PCV according to the Subheading 3.1.1. The expected pH after 24 h incubation is about 6.8 ± 0.2 and PCV >5%.
4. Transfer the total culture broth to the 5 L fermentor.

3.1.3 *Mycobacterium neoaurum B-3805 Third Vegetative Stage (5 L)*

The bioconversion process done in 20 L fermentor requires a third vegetative stage to get enough biomass to inoculate the productive stage. This culture can be done in a fermentor as an alternative to using several 2 L flasks, thus allowing to increase the parameter controls and to lower the contamination risk during the transfer of the cultures from several flasks to the fermentor.

Media and Fermentor Preparation

1. Weigh the raw materials of the growth medium and dissolve them sequentially in warm water (final volume 3 L).
2. Adjust to pH 7.0 using 20% NaOH.
3. Place the medium inside the fermentor, assemble all fermentor parts and autoclave at 121 °C for 70 min.
4. After sterilization, place the fermentor on its control console, connect all accessories and set these fermentation parameters: temperature 31 °C, air flow rate 2.4 L/min, and agitation 450 rpm, and dissolved oxygen >30% (*see* **Note 5**).
5. Calibrate the DO_2 probe assigning a theoretical value of 100% before inoculating (maximal oxygen supply at the initial aeration, pressure, and agitation). A value of 0% is obtained by briefly disconnecting the lead cable from the DO electrode and setting the zero.

Incubation

1. Transfer 300 mL of a well-grown *M. neoaurum* B-3805 inoculum obtained according to Subheading 3.1.2 to a sterile inoculation flask in sterile conditions (*see* **Note 2**).
2. Inoculate the 5 L fermentor by spiking the needle in the septum previously sterilized with a sporicidal solution.
3. Keep the incubation parameters defined in Subheading "Media and Fermentor Preparation".
4. Take samples every 24 h in order to monitor *M. neoaurum* growth according to the procedure described in Subheading 3.1.1.
5. The culture is ready to inoculate the bioconversion stage when pH is about 6.8 ± 0.2 and PCV > 5%.

3.2 AD Bioconversion at 5 L Scale

3.2.1 Media and Fermentor Preparation

1. Weigh the raw materials of the bioconversion medium, except phytosterol and oil, and dissolve them sequentially in warm water (*see* **Note 6**).
2. Adjust to pH 7.0 with 20% NaOH.
3. Weigh oil and phytosterol separately, add them to the vessel of the fermentor and mix (*see* **Note 7**).
4. Add then the bioconversion medium previously prepared over the oil–phytosterol mixture to achieve a total volume of 3 L.
5. Assemble all fermentor parts and autoclave at 121 °C for 70 min.
6. After sterilization, place the fermentor in its control console (*see* **Note 8**). Connect all accessories and set these fermentation parameters: temperature 31 °C, air flow rate 2.4 L/min, and agitation 800 rpm.
7. Calibrate the DO_2 probe according to Subheading "Media and Fermentor Preparation".

3.2.2 Bioconversion of Phytosterol to AD

1. Inoculate the 5 L fermentor with 300 mL of a well-grown *M. neoaurum* B-3805 inoculum (optimal pH and PCV parameters are 6.8 ± 0.2 and >5%, respectively) according to the procedure described in the Subheading 3.1.2.
2. Incubate at these fermentation parameters: temperature 31 °C, air flow rate 2.4 L/min, and agitation 800 rpm (*see* **Note 9**).
3. Take samples every 8 h to check the absence of microbial contamination (*see* **Note 2**), pH (*see* **Note 10**), growth rate measured as PCV (Subheading 3.1.1) (*see* **Note 11**), phytosterol by TLC, and AD by TLC and HPLC (Subheading 3.4).
4. Keep the bioconversion process until the residual concentration of phytosterol in the broth is under 1 g/L (analyzed by TLC). The bioconversion should be completed in about 120 h (*see* **Note 12**).

3.3 AD Bioconversion at 20 L Scale

3.3.1 Media and Fermentor Preparation

1. Weigh the raw materials of the bioconversion medium, except phytosterol and oil, and dissolve them sequentially by stirring in 4–5 L of warm water. Adjust to pH 7.0 with 20% NaOH.
2. Place the pH and pO_2 probes in the corresponding inputs of the fermentation vessel and calibrate the pH probe with the appropriate buffer standard solutions (7.01 and 4.01 buffers).
3. Check that the air filter is clean, and change the position of the air entry to "Sterilization".
4. Place the dissolved culture medium inside the fermentation vessel, and check the pH with the fermentor probe.
5. Complete to 12.5 L with distilled water and begin stirring at 200 rpm.

6. Add the oil and phytosterol inside the fermentor to get a final volume of 15 L (*see* **Note 13**).
7. Close all the ports of the fermentation vessel and do a pressure test to check that the vessel is hermetically sealed (*see* **Note 14**). After that, open the air output to release pressure.
8. Start the sterilization cycle according to the following sterilization parameters: temperature 121 °C, agitation 200 rpm, time 30 min, and final temperature 31 °C (*see* **Note 15**).
9. After finishing the sterilization process, set the following fermentation parameters: temperature 31 °C, agitation 450 rpm, air flow rate 15 L/min, backpressure 0.8 bars, and DO_2 >30%.
10. Calibrate the DO_2 probe according to Subheading "Media and Fermentor Preparation".

3.3.2 Bioconversion of Phytosterol to AD

1. Inoculate the bioconversion medium with 1.5 L of *M. neoaurum* B-3805 third vegetative (Subheading 3.1.3).
2. Incubate according to the fermentation parameters defined in Subheading 3.3.1 (*see* **Note 16**).
3. Take samples every 8 h to check the growth level and AD production (Subheading 3.2.2) (*see* **Note 17**).
4. Keep the bioconversion process until the residual concentration of phytosterol in the broth is under 1 g/L (analyzed by TLC).
5. Harvest the bioconversion broth and begin the downstream process to purify AD.

3.4 Quantification of Sterols and AD

3.4.1 Analysis by HPLC Method

Sample Extraction

1. Take 20 mL of the bioconversion broth and mix with a magnetic stirrer for 10 min.
2. Take 100 μL of the stirred sample and mix with 900 μL of butyl acetate in a 2.2 mL microtube.
3. Stir at 50 °C for 20 min.
4. Centrifuge at 10,000 × *g* for 10 min.
5. Take a sample of 200 μL of supernatant and make a second dilution in 800 μL of acetonitrile HPLC grade. Final dilution of the broth sample is 1/50.

Standard Preparation

1. Weigh 10 mg of AD standard in a 10 mL volumetric flask and bring up to the 70% of the final volume with acetonitrile HPLC grade.
2. Sonicate until complete dissolution and bring up to 10 mL to achieve a concentration of 1 mg/mL.
3. Dilute 1 mL in a 50 mL volumetric flask with acetonitrile HPLC grade to achieve a final concentration of 20 mg/L.

Parameter Analysis

1. Analyze 10 μL of the sample according to the following HPLC conditions: temperature 50 °C, flow 1 mL/min, and wavelength 240 nm.
2. Check the products according to their retention time (RT): ADD 5.4 min, testosterone 6.5 min, and AD 6.7 min (*see* **Note 18**).

3.4.2 Analysis by TLC Method

Sample Extraction

1. Take 20 mL of the bioconversion broth and mix with a magnetic stirrer for 10 min.
2. Take 100 μL of the stirred sample and mix with 900 μL of ethyl acetate in a 2.2 mL microtube (*see* **Note 19**).
3. Stir at 50 °C for 20 min.
4. Centrifuge at 10,000 × *g* for 10 min.
5. Transfer the supernatant to another microtube and discard the pellet.

TLC Development

1. Cut a TLC sheet of 8–10 cm in length and draw a line at about 1 cm from the bottom with a pencil.
2. Put 10 μL of the supernatant sample. If several samples are running, keep a 1 cm distance between samples and on each side. Let the solvent evaporate for 5 min (*see* **Note 20**).
3. Place the TLC sheet into the tilt chamber (at about a 45° angle in relation to the wall chamber).
4. Fill the TLC chamber up to 0.5 cm with mobile phase and close the chamber.
5. Run the TLC until the mobile phase reaches to the upper side of the TLC sheet (0.5 cm before the edge).
6. Dry the sheet for 5–10 min at room temperature.
7. Develop the TLC by spreading 20% H_2SO_4 onto the surface with a cotton swab or spraying it.
8. Let it dry and then heat it on a heating plate or in an oven at 100 °C until color development (*see* **Note 21**).

4 Notes

1. The preparation of a Working Cell Bank (WCB) kept at −80 °C is desirable to adequately preserve the strain and standardize the seeding procedure.
2. Seed a TSA Petri plate and incubate at 31 °C for 24–48 h to detect potential microbial contaminations.
3. If necessary, the incubation time can be extended up to 80 h without any negative effect.

4. It is important to avoid wetting the plugs, so you can incubate at 180 rpm if necessary.
5. It is important to keep the DO_2 over 30% to guarantee an optimal oxygen supply for growth. An agitation cascade associated to DO_2 level is useful; when DO_2 goes down, the agitation goes up to keep DO_2 over 30%.
6. To calculate the volume of this solution, subtract the volume of the oil from the final volume (3 L).
7. The use of a blender will facilitate the breakdown of the phytosterol crystals, improving the suspension of oil–phytosterol. Place the oil and phytosterol inside the glass vessel and mix with the blender for 2–5 min.
8. Once the autoclave sterilization occurs, let the fermentor cool down while applying strong agitation; otherwise the oil–phytosterol mixture can solidify at the bottom and it will not be available for *M. neoaurum* during the bioconversion process.
9. Up until 60–70 h, the broth has a heterogeneous aspect because of the presence of phytosterol globules. From 60–70 h onwards the broth acquires a more homogeneous appearance and the DO_2 rises to reach values of about 100% at the end of the bioconversion.
10. If there is a significant difference between the pH values measured offline and online, you must recalibrate the online pH.
11. In this case, the packed cells are located in the oily phase at the top of the tube due to the hydrophobic nature of the mycobacterial cellular surface.
12. Phytosterols are not soluble in water and it is usual to find some phytosterol stuck to the vessel wall or baffles, which is not available for the bioconversion, decreasing the AD bioconversion yield. To minimize this effect, it is advisable to disconnect the engine daily and manually shake the glass vessel, trying to break the unavailable phytosterol free.
13. In this case it is not necessary to mix previously the oil and phytosterol with a blender, because the medium sterilization process occurs with agitation.
14. Apply the pressure test, close the air output and start to put air inside (5 L/min) until it reaches a pressure of 1 bar, then stop the air flow and wait at least 5 min. The pressure should be maintained during that time.
15. During the cooling phase of the sterilization process, beware when changing the filter position from “Sterilization” to “Fermentation”, as some foaming may occur. To minimize it, change the filter position when the temperature reaches 60 °C. If temperature is lower, the phytosterol may crystallize inside the air sparger and block the air entry.

16. The bioconversion process at 20 L scale is similar to the one described at 5 L scale, but the third vegetative used to inoculate the 15 L broth has normally more biomass content, and the physical parameters can be slightly different (impellers size, backpressure, oxygen transfer rate, etc.) [22]. At 20 L scale, the heterogeneous aspect of the bioconversion broth changes at 50–60 h (about 10–20 h earlier than at 5 L scale) and the bioconversion is completed after 96–100 h.
17. To take a sample, sterilize with steam the sample taker for 15 min. Then, discard the first 50 mL of sample and place 50–60 mL in a sterile container. Finally, sterilize the sample taker again for 15 min with steam.
18. Butyl acetate retention time is 4.5 min. It could cause confusion and hide some other HPLC peaks.
19. The use of organic solvents with low boiling point, such as ethyl acetate, is highly recommendable for the sample extraction to be analyzed by TLC. These solvents evaporate quickly once the sample on the TLC sheet is uploaded, avoiding thus the sample spreading.
20. For a proper TLC development, put 3 μl of the sample as a small drop on the line and let the solvent evaporate. Repeat the operation over the same drop until completing the sample loading.
21. It is advisable to include standards of phytosterol and AD of known concentration in the TLC sheet to help identify and quantify the product bands [23].

Acknowledgments

This work was fully supported by a grant of the European Union program ERA-IB [MySterI (EIB.12.010)] through the APCIN call of the Spanish Ministry of Economy and Competitiveness (MINECO, Spain) (PCIN-2013-024-C02-01). The authors want to thank the European Union program ERA-IB; the Spanish Ministry of Economy and Competitiveness (MINECO, Spain) and the MySterI Consortium (INBIOTEC, Pharmins Ltd., University of York, SINTEF, Technische Universität Dortmund, and Gadea Biopharma S.L.).

References

1. Mahato SB, Garai S (1997) Advances in microbial steroid biotransformation. Steroids 62:332–345
2. Fernandes P, Cruz A, Angelova B, Pinheiro HM, Cabral JMS (2003) Microbial conversion of steroid compounds: recent developments. Enzym Microb Technol 32:688–705
3. Batthi H, Khera RA (2012) Biological transformations of steroidal compounds: a review. Steroids 77:1267–1290

4. Al Jasem Y, Khan M, Taha A, Thiemann T (2014) Preparation of steroidal hormones with an emphasis on transformations of phytosterols and cholesterol. Med J Chem 3:796–830
5. Marques MPC, Carvalho F, de Carvalho CCCR, Cabral JMS, Fernandes P (2010) Steroid bioconversion: towards green processes. Food Bioprod Process 88:12–20
6. Moreau RA, Whitaker BD, Hicks KB (2002) Phytosterols, phytostanols, and their conjugates in foods: structural diversity, quantitative analysis, and health-promoting uses. Prog Lipid Res 41:457–500
7. Fernandes P, Cabral JMS (2007) Phytosterols: applications and recovery methods. Bioresour Technol 98:2335–2350
8. Kutney JP, Milanova RK, Vassilev CD, Stefanov S, Nedelcheva NV (2000) Process for the microbial conversion of phytosterol to androstenedione and androstadienedione. US Patent 6,071,714
9. Donova MV (2007) Transformation of steroids by actinobacteria: a review. Appl Biochem Microbiol 43:1–14
10. Rodina NV, Molchanova MA, Voishvillo NE, Andryushina VA, Stytsenko TS (2008) Conversion of phytosterols into androstenedione by *Mycobacterium neoaurum*. Appl Biochem Microbiol 44:48–54
11. Pérez C, Falero A, Duc HL, Balcinde Y, Hung BR (2006) A very efficient bioconversion of soybean phytosterols mixtures to androstanes by mycobacteria. J Ind Microbiol Biotechnol 33:719–723
12. Egorova OV, Gulevskaya SA, Puntus IF, Filonov AE, Donova MV (2002) Production of androstenedione using mutants of *Mycobacterium* sp. J Chem Technol Biotechnol 77:141–147
13. Zhang XY, Peng Y, Su ZR, Chen QH, Ruan H, He GQ (2013) Optimization of biotransformation from phytosterol to androstenedione by a mutant *Mycobacterium neoaurum* ZJUVN-08. J Zhejiang Univ Sci B 14:132–143
14. Marsheck WJ, Kraychy S, Muir RD (1972) Microbial degradation of sterols. Appl Microbiol 23(1):72–77
15. Marsheck WJ, Heights A, Kraychy S (1973) Selective microbiological preparation of androst-4-ene-3,17-dione. US Patent 3,759,791
16. Rodríguez-García A, Fernández-Alegre E, Morales A, Sola-Landa A, Lorraine J, Macdonald S, Dovbnya D, Smith MC, Donova M, Barreiro C (2016) Complete genome sequence of '*Mycobacterium neoaurum*' NRRL B-3805, an androstenedione (AD) producer for industrial biotransformation of sterols. J Biotechnol 224:64–65
17. Malaviya A, Gomes J (2008) Androstenedione production by biotransformation of phytosterols. Bioresour Technol 99:6725–6737
18. Hesselink PGM, Vliet SV, Vries HD, Witholt B (1989) Optimization of steroid side chain cleavage by *Mycobacterium* sp. in the presence of cyclodextrins. Enzym Microb Technol 11:398–404
19. Carvalho F, Marques MPC, de Carvalho CCCR, Cabral JMS, Fernandes P (2009) Sitosterol bioconversion with resting cells in liquid polymer based systems. Bioresour Technol 100:4050–4053
20. Cruz A, Fernandes P, Cabral JMS, Pinheiro HM (2001) Whole-cell bioconversion of β-sitosterol in aqueous-oganic two-phase systems. J Mol Catal B Enzym 11:579–585
21. Stefanov S, Yankov D, Beschkov V (2006) Biotransformation of phytosterols to androstenedione in two phase water-oil systems. Chem Biochem Eng Q 20:421–427
22. García-Ochoa F, Gómez E (2009) Fermentor scale-up and oxygen transfer rate in microbial processes: an overview. Biotechnol Adv 27(2):153–176
23. Gulla V, Banerjee T, Patil S (2008) Quantitative TLC analysis of steroid drug intermediates formed during bioconversion of soysterols. Chromatographia 68:663–667

Chapter 15

Bioconversion of Phytosterols into Androstadienedione by *Mycobacterium smegmatis* CECT 8331

Julia García-Fernández, Igor Martínez, Lorena Fernández-Cabezón, Carmen Felpeto-Santero, José-Luis García, and Beatriz Galán

Abstract

The C19 steroid 1,4-androstadiene-3,17-dione (androstadienedione, ADD) is an added value product used as a synthon in the pharmaceutical industry for the commercial production of corticosteroids, mineralocorticoids, oral contraceptives, and other pharmaceutical steroids. Phytosterol biotransformation catalyzed by microbial whole cells is actually a very well-established research area in white biotechnology. The protocol below provides detailed information on ADD production by the mutant CECT 8331 of *Mycobacterium smegmatis* mc^{2}155 using phytosterols as raw material in a lab scale. This protocol describes the bioconversion of phytosterols into ADD in a single fermentation step.

Key words *Mycobacterium smegmatis*, Steroids, Androstadienedione, ADD, Phytosterols, Biotransformation

1 Introduction

Steroids represent one of the largest sectors in the pharmaceutical industry with huge production exceeding 1,000,000 tons/year [1]. Androstenedione (4-androstene-3,17-dione; AD) and androstadienedione (1,4-androstadiene-3,17-dione; ADD) are key intermediates of microbial sterol metabolism [2]. These compounds belong to the 17-keto steroid family and are used as the starting materials (synthons) for the preparation of different pharmaceutical steroids [2–4]. They can be produced by microbial side chain cleavage of cholesterol or phytosterols as an alternative to conventional multistep chemical synthesis based on diosgenin, a steroid found exclusively in the flowers and leaves of *Digitalis* plants, as the starting material. Industrially, AD and ADD are currently produced from phytosterols as a raw material through fermentative processes using wild-type microorganisms that have been subsequently modified and optimized by conventional

José-Luis Barredo, Ignacio Herráiz (eds.), *Microbial Steroids: Methods and Protocols*, Methods in Molecular Biology vol. 1645, DOI 10.1007/978-1-4939-7183-1_15, © Springer Science+Business Media LLC 2017

random mutagenic procedures or by recombinant metabolic engineering. In principle, rational metabolic engineering provides a more effective and well-defined alternative to strain development than random mutagenesis. The strategy for developing an ADD producer organism by metabolic engineering consists in blocking the B-ring cleavage by inactivating the 9α-hydroxylase activity (KshAB), whereas the strategy for developing an AD producer must include and additional inactivating mutation in the 3-ketosteroid Δ^1-dehydrogenase activity (KstD) [2, 4–6]. In recent years, microbial selective cleavages of phytosterols to AD and ADD have been reported by different researchers [3, 4].

Bacterial strains belonging to *Mycobacterium* genera are known to degrade a range of naturally occurring steroids. In this sense, *Mycobacterium smegmatis* mc^{2}155 has recently been engineered in our laboratory to modify the catabolic pathway of sterols to redirect the major metabolic flux to the targeted product ADD by generating a *kshB* mutant deposited as CECT 8331 (ES P201430272) [7]. This strain has the ability to degrade the 17-alkyl side chain of natural sterols such as cholesterol and phytosterols to produce ADD without appreciable degradation of the steroid nucleus.

The main bottleneck to achieve sterol biotransformation is the low aqueous solubility of these highly hydrophobic substrates (bellow 1 μM). This means that in water solutions the total amount of soluble steroid accessible to the microorganism for its industrial biotransformation is very limited. Therefore, solubilization of the lipophilic steroid with different agents, like cell envelop-permeabilizing antibiotics, surfactants, cyclodextrins, and organic solvents, has been used to overcome this problem.

In the protocol presented below we describe two different methods to produce ADD by microbial biotransformation from phytosterols. The first method describes a fermentation performed in shaken flasks and the second describes a fermentation performed in 2 L stirred tank bioreactor (STBR) to partially scale up the process.

2 Materials

2.1 Bacterial Strain and Culture Media

Chemical substances for preparing bacterial culture media can be individually purchased from any supplier of common bacterial growth components such as Sigma (https://www.sigmaaldrich.com) or as preprepared media from suppliers such as Difco (http://www.bd.com).

For solid cultures in petri agar plates we use Middlebrook 7H10 Broth. For liquid cultures we use: (1) 7H9-GAT Broth (*Glycerol, ADC, Tween*) for the precultures in Falcon tubes (https://www.fishersci.com); (2) 7H9-Phy Broth for the phytosterol biotransformation in flasks; (3) STBR broth for phytosterol biotransformation in STBR [8]. The components and amounts per liter of the culture media are given bellow.

1. *Mycobacterium smegmatis* CECT 8331 (Colección Española de Cultivos Tipo, CECT collection: http://www.uv.es/cect) (*see* **Note 1**).
2. Middlebrook 7H10 Broth: Dipotassium phosphate (1.5 g), monopotassium phosphate (1.5 g), ammonium sulfate (0.5 g), monosodium glutamate (0.5 g), sodium citrate (0.4 g), ferric ammonium citrate (0.04 g), magnesium sulfate (25.0 mg), copper sulfate (1 mg), pyridoxine hydrochloride (1 mg), zinc sulfate (1 mg), biotin (0.5 mg), calcium chloride (0.5 mg), malachite green (0.25 mg), agar (15 g), glycerol (5 mL), ADC enrichment (100 mL) (*see* **Note 2**), and demineralized water (900 mL).
3. Middlebrook ADC Enrichment: sodium chloride (8.5 g), bovine albumin (50 g), dextrose (20 g), and catalase (30 mg) (*see* **Note 3**).
4. 7H9-GAT Broth: Ammonium sulfate (0.5 g), l-glutamic acid (0.5 g), sodium citrate (0.1 g), pyridoxine (1.0 mg), biotin (0.5 mg), disodium phosphate (2.5 g), monopotassium phosphate (1.0 g), ferric ammonium citrate (40 mg), magnesium sulfate (50 mg), calcium chloride (0.5 mg), zinc sulfate (1.0 mg), copper sulfate (1.0 mg), glycerol (2 mL), 0.05% Tween 80 (v/v) (*see* **Note 4**), ADC enrichment (100 mL), and demineralized water (900 mL).
5. 7H9-Phy Broth: Ammonium sulfate (0.5 g), l-glutamic acid (0.5 g), sodium citrate (0.1 g), pyridoxine (1.0 mg), biotin (0.5 mg), disodium phosphate (2.5 g), monopotassium phosphate (1.0 g), ferric ammonium citrate (40 mg), magnesium sulfate (50 mg), calcium chloride (0.5 mg), zinc sulfate (1.0 mg), copper sulfate (1.0 mg), 3.6% tyloxapol (v/v) (Sigma; https://www.sigmaaldrich.com) (*see* **Note 5**), glycerol 18 mM, phytosterols (0.4 g/L), and demineralized water up to 1000 mL.
6. STBR broth: Corn steep liquor (20 g), sodium nitrate (5.4 g), ammonium chloride (0.28 g), sodium phosphate monobasic (0.72 g), and demineralized water up to 1000 mL.
7. NaCl–Tween 80 solution: 0.9% NaCl and 0.05% Tween 80.

2.2 Phytosterol Solubilization

1. Phytosterols.
2. Tyloxapol (v/v) (Sigma; https://www.sigmaaldrich.com) (*see* **Note 5**).
3. Sunflower oil (*see* **Note 6**).

2.3 Biotransformation

1. Bench centrifuge.
2. 0.9% NaCl solution in 0.05% Tween 80.
3. Orbital shaker.
4. Stirred tank bioreactor (STBR).

2.4 Steroid Extraction

1. Chloroform.
2. Vortex.
3. Centrifuge Mini Spin.
4. Ethyl acetate.
5. Thermoblock (80 ° C).

2.5 Monitoring Phytosterol Biotransformation by TLC Analysis

1. Chromatogram development: CAMAG® Flat Bottom Chamber for 20 × 20 cm plates, with stainless steel lid (http://www.camag.com).
2. TLC plates: TLC Silica gel 60 F_{254} aluminum sheets 20 × 20 cm (Merck; http://www.merckmillipore.com).
3. Solvent mixture for mobile phase: hexane/ethyl acetate (10/4).
4. TLC/HPTLC Sprayer (CAMAG; http://www.camag.com).
5. Development solution: 20% H_2SO_4 (v/v) prepared in water.

2.6 Steroid Quantification: Gas Chromatography–Mass Spectrometry

1. Androstadienedione (1,4-androstadiene-3,17-dione; ADD) (Sigma; https://www.sigmaaldrich.com).
2. Androstenedione (4-androstene-3,17-dione; AD) (Sigma; https://www.sigmaaldrich.com).
3. Phytosterols (Arboris, USA).
4. N,O-bis(trimethylsilyl) trifluoroacetamide (BSTFA) (Merck; http://www.merckmillipore.com).
5. Pyridine.
6. Gas chromatography (GC) equipment, with split/splitless injector and quadrupole mass detector (MS) equipped with a HP-5MS capillary column (30 m × 0.25 mm × 0.25 μm). Our system is the Agilent 7890A/5975C, and data are acquired and processed with the software MSD ChemStation E.02.00.493 (Agilent Technologies Inc.; http://www.chem.agilent.com).

2.7 Steroid Quantification: Liquid Chromatography–Mass Spectrometry

1. Androstadienedione (1,4-androstadiene-3,17-dione; ADD) (Sigma; https://www.sigmaaldrich.com).
2. Androstenedione (4-androstene-3,17-dione; AD) (Sigma; https://www.sigmaaldrich.com).
3. 4-HBC (22-hydroxy-23,24-bisnorchol-4-en-3-one) (Sigma; https://www.sigmaaldrich.com).

4. 1,4-HBC (3-oxo-23,24-bisnorchol-1,4-dien-22-oic acid) (Sigma; https://www.sigmaaldrich.com).
5. Phytosterols (Arboris, USA).
6. Liquid chromatography (LC-MS) equipment: Our system is a Surveyor Plus LC equipped with automatic injector coupled to ion trap (LXQ) equipped with atmospheric pressure chemical ionization source (APCI) and column Teknokroma Tracer Escel 120 ODSB (5 μm, 15 × 0.46 cm). Data are processed using Xcalibur software.
7. Acetonitrile.
8. Isopropanol LC-MaScan (LAB-SCAN, Analytical sciences).
9. Mobile phase A: water plus 0.01% (v/) formic acid.
10. Mobile phase B: acetonitrile plus 0.01% (v/) formic acid.
11. Mobile phase C: isopropanol plus 0.01% (v/) formic acid.

2.8 General Equipment

1. Spectrophotometer.
2. Filter 0.4 μm pore size.
3. Vacuum pump device.

3 Methods

3.1 Culture Medium Preparation

3.1.1 Preparation of Middlebrook 7H10 Broth

1. Dissolve all the components; bring the volume up to 900 mL and autoclave at 121 °C for 20 min.
2. Add the preheated (45 °C) filtered ADC (100 mL/L).
3. Mix all the components and pour the medium into Petri plates (*see* **Note 2**).

3.1.2 Preparation of ADC Enrichment Solution

1. Dissolve all the ingredients and bring the volume up to 1 L.
2. Filter the solution to be sterilized. The ADC solution contains heat-sensitive components has to be sterilized by filtration (*see* **Note 3**).

3.1.3 Preparation of Middlebrook 7H9-GAT Broth

1. Dissolve all the components; bring the volume up to 900 mL and autoclave at 121 °C for 20 min.
2. Add sterile glycerol 2 mL/L, 0.05% Tween 80, and ADC (100 mL/L).
3. Mix all the components and keep the mixture at 4 °C.

3.1.4 Preparation of Middlebrook 7H9-Phy Broth

1. Dissolve all the components, bring the volume up to 900 mL and autoclave at 121 °C for 20 min.
2. Add glycerol 18 mM, 3.6% tyloxapol (v/v) (*see* **Note 5**), and 0.4 g/L of phytosterols (*see* **Note 6**) (Fig. 1).

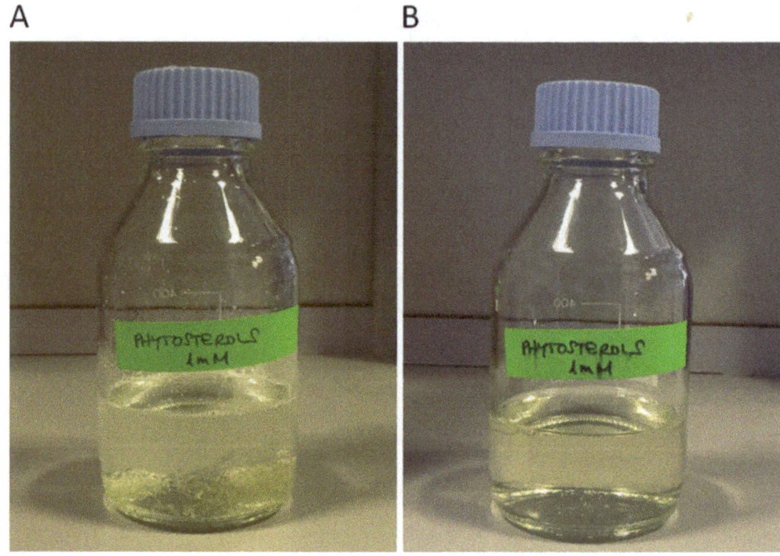

Fig. 1 Pictures of 7H9-Phy Broth before (**a**) and after (**b**) autoclaving

3.1.5 Preparation of STBR Broth

1. Prepare 1480 mL of the STBR broth (all components except the phytosterols in sunflower oil) and adjust the pH at 8.1.
2. Meanwhile, 20 g of phytosterols is solubilized in 320 mL of sunflower oil at 120 °C by stirring the mixture until it becomes clear.
3. Mix the STBR broth and the solubilized phytosterols in the bioreactor and autoclave the mixture (*see* **Notes** 7 and **8**) (Fig. 2).
4. The 200 mL remaining to reach the total 2 L volume corresponds to the bacterial inoculum that will be added just to start the fermentation process.

3.2 Method 1: Shaken Flask Biotransformation

3.2.1 Preculture Preparation

The bacterial preculture is necessary to obtain the biomass required to inoculate the culture medium at an appropriate OD_{600} in order to achieve an efficient phytosterol biotransformation.

Inoculate an isolated colony of CECT 8331 from a 7H10 agar plate in 10 mL of 7H9-GAT Broth placed in a sterile 50 mL tube and place it in an orbital shaker at 37 °C and 250 rpm to allow bacterial growth for 24–30 h.

3.2.2 Biotransformation

Phytosterol biotransformation in shaken flask is carried out using CECT 8331 growing cells in 100 mL flasks containing 25 mL of 7H9-Phy Broth (*see* Subheading 3.1).

1. Take a well-grown (OD_{600} = 2–2.5) preculture of CECT 8331 in 7H9-GAT Broth. Prior to its inoculation, centrifuge and wash out the medium with one volume of NaCl–Tween

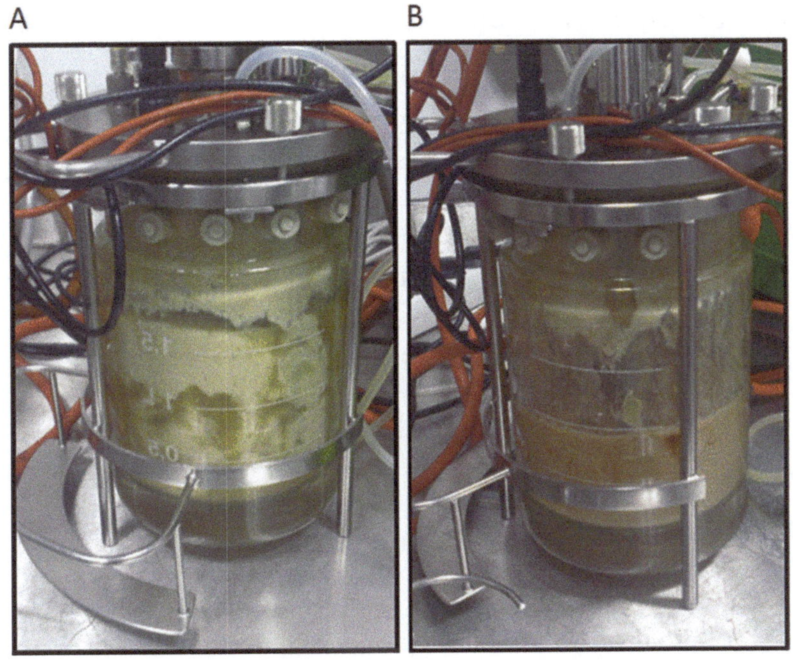

Fig. 2 Pictures of the 2 L bioreactor containing the STBR Broth and the bacterial inoculum before (**a**) and after (**b**) cultivation

80 solution and finally resuspend the pellet in 0.5 mL of the same solution to measure the OD_{600}.

2. To start the biotransformation procedure, inoculate the flasks containing 25 mL of 7H9-Phy Broth to an initial OD_{600} of 0.05.
3. Grow shaking (250 rpm) at 37 °C for 96 h.
4. Take culture samples every 24 h (0, 24, 48, 72, and 96 h) to monitor the phytosterol biotransformation: 1 mL to measure OD_{600}, 0.8 mL for the TLC analysis, and 0.8 mL for GC-MS analysis.

3.3 Method 2: Stirred Tank Bioreactor Biotransformation (STBR)

3.3.1 Preculture Preparation

1. Inoculate an isolated colony of CECT 8331 from a 7H10 agar plate in 10 mL of 7H9-GAT Broth placed in a sterile 50 mL tube.
2. Incubate in an orbital shaker at 37 °C and 250 rpm to allow bacterial growth for 24–30 h.
3. Use this culture to inoculate 200 mL of fresh 7H9-GAT Broth in a 1 L Erlenmeyer flask at an initial OD_{600} of 0.1.
4. Incubate this second inoculum stage for 24–30 h at 37 °C and 250 rpm in an orbital shaker.

3.3.2 Biotransformation

The process is carried out by growing CECT 8331 cells in a 2 L STBR containing 1800 mL of STBR broth and phytosterols (20–40 g) solubilized in sunflower oil (*see* Subheading 3.1) (Fig. 2).

1. Inoculate the autoclaved STBR with 200 mL of well-grown (OD_{600} = 2–2.5) 7H9-GAT CECT 8331 cells.
2. Keep the culture conditions as follows: Temperature 37 °C, initial pH 7.4 (it is not under control during the biotransformation), initial stirred speed 600 rpm, and air flow 2 L/min.
3. Maintain the dissolved oxygen over 20% of air saturation by increasing the stirrer speed.
4. Take culture samples every 24 h (0, 24, 48, 72, and 96 h) to monitor the biotransformation. 1 mL for the TLC and GC-MS analysis (*see* **Note 9**).

3.4 Steroid Extraction

Previous to TLC or GC-MS analyses, steroids must be extracted in an organic phase by duplicates.

3.4.1 Extraction of Shaken Flask Biotransformations

1. Take duplicate samples of 0.2 mL from each culture (*see* Subheading 3.2).
2. Add 0.4 mL of chloroform to each sample.
3. Vortex for 1 min.
4. Freeze the tubes at −80 °C for 30 min to separate the organic phase from the aqueous phase.
5. Discard the aqueous phase (*see* **Note 10**).
6. Place the tubes in a thermoblock at 60 °C until the samples are dry.

3.4.2 Extraction of STBR Biotransformations

1. Take one sample of 1 mL from the culture (*see* Subheading 3.3).
2. Centrifuge the sample by using a microfuge to separate oil and aqueous phases.
3. Take 50 μL of clean oil of each sample (upper phase).
4. Add 200 μL of ethyl acetate and vortex for 1 min.

3.5 Monitoring Phytosterol Biotransformation by TLC Analysis

The biotransformation of phytosterols into ADD is monitored every 24 h during the process (96 h) (Fig. 3).

1. Extract the steroid fraction as described in Subheading 3.4.
2. Cut TLC plates leaving 1 cm margin on each side, 8 cm in length to allow the sample running, and 1 cm separation to upload the different samples.
3. Dissolve the samples following method 1 (*see* Subheading 3.4) in 20–100 μL of chloroform depending on the expected steroid concentration and always using the same volume for all

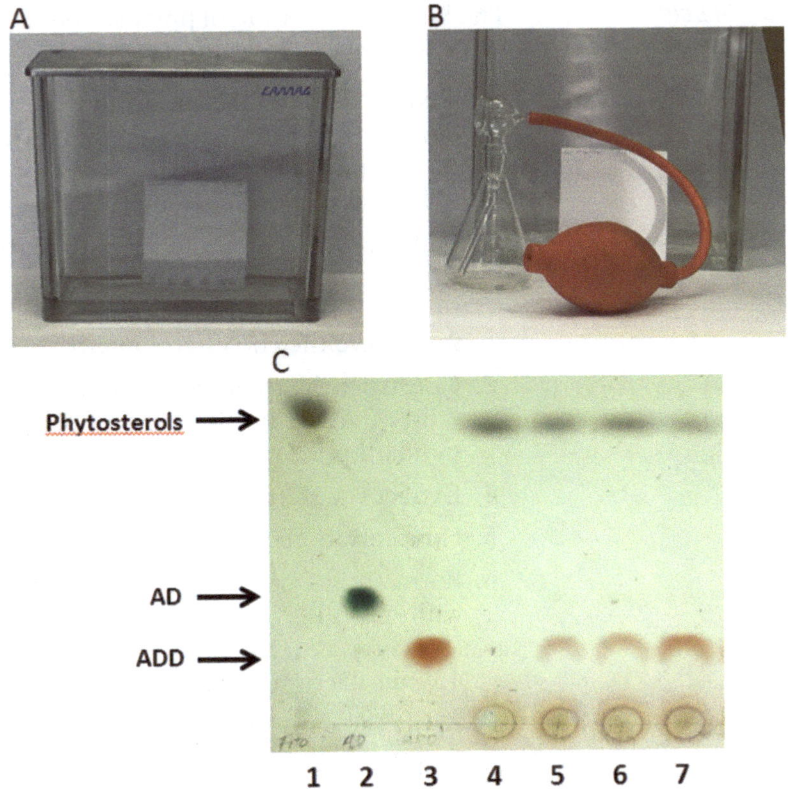

Fig. 3 Biotransformation monitoring by TLC. (**a**) TLC chamber device. (**b**) Sprayer for developing. (**c**) H_2SO_4 developed TLC; controls for phytosterols, AD, and ADD (*lines 1*, *2*, and *3*, respectively); phytosterol biotransformation at 0 h (*line 4*), 24 h (*line 5*), 48 h (*line 6*), and 96 h (*line 7*)

the samples. Apply the samples obtained following method 2 (*see* Subheading 3.4) directly on the TLC plates.

4. Apply 10 μL of each sample on to the TLC plates and let it dry for 5 min at room temperature. It is recommended to load standards of the substrate and expected products in the same TLC plate to better interpret the results of the analysis.
5. Place the loaded plate into the TLC chamber containing 50 mL of mobile phase (*see* Subheading 2.5).
6. Run the TLC until the mobile phase gets to the delimited frame on the TLC plate (approximately 1 cm from the end).
7. Dry the plate for 10 min at room temperature.
8. Develop the plate by spraying the surface with a solution containing 20% H_2SO_4 in water (v/v) and heat the developed plate at 100 °C in an oven until color is maximal (approximately 10 min) (*see* **Note 11**).

3.6 Steroid Quantification by GC-MS Analysis

The biotransformation of phytosterols into ADD is also monitored by GC-MS every 24 h during 96 h (Fig. 4).

1. Take a sample of 0.2 mL from the bacterial culture every 24 h during 96 h in duplicate and freeze the cells at −80 °C until all the samples are collected. Skip this step with samples obtained following method 2.
2. Unfreeze the samples and add 50 μL of cholesterol dissolved in chloroform (≈20 μg) as internal standard to each one prior to the steroid extraction. For samples obtained following method 2, dissolve the cholesterol in ethyl acetate.
3. Construct the standard curve from 56 to 7 μg of ADD, 56 to 7 μg of AD, and 80–10 μg of phytosterols. Add the internal standard prior to extraction.
4. Extract the steroid fraction as described in Subheading 3.4.
5. Concentrate the chloroform fraction by evaporation.
6. Form the trimethylsilyl (TMT) ether derivatives by reaction with 50 μL of N,O-bis (trimethylsilyl) trifluoroacetamide (BSTFA) and 50 μL of pyridine, and heating at 60 °C for 45 min. Derivatize the calibration standards in the same way.
7. Apply the derivatized samples to the GC-MS.
8. Use the following working conditions: split ratio (20:1), injector temperature 320 °C; column temperature 240 °C for 3 min and then heat to 320 °C at 5 °C/min. Record the mass spectra in electron impact (EI) mode at 70 eV within the *m/z* range 50–550.
9. For quantification of the peak area, use these quantitative masses: 329 + 458 *m/z* for cholesterol; 343 + 384 *m/z* for cholestenone; 382 + 472 *m/z* for campesterol; 394 + 484 *m/z* for stigmasterol; 357 + 486 *m/z* for β-sitosterol; 244 + 286 *m/z* for AD, and 122 + 284 *m/z* for ADD in selected ions of monitoring (SIM). Use EI mass spectra and retention data to assess the identity of compounds by comparing them with those of standards in the NIST Mass Spectral Database and commercial standards (NIST 2011).

3.7 Steroid Quantification by LC-MS Analysis

This analytical method can be used instead of GC-MS to quantify the steroids produced in the biotransformation (Fig. 5).

1. Take 0.2 mL from the culture every 24 h during 96 h in duplicate and freeze the cells at −80 °C until all the samples are collected.
2. Unfreeze the samples and add 20 μL of 5 mM testosterone dissolved in 10% tyloxapol (v/v) as internal standard (ISTD) to each one prior to the steroid extraction.

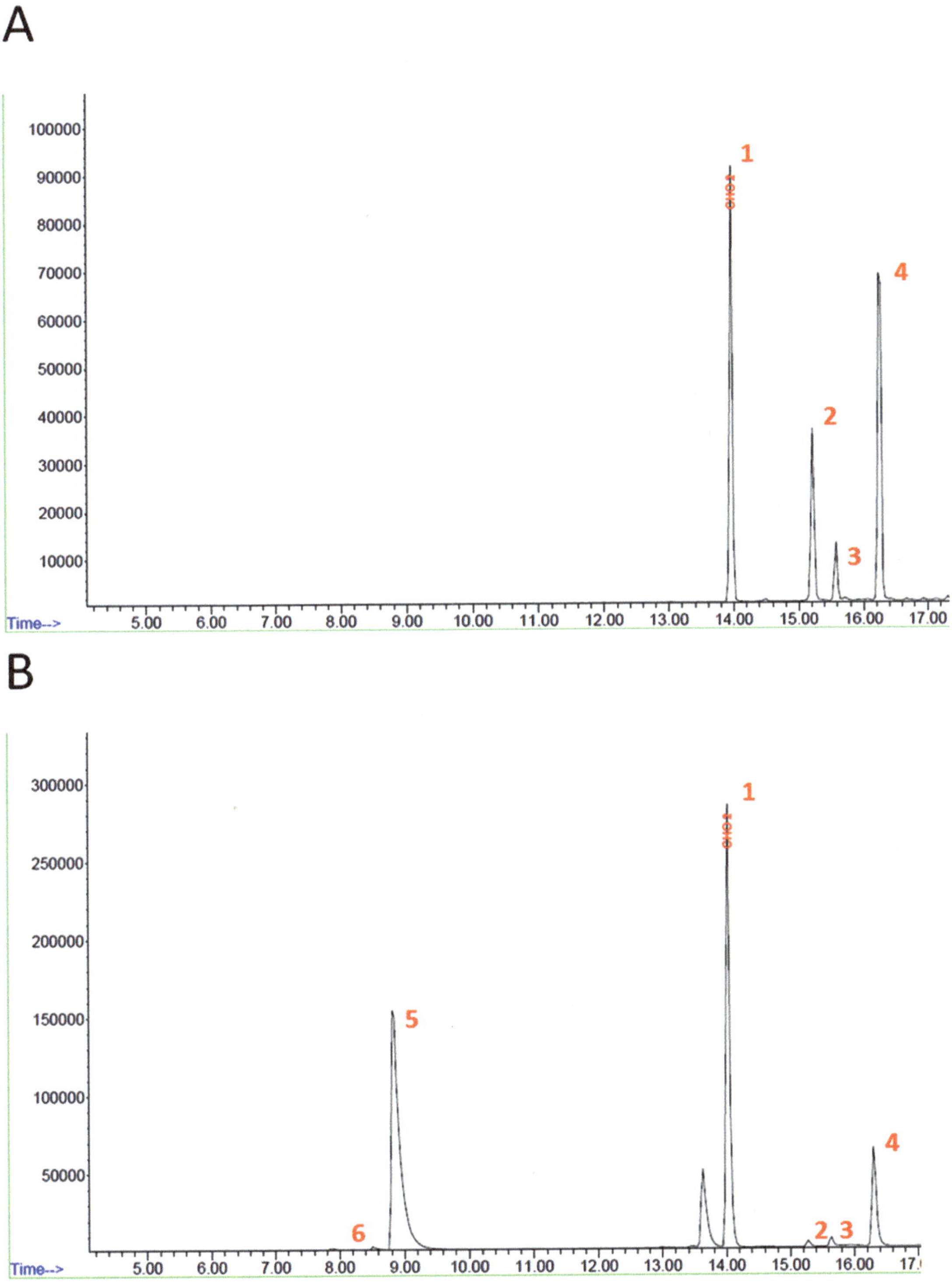

Fig. 4 Analysis by GC-MS of the ADD production by *M. smegmatis* CECT 8331 strain in shaken flasks after 96 h of incubation of phytosterols. (**a**) Starting time and (**b**) 96 h. (*1*) Cholesterol, (*2*) campesterol, (*3*) stigmasterol, (*4*) β-sitosterol, (*5*) ADD, (*6*) AD

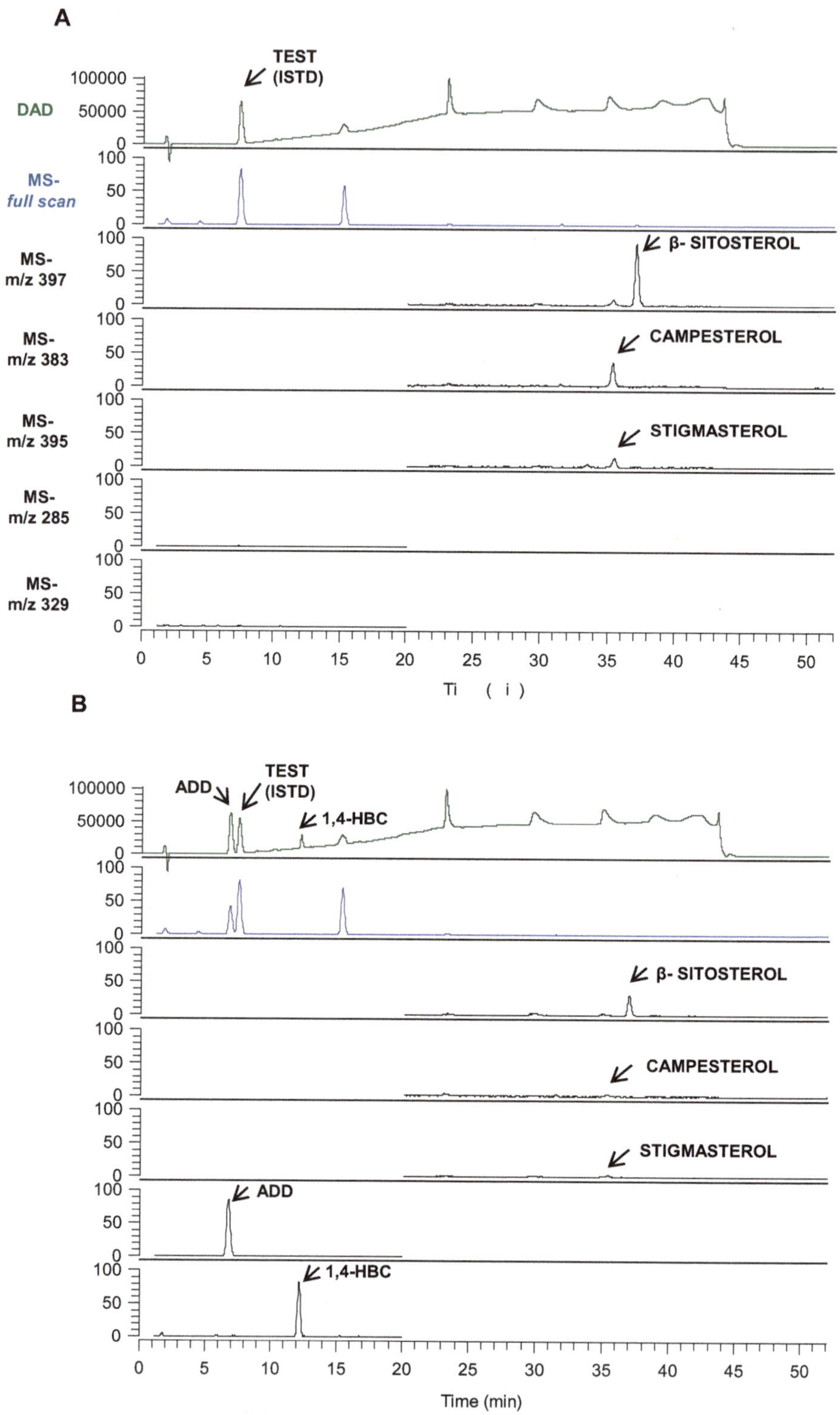

Fig. 5 Analysis by HPLC-DAD-MS of the ADD production by *M. smegmatis* CECT 8331 strain in shaken flasks after 96 h of incubation on phytosterols. (**a**) Starting time and (**b**) 96 h

3. Construct the standard curve from 1,000,000 μM to 15.625 μM (0.4 g/L) of phytosterols, AD, and ADD. Add the internal standard prior to extraction. Use 7H9-Phy Broth without phytosterols as a solvent to make the dilutions (*see* **Note 12**).
4. Extract the steroid fraction as described in Subheading 3.4.
5. Concentrate the chloroform fraction by evaporation and resuspend in 1.5 mL of acetonitrile LC-MaScan.
6. Use the following equipment working conditions: volume to inject 25 μL, capillary temperature 275 °C, vaporizing temperature 425 °C, capillary voltage 39 V, corona discharge 6 kV, source power 6 μA, and dissociation by collision energy 15 eV. Use high pure nitrogen as an auxiliary gas and sprayer.
7. Perform the chromatographic separation at a flow of 1 mL/L in gradient using three mobile phases (A, B, and C) as follows:

Time (min)	A%	B%	C%
0	50	50	0
5	50	50	0
15	20	71	9
20	0	91	9
40	0	70	30
41	0	85	15
42	50	50	0
52	50	50	0

8. For quantification of the peak area, use the following quantitative masses: 289 *m/z* for testosterone (ISTD); 269 *m/z* for cholesterol, 383 *m/z* for campesterol; 395 *m/z* for stigmasterol; 397 *m/z* for β-sitosterol; 287 *m/z* for AD and 285 *m/z* for ADD, 331 *m/z* for 4-HBC (22-hydroxy-23,24-bisnorchol-4-en-3-one), and 329 for 1,4-HBC (3-oxo-23,24-bisnorchol-1,4-dien-22-oic acid) in selected ions of monitoring (SIM). EI mass spectra and retention data are used to assess the identity of compounds by comparing them with those of standards.

4 Notes

1. CECT 8331 can be stored at −80 °C in 50% glycerol (v/v) (viable >5 years) or at 4 °C on 7H10 agar plates (viable for 1 month).

2. To prepare 7H10 agar plates preheat the ADC (enrichment solution) at 45 °C for 30 min before adding to the medium. Mix both components carefully because albumin generates air bubbles that must be avoided.
3. The commercial Middlebrook ADC Enrichment contains catalase, which is absolutely required for *M. tuberculosis* growth but not for *M. smegmatis*. The ADC used to culture *M. smegmatis* can be prepared in the lab making 100 mL aliquots without adding catalase and keeping them at 4 °C once prepared to avoid contamination.
4. Add 0.05% Tween 80 (v/v) from a 20% sterile stock to avoid the CECT 8331 clumping in the 7H9 liquid medium when tyloxapol is not present.
5. Tyloxapol is a nonionic polymeric surfactant not used as carbon source by CECT 8331 that allows the dissolution of phytosterols. Solubilizing the phytosterols enables the monitoring of the cultures following its DO at 600 nm. To pipette correctly the tyloxapol preheat it using a microwave to reduce the thickness.
6. Place the biotransformation medium in a sonication bath at 80 °C for 1–4 h to enhance the phytosterol solubilization before autoclaving.
7. Sunflower oil is used to dissolve high amounts of phytosterols that are insoluble using tyloxapol. When the dissolved phytosterols are added to the aqueous phase a thick emulsion is formed. It is needed to perform the biotransformation in a STBR in order to obtain a good gas and mass transfer.
8. It is important to stir the STBR immediately after autoclaved to avoid the formation of a solid phase in the organic fraction.
9. Due to the characteristics of the emulsion, it is not possible to monitor the culture by measuring the OD_{600}.
10. Place the samples at room temperature for 5 min and recover the organic phase that is not frozen with a pipette before the ice melts.
11. Before spraying the TLC with 20% H_2SO_4 those compounds that contain conjugated double bonds like ADD can be observed under UV light.
12. A calibration curve should be generated for each analyte in the sample and the concentration of the standards should be chosen on the basis of the concentration range expected in the study. To prepare a good calibration curve, a blank and at least six samples covering the expected range must be prepared and analyzed.

References

1. Znidarssic-Plazl P, Plazl I (2010) Development of a continuous steroid biotransformation process and product extraction within microchannel system. Catal Today 157:315–320
2. García JL, Uhía I, Galán B (2012) Catabolism and biotechnological applications of cholesterol degrading bacteria. Microb Biotechnol 5:679–699
3. Malaviya A, Gomes J (2008) Androstenedione production by biotransformation of phytosterols. Bioresour Technol 99:6725–6737
4. Donova MV, Egorova OV (2012) Microbial steroid transformations: current state and prospects. Appl Microbiol Biotechnol 94:1423–1447
5. Donova MV, Gulevskaya SA, Dovbnya DV, Puntus IF (2005) *Mycobacterium* sp. mutant strain producing 9α-hydroxyandrostenedione from sitosterol. Appl Microbiol Biotechnol 67:671–678
6. Andor A, Jekkel A, Hopwood DA, Jeanplong F, Ilkoy E, Kónya A, Kurucz I, Ambrus G (2006) Generation of useful insertionally blocked sterol degradation pathway mutants of fast-growing mycobacteria and cloning, characterization, and expression of the terminal oxygenase of the 3-ketosteroid 9α-hydroxylase in *Mycobacterium smegmatis* mc^{2}155. Appl Environ Microbiol 72:6554–6559
7. García JL, Uhía I, Galán B (2014) Mutantes recombinantes selectivos de *Mycobacterium smegmatis* mc2155 y su uso para la producción de 1,4-androsten-3,17-diona o 4-androsten-3,17-diona a partir de esteroles naturales. Spanish Patent N° P201430272.
8. Kutney JP (2003) Process for fermentation of phytosterols to androstadienedione. WO 2003/064674

Chapter 16

Process for Biotransformation of Androsta-4-ene-3, 17-Dione (4-AD) to Androsta-1,4-Diene-3,17-Dione (ADD)

Surya Prakash and Abhay Bajaj

Abstract

Androsta-1,4-diene-3,17-dione (androstadienedione, ADD) is key intermediate for the organic synthesis of a variety of female sex hormones such as estrone, estradiol, estriol and other related derivatives. De novo synthesis of this molecule is not yet reported in any form of living system, i.e., microbial, plant, and animal. The structural complexities due to presence of several chiral carbon centers create significant hurdles in chemical synthesis of such molecules. Microbe-mediated biotransformation offer a highly reliable, cost-effective, and relatively non hazardous way for commercial manufacturing of steroidal key intermediates. Currently microbial biotransformations are extensively being exploited for large-scale production of basic intermediates such as androstenedione (AD), ADD, and several types of hydroxylated derivatives of androstane compounds. In this chapter several aspects of microbial biotransformation process of AD to ADD are discussed.

Key words Androstenedione (AD), Androstadienedione (ADD), Biotransformation, *Arthrobacter simplex*, Fermentor

Abbreviations

AD	Androstenedione
ADD	Androstadienedione
CSL	Corn steep liquor
DM	Demineralized
FAD	Flavin adenine dinucleotide
HPLC	High performance liquid chromatography
kg/cm^2	Kilogram per centimeter square
KSTD	Keto steroid dehydrogenase
mpa	Megapascal
rpm	Revolutions per minute
TLC	Thin layer chromatography

José-Luis Barredo, Ignacio Herráiz (eds.), *Microbial Steroids: Methods and Protocols*, Methods in Molecular Biology vol. 1645, DOI 10.1007/978-1-4939-7183-1_16, © Springer Science+Business Media LLC 2017

1 Introduction

Microbial dehydrogenation involving C-1,2 positions is one the most important reactions of the steroid biotransformation. Such type of reaction is prevalent among different bacterial and fungal isolates under variety of conditions. Industrial steroid biotransformations employing dehydrogenation are efficient and economically viable, thus holding commercial significance.

C-1,2 dehydrogenation reaction is reported to be catalyzed by a 3-oxosteroid 1-dehydrogenase, which is a type of keto steroid dehydrogenase enzyme (KSTD) with E.C. number 1.3.99.4. This enzyme is reported from a variety of microbial sources like *Actinobacteria* [1] and *Septomyxa* sp. [2] involved in secondary metabolism in the organism. The reaction is dependent on FAD as an acceptor. The enzyme is known to be localized in cytoplasm in *Aspergillus* [3], membrane bound in *Rhodococcus rhodochrous* [4], or particulate bound in *Comamonas testosteroni* [5]. A recent study on a 3-ketosteroid-Δ (1)-dehydrogenase (kstD (F)) involved in the biosynthesis of the antibiotic *Fusidane* in *Aspergillus fumigatus* CICC 40167 reported that kstDF is a promising enzyme in steroid delta-1-dehydrogenation and steroid reduction, which is favorable to construct genetically engineered steroid-transforming recombinants by heterologous overexpression [3].

A general reaction scheme involving KSTD can be represented as follows:

3-oxosteroid + acceptor → 3-oxo-DELTA1-steroid + reduced acceptor.

FAD is the electron acceptor and $FADH_2$ is the reduced form of acceptor.

One of the major biotransformation products of this reaction is androsta-1,4-diene-3,17-dione (ADD). The substrate for the reaction is androst-4-ene-3,17-dione (AD), a key intermediate formed during microbial degradation of side chain of sterols. Both AD and ADD belongs to the 17-keto family of C-19 steroids and serve as starting material for a variety of steroids with high biological activity and commercial value. ADD serves as a key starting material for entire estrone series female sex hormones [6]. Though microbial biotransformation of AD into ADD is commercially being exploited, there are several factors which limit its yield, the major being the low aqueous solubility of the substrate (AD), which limits the bioavailability of AD for the enzymatic reaction. Another factor which governs the rate of enzymatic reaction includes the tolerance of the microorganism toward the toxicity of AD and ADD. Apart from this, routine process parameters of physiochemical nature such as temperature, pH, and aeration also affect the efficiency of biotransformation.

2 Materials

2.1 Culture Media and Strain

All the components of media like glucose, peptone, sodium chloride, yeast extract, beef extract, agar-agar, CSL (corn steep liquor), and sodium hydroxide are of lab grade. The hardness of water used for preparation of media should be about 300 ppm.

1. Nutrient Broth (NB): D(+)glucose 1 g/L, peptone 15 g/L, sodium chloride 6 g/L, and yeast extract 3 g/L. Adjust to pH 7.0 with NaOH (1 N).
2. Nutrient Agar (NA): Agar-agar (20 g/L) in NB. Add agar-agar after pH adjustment.
3. Seed Culture Medium: Beef extract 2 g/L, yeast extract 1 g/L, peptone 5 g/L, glucose 16 g/L, and NaCl 5 g/L. Adjust to pH 6.8 with NaOH (1 N).
4. Biotransformation Medium: Glucose 10 g/L, CSL 10 g/L, and peptone 2.5 g/L. Adjust to pH 6.8 with NaOH (1 N). Add antifoam 0.1 mL/L and Tween 80 1 g/L.
5. *Arthrobacter simplex* ATCC 6946 procured from the American Type Culture Collection (USA). Store the lyophilized culture under recommended conditions to maintain its viability and sterility. Do the subculturing using the standard procedure described in culture maintenance sheet provided by the collection center.

2.2 Substrate and Solvents

1. ADD (Sigma).
2. 4-AD (Tianjin Tianyao Pharmaceutical Co. Ltd.).
3. Methanol, analytical grade for HPLC and lab grade for DSP.
4. n-Hexane, lab grade.
5. Cyclohexane, lab grade.
6. Ethyl acetate, lab grade.
7. Acetone, lab grade.
8. Dichloromethane (CH_2Cl_2), lab grade.
9. Ammonium sulfate.
10. Sulfuric acid.

2.3 Equipment Used for Laboratory Activities

1. Weighing balances, precision 0.01 g.
2. pH meter.
3. Magnetic stirrer.
4. Autoclave.
5. Biosafety cabinet—Class II, A2.
6. Incubator.

7. Orbital shaker.
8. Refrigerator that can achieve a temperature up to 2–4 °C.
9. Deep freezer that can achieve a temperature up to −60 °C.
10. Hot air oven.
11. Microscope.

2.4 Glassware

1. Petri plates (90 mm diameter).
2. Tubes (8/180 mm).
3. Tubes (40/200 or 18/180 mm).
4. Pipettes (1 and 25 mL).
5. Flasks (150, 1000, and 5000 mL).
6. Inoculation needles.
7. Scrapping tools and inoculation loops.
8. TLC plates (precoated activated silica plates; e.g., Merck 1.05554.007).

2.5 Equipment for Fermentation and Downstream Processing

1. Fermentors: In situ sterilizable fermenters with 50 L vessel capacity (e.g., Shanghai Baoxing biological equipment Engineering Co. Ltd., model Biotech 50 JS).
2. Jacketed glass reactors of 20 L capacity (e.g., Shanghai Yuhua Equipment Co. Ltd., model S212).
3. Vacuum filtration assembly.
4. Vacuum pump.
5. Vacuum oven.
6. Rotary evaporator (e.g., Shanghai Yuhua Equipment Co. Ltd., model RE 205).
7. Filter cloth 768.
8. Process utilities, including heating (steam), cooling (brine), and aeration (compressed air).

2.6 Equipment for Analysis

1. High Performance Liquid Chromatography (HPLC) equipment (e.g., Agilent Technology, model 1260).
2. Analytical balance with an accuracy to 0.001 g.
3. Ultrasonic sonicator.
4. Chromatography column Hypersil BDSC18, Sum, 4.0 × 125 mm or equivalent.
5. Chromatography column Hypersil C18 (4.6200 mm) 10 μm or equivalent.
6. Chromatography column Diamonsil C18, 5 μm particles, 250 mm × 4.6 mm or equivalent.
7. Detector: Dual wavelength UV/Visible detector.

8. Filters 8–10 μm (Whatman).
9. Filters 0.45 μm (Millipore-type HV).
10. Mobile phase (Isocratic): Methanol–DM water (60:40).

3 Methods

Details of methods involved in flask level and fermentor level experiments are described below and are performed sequentially.

3.1 Inoculum Preparation

3.1.1 Preparation of Seed Culture Slants

Revival and preparation of slants from lyophilized culture of *Arthrobacter simplex* ATCC 6946 is performed by following steps under strict aseptic condition using a biosafety cabinet (Fig. 1):

1. Add 2 mL sterilized NB medium in the lyophilized vial containing the bacterial culture.
2. Transfer the suspension into sterile 30 mL nutrient broth (NB) medium in a 150 mL flask.
3. Incubate at 31 °C and 150–180 rpm for 2–5 days in an incubator shaker.
4. Set up 1/10 serial dilution of the culture in 5 mL NB tubes (*see* **Note 1**).
5. Plate the appropriate serial dilution on nutrient agar plate, and incubate at 31 °C for 2–4 days till colonies appear.
6. Select the single colony based on marked phenotypic characters and streak it on fresh NB agar plate. Incubate at 31 °C for 2–4 days till colonies appear.
7. Prepare the master slants using the culture grown onto nutrient agar medium containing Petri plate.
8. Make multiple aliquots of working culture from this master slant (*see* **Note 2**).
9. Store master slants under sterile condition at −20 °C.
10. Using this working culture, prepare inoculum flask separately as and when required.

3.1.2 Preparation of Seed Culture in Flask

1. Prepare 100 mL seed culture medium, pH 6.8 in 500 mL capacity conical flasks and sterilize by autoclaving at 121 °C for 30 min.
2. Inoculate the flasks having sterilized seed culture medium with working culture slant under aseptic condition using the biosafety cabinet.
3. Incubate the seed culture on rotary shaker at 150 rpm and 31 °C for 24 h (*see* **Note 3**).

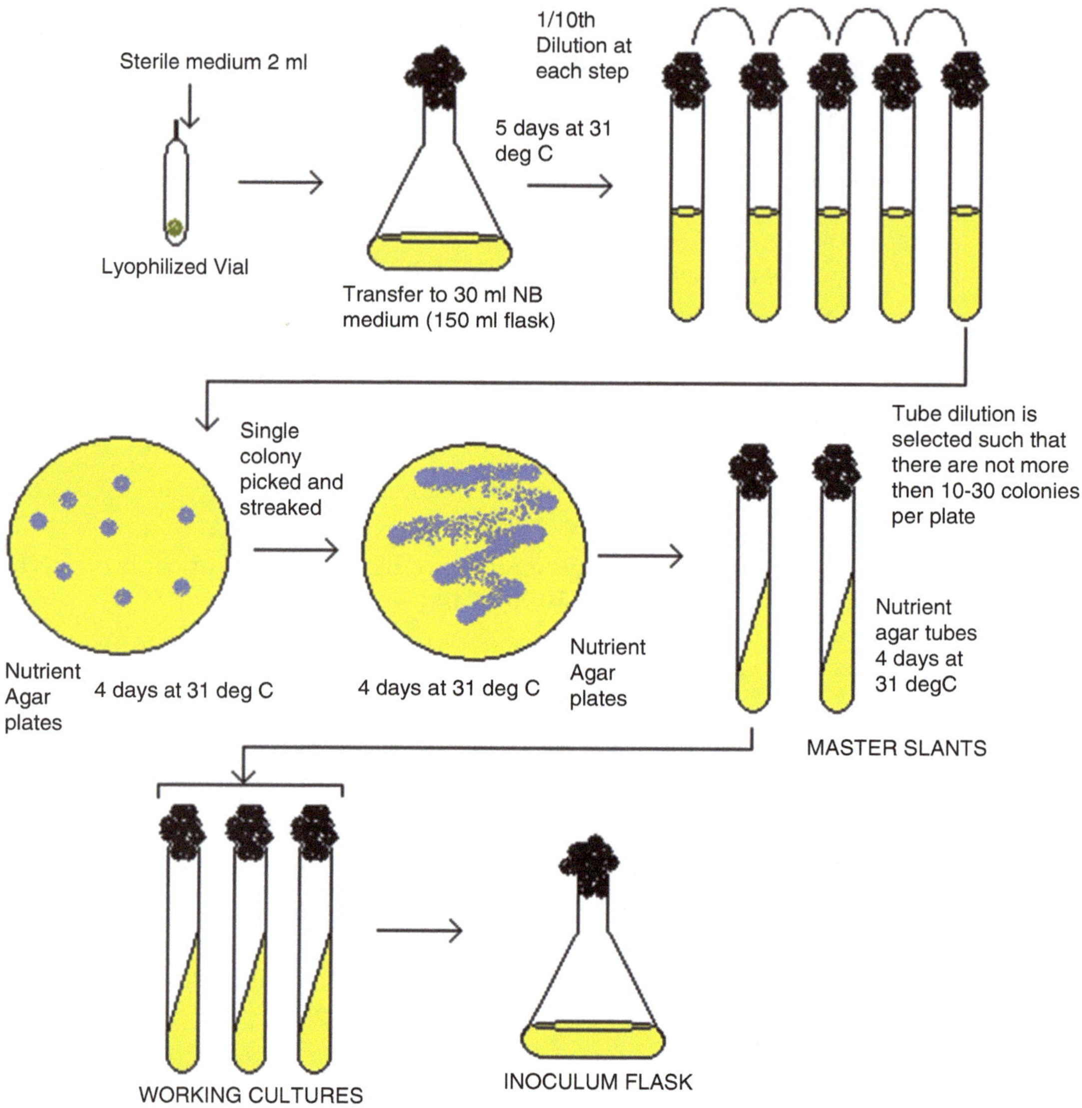

Fig. 1 Schematic representation of the process for the preparation of slants from a lyophilized vial

Upon observation of profuse microbial growth in flask, record pH and check under microscope to validate the monoseptic growth (*see* **Note 4**).

3.2 Strain Activity Test

Laboratory test for confirmation of strain activity is performed under aseptic conditions using biosafety cabinet by following steps:

1. Prepare 100 mL biotransformation medium in 500 mL capacity conical flasks (in triplicate) and sterilize at 121 °C for 30 min in an autoclave.
2. Prepare 100 mL of seed medium, in 500 mL flask, and take 50 mL DM water in separate 250 mL flask and sterilize both

the seed medium and DM water containing flasks at 121 °C for 30 min in an autoclave.

3. Pour 7–8 mL sterilized DM water in the working culture slant and prepare microbial suspension aseptically using biosafety cabinet.
4. Inoculate the 500 mL flask containing 100 mL seed medium using microbial suspension prepared in **step 2** under same aseptic condition.
5. Incubate the seed culture flasks in a shaker at 150 rpm and 31 °C for 24 h.
6. Take a sample from the growing seed culture under aseptic condition and check culture purity by microscopy (*see* **Note 4**).
7. Record pH of culture and take pictures of microscopic slides for future reference.
8. Inoculate the conical flasks (in triplicate) containing 100 mL biotransformation medium under aseptic condition with 20 mL of seed culture prepared in the previous steps.
9. Incubate the flasks in an incubator shaker at 150 rpm and 32 °C for 12 h. Once growth is observed, the substrate, i.e., AD (0.2% w/v) solubilized in a minimum volume of acetone is added to all three flasks and incubation is continued for 2 days. Take 10 mL samples from each biotransformation flask aseptically in biosafety cabinet at regular interval of 12 h. Make observation of the pH trend and the biotransformation of substrate to product by TLC and HPLC (*see* Subheading 3.4).

3.3 Biotransformation Process

The fermentation facility is prepared prior to the trial batches. The activities include:

3.3.1 Preparation of Fermentation Facility

1. Cleaning: The 50 L fermentor and stainless steel piping encasing the main filter, air inlet line, and steam inlet line should be thoroughly cleaned prior to use.
2. Leak Test: Leak test should be performed for fermentor and main filters prior to use.
3. Utility Validation: All utilities including boiler, chiller, air compressor and electricity should be validated for its operational efficiencies prior to use.
4. Blank Sterility Trial: If all the above results are satisfactory, then blank trials are undertaken in fermentor and sterility of the batch is assessed.

3.3.2 Biotransformation Process in 50 L Fermentor

Prior to preparation of biotransformation media in the 50 L fermentor, an adequate volume (3 L) of fresh seed culture is prepared in flasks (*see* Subheading 3.1.2) which is used as inoculum for fermentor.

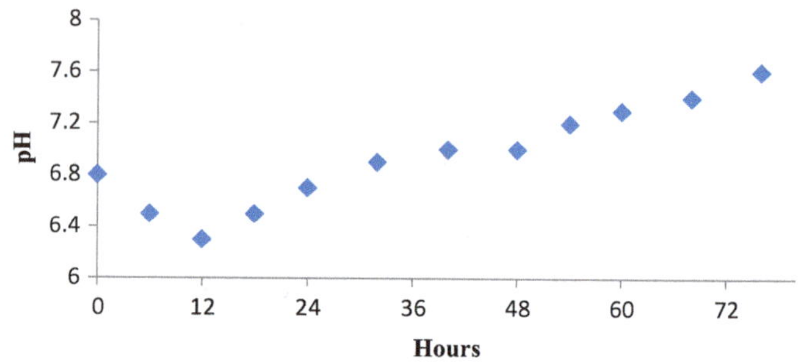

Fig. 2 Typical pH profile observed during biotransformation of AD to ADD

1. To prepare the biotransformation medium, mix accurately weighed ingredients for a working volume of 30 L in the defined volume of water (20 L), then charge it into the fermentor.
2. Adjust the pH to 6.8 by adding 1 N NaOH slowly with continuous mixing and make up the final volume to 27 L.
3. Sterilize the whole fermentor vessel containing the biotransformation medium at 121 °C for 45 min. After sterilization, cool the medium to 32 °C, maintaining the vessel pressure of 0.5 kg/cm^2 and continuous agitation (100 rpm).
4. Transfer the seed culture from flask to fermentor using a peristaltic pump. Minimize the air flow and the pressure of vessel to get complete culture transfer without hindrance.
5. Set the following fermentation parameters: temperature 32 ± 0.5 °C, agitation from 0 to 6 h at 200 rpm and from 6 to 48 h at 300 rpm, aeration 0.5–1.0 vvm, and vessel pressure between 0.2 and 0.3 kg/cm^2 (*see* **Note 5**).
6. Ideally after 12 h of seed culture inoculation, sufficient growth appears (*see* **Note 4**).
7. Then, feed the fermentor a defined quantity (1% w/v, i.e., 300 g) of substrate (4-AD) using a peristaltic pump (*see* **Note 6**).

   During the biotransformation process, the pH is observed and recorded for future references. The pH trend should be downward in first 0–8 h and it should move gradually upward till the completion of biotransformation (Fig. 2).
8. Take a 250 mL sample of the biotransformation broth every 12 h and analyze them by HPLC until the completion of biotransformation reaction. When biotransformation is complete, normally the pH in fermentor reaches above 7.5 and $DO_2 \geq$ 20% (*see* **Note 7**).
9. Terminate the biotransformation process in fermentor by stopping air, followed by in situ sterilization of the fermentation broth at 121 °C for 15 min.

3.4 Analytical Procedure

High performance liquid chromatography (HPLC) analysis is used for quantitative determination of AD and ADD, and simultaneously provides approximate quantitative determination of major by-products of the reaction, i.e., boldenone and testosterone. Thin layer chromatography (TLC) can also be an alternative for the assessment of product formation and biotransformation process [7].

3.4.1 Preparation of a Calibration Curve for Product Quantification

For quantification of the product from biotransformation process, an absorbance curve using known concentration of authentic ADD is drawn. To achieve this, four known concentrations of ADD are used. The steps followed for this purpose are:

1. Weight accurately 3, 6, 9, and 12 mg of ADD and transfer in separate volumetric flasks of 100 mL.
2. Add 20 mL of methanol.
3. Add 25 mL of mobile phase.
4. Keep the volumetric flask in an ultrasonic bath for 5 min.
5. Fill up to the measure with mobile phase.
6. Filter an aliquot of solution from each volumetric flask using Millipore filter type HV (0.45 μm) and analyze them using HPLC equipped with a Diamonsil C18 column and a UV/Visible detector. Monitor the process at 254 nm with a flow rate of 1 mL/min using methanol and water (60:40 v/v) as a mobile phase at 30 °C for 30 min.
7. The absorbance value obtained at various concentration of authentic ADD is used to prepare the calibration curve. The linear regression equation is generated and for standard curve use ADD (mg) = (7.43 Peak Area + 72.22) × 10−4.
8. The retention time for ADD is nearly at 13.20 min and AD is 8.73 min (±10%).

3.4.2 Product Quantitation in Fermentor Samples Using HPLC

1. For routine analysis of the product formed during the biotransformation process, take an aliquot (50 mL) of bioconversion media and extract with 2 volumes of dichloromethane (CH_2Cl_2).
2. Filter the solvent layer (CH_2Cl_2) through Whatman filter paper (10 μm) to remove any particulate matter and dry using rotavapor; nearly 700 mg of crude product is obtained.
3. Resuspend the crude product in 100 mL methanol (HPLC grade) and filter using Millipore filter type HV (0.45 μm) prior to placing in an appropriate vial for HPLC analysis.
4. Analyze AD and ADD in a HPLC instrument equipped with a Diamonsil C18 column and a UV/Visible detector. Monitor the process at 254 nm with a flow rate of 1 mL/min using methanol and water (60:40 v/v) as mobile phase at 30 °C for 30 min [8].

5. Identify the products by comparison of their spectral properties with that of the authentic standards.
6. Quantify the product using the calibration curve drawn from known concentration of authentic standards (*see* Subheading 3.4.1).

3.4.3 Thin Layer Chromatography

For quick semi-qualitative assessment of product formation and monitoring of the biotransformation process, this method can be used.

1. Extract the biotransformation broth sample with 4 volumes of dichloromethane or alternatively suspend solid samples in dichloromethane (1:4 w/v).
2. Spot few drops of dichloromethane extract along with a standard ADD solution (prepared in dichloromethane) on pre-coated thin-layer silica gel plates (10 cm × 4 cm) using ordinary capillary tubes.
3. Develop the plates with cyclohexane–ethyl acetate (3:2 v/v).
4. Observe the developed plates under ultraviolet light.
5. Visualize the plates by spraying with 2% ceric ammonium sulfate/1 M H_2SO_4 and heating the plate at 160 °C until color development.
6. Use the disappearance and appearance of spots equivalent to authentic standards as a reference to assess the product formation and monitor the biotransformation process.

3.5 Downstream Process

3.5.1 Filtration

1. Upon completion of the biotransformation as assessed by HPLC or TLC or both, heat the broth to 121 °C with stirring at 100 rpm for 15 min, and then allow it to cool down to 90 °C.
2. Feed the broth into the filtration system where a clear aqueous layer is separated from the biomass together with ADD and impurities like residual AD.
3. Dry the filtered cake overnight at 50 °C in a hot air oven. The net quantity of filter cake varies batch to batch after drying.

3.5.2 Solvent Extractions

1. Extract the dried filtrate cake with methanol (1:3 w/v) in a jacketed glass reactor of 20 L at room temperature.
2. Mix thoroughly and separate using a filtration assembly.
3. Repeat this process three times sequentially until the complete product is extracted from dried cake.
4. Check the residual product (ADD) by HPLC/TLC in the cake solids after the third extraction. If there is residual ADD, more extraction steps can be done using 1:1 (w/v) cake: methanol ratio.

3.5.3 First Concentration

1. Combine the methanol extracts in a 20 L jacketed glass reactor.

2. Concentrate until methanol is reduced to 10% of initial volume by maintaining the following process conditions: vacuum 00.08 mpa, temperature 60 °C, and agitation 50 rpm.

3.5.4 Crude ADD Isolation

1. Cool the concentrated methanol up to 10 °C by brine circulation in the reactor jacket, keeping other process conditions (i.e., agitation and vacuum) as in Subheading 3.5.3 for 10–12 h.
2. Check the formation of visible crystals in the reactor. The reaction mixture should appear viscous.
3. Filter the crude ADD through a Nutsche filter and take back the clear methanol to the same reactor to concentrate it further to get additional crude ADD.

3.5.5 Crude ADD Purification Process

1. Reflux the crude ADD with hexane (1:1 w/v) at 40 °C with agitation (50 rpm) for 1 h without vacuum in the reactor, which is used for crystallization. Filter the product–hexane suspension through a Nutsche filter and dry under vacuum for 2 h.
2. Decolorization: Use methanol to solubilize the refluxed ADD and add active carbon (1:0.16 w/w) to remove color impurities. Reflux the mixture at 50 rpm for 1 h at 60 °C and then allow it to be cool to room temperature before filtration. Filter the active carbon under vacuum using a Nutsche filter. Wash the active carbon with a small volume of methanol on the same filter.
3. Second concentration: Transfer the decolorized methanol to the crystallization reactor and concentrate under the following process conditions till the volume becomes 10% of original volume: vacuum 0.08 mpa, temperature 60 °C, and agitation 50 rpm.
4. Crystallization: Perform crystallization of the product from the saturated methanol solution in the crystallization reactor by maintaining the following parameters: vacuum 0.08 mpa, temperature 10 °C, and agitation 50 rpm. Maintain the parameters for 10–12 h to allow slow crystallization.
5. Filtration: Filter the crystals using filtration cloth 768 using vacuum.
6. Drying: Dry the filtered crystals under the following conditions: vacuum 0.09 mpa, temperature 50 °C, and time 12 h.

3.6 Specifications of Final Product

The chemical formula of ADD is $C_{19}H_{24}O_2$, with a molecular mass of 284.39 g/mol (Fig. 3). The specifications of the final product arc: white crystalline powder, purity ≥98% as per HPLC analysis, and loss on drying ≤0.5%.

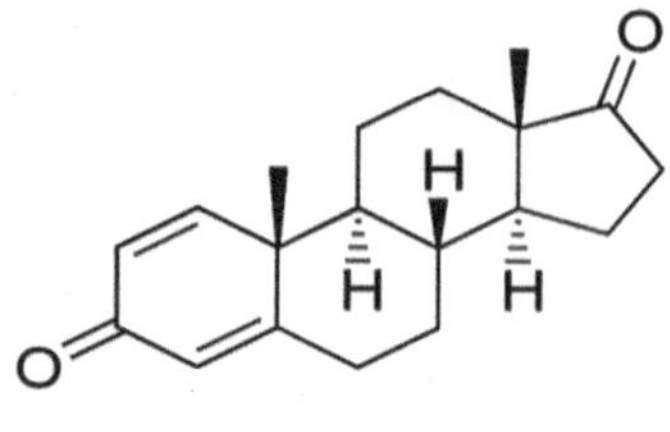

Fig. 3 Molecular structure of ADD

4 Notes

1. Serial dilution for plating is selected to obtain not more than 10–30 colonies per Petri plate.
2. The use of slants more than 2 weeks old is not suggested.
3. The longevity of the seed material is around 24 h in a refrigerator at 0–4 °C. Longer storage time is not recommended.
4. Use Gram staining technique for identification.
5. If needed, add silicone oil-based antifoam of polydimethylsiloxane class (0.1 mL/L) to the biotransformation medium.
6. Substrate preparation method: AD (300 g) is accurately weighed and solubilized in 1.2 L of acetone in a 5 L capacity conical flask; this solution is prepared immediately before use.
7. Sampling method from fermentor: Sterilize the sample port of fermentor by flowing pressurized steam for 30 min. Then close the steam valve and open the sample valve to collect a sample in a clean glassware. After collecting an adequate volume of sample, close the sample valve of the fermentor and again open the steam valve to pass pressurized steam for another 30 min.

References

1. Donova MV (2007) Transformation steroid compounds by actinobacteria. Appl Biochem Microbiol 43:1–14
2. Singh K, Sehgal SN, Vezina C (1968) Large-scale transformation of steroids by fungal spores. Appl Microbiol 16:393–400
3. Chen MM, Wang FQ, Lin LC, Yao K, Wei DZ (2012) Characterization and application of fusidane antibiotic biosynethsis enzyme 3-ketosteroid-Δ1-dehydrogenase in steroid transformation. Appl Microbiol Biotechnol 96:133–142
4. Morii S, Fujii C, Miyoshi T, Iwami M, Itagaki E (1998) 3-Ketosteroid-DELTA1-dehydrogenase of *Rhodococcus rhodochrous*: sequencing of the genomic DNA and hyperexpression, purification, and characterization of the recombinant enzyme. J Biochem 124:1026–1032
5. Levy HR, Talalay P (1959) Bacterial oxidation of steroids: II. Studies on the enzymatic mechanism of ring A dehydrogenation. J Biol Chem 234:2014–2021
6. Malaviya A, James G (2008) Androstenedione production by biotransformation of phytosterols. Bioresour Technol 99:6725–6737
7. Gulla V, Banerjee T, Patil S (2008) Quantitative TLC analysis of steroid drug intermediates formed during bioconversion of soysterols. Chromatographia 68(7):663–667
8. Shao M, Zhang X, Rao Z, Xu M, Yang T, Li H, Xu Z (2015) Enhanced production of androst-1,4-diene-3,17-dione by *Mycobacterium neoaurum* JC-12 using three-stage fermentation strategy. PLoS One 10(9):e0137658

Chapter 17

Laboratory-Scale Hydroxylation of Steroids by $P450_{BM3}$ Variants

Sabrina Hoebenreich, Martin Spinck, and Nathalie Nett

Abstract

This chapter describes the asymmetric hydroxylation of steroids on laboratory preparative scale, using engineered variants of $P450_{BM3}$ (CYP102A1) as enzyme catalyst. The following protocol covers the creation of an *Escherichia coli* BL21-Gold (DE3) expression strain, including necessary control experiments like plasmid preparation, test expression, and creation of storage cultures, to verify successful experimental access to recombinant expressed $P450_{BM3}$ variants. The recombinant expressed $P450_{BM3}$ variants are obtained as cleared cell lysate and used in a biotransformation setup to hydroxylate 2.8 mg and up to 15 mg testosterone in the presented protocol. Since $P450_{BM3}$ depends on NADPH as an electron source for the reaction, a glucose and glucose dehydrogenate based recycling system is added to the reaction. The protocol further includes liquid–liquid extraction of hydroxytestosterone and directs the experimenter to compound purification via column chromatography.

Key words Preparative scale, Steroid hydroxylation, CYP102A1, Testosterone, Biocatalysis, Steroid biotransformation, Bacillus megaterium

1 Introduction

Achieving asymmetric C-H activation in steroids is a challenging chemical transformation, where chemical synthesis methods are sparse. Steroid hydroxylation, among other steroidal transformations, has always been an area of biotechnology using either whole cells or recombinant expressed enzymes as biocatalysts [1–8]. Among them, cytochrome $P450_{BM3}$ or CYP102A1 from *Bacillus megaterium* [9] has a high synthetic value for laboratory scale synthesis of functionalized steroids [10–13]. It is one of the benchmark systems in P450 research, where a huge quantity of data on protein engineering and evolution is available [14, 15].

Most importantly for this chapter, $P450_{BM3}$ has several benefits from a synthetical point of view. First, it is a natural fusion protein with its associated FMN and FAD containing reductase domain, and therefore, researchers have to express only a single enzyme [9].

José-Luis Barredo, Ignacio Herráiz (eds.), *Microbial Steroids: Methods and Protocols*, Methods in Molecular Biology vol. 1645, DOI 10.1007/978-1-4939-7183-1_17, © Springer Science+Business Media LLC 2017

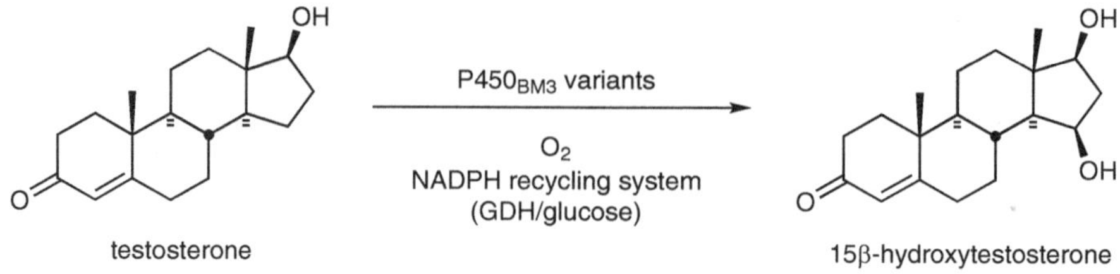

Fig. 1 Testosterone conversion obtained by $P450_{BM3}$ variants

Second, it shows a high expression rate in *E. coli* laboratory strains, yielding up to 1.5 g/L soluble enzyme [16], and third, $P450_{BM3}$ and its cofactors are air tolerant during purification, handling and over longer reaction times, in contrast to other nonheme iron oxygenases.

Cytochrome $P450_{BM3}$ wild type does not hydroxylate steroids [17, 18]. Several research groups have therefore expanded the substrate scope of $P450_{BM3}$ by protein engineering and directed evolution to accept steroids like testosterone [17–24], testosterone acetate [25], progesterone [19], 11β-hydroxyprogesterone [25], estradiol [18, 26], trimethyl estriol [25], norandrostenedione [27], and norethisterone [24, 28, 29]. Hydroxylation and dihydroxylation was observed mainly at positions 1, 2, 4, 15, 16, and 19 of the steroid scaffold, depending on the variant and substrate. The general chemical equation of this reaction involves molecular oxygen and NADPH to achieve steroid hydroxylation (Fig. 1). It is therefore necessary to couple the steroid transformation to the energy metabolism of whole cells, by performing whole cell biotransformation, or to provide an NADPH regeneration system into the reaction mixture [30].

This chapter describes the general procedure to synthesize hydroxylated steroids using the example of testosterone by $P450_{BM3}$ variants. The overall workflow includes recombinant expression, cell rupture and cleared cell lysate preparation, as well as the biocatalytical transformation and product purification.

2 Materials

This protocol is a mixture of molecular biology/biochemistry and chemical synthesis techniques (*see* **Note 1**). Especially the biology part is highly dependent on a longer list of special instruments.

2.1 Instruments

1. Autoclave, e.g., a FVA3/A1 (Fedegari).
2. Shaker, e.g., a Multitron, tray with 25 mm shaking diameters (HT-Infors, Einsbach, Germany).
3. Cell incubator, e.g., an isotherm incubator (Fisher Scientific).

4. Centrifuge, e.g., a Sorvall RC 6 Plus centrifuge equipped with F21S rotor (Thermo Scientific).
5. pH-meter, e.g., UltraBasic Benchtop Meters (Denver Instrument).
6. Hot-plate magnetic-stirrer device, e.g., a Hei-Tec magnetic stirrer device (Heidolph).
7. Heat gun, e.g., PHG 630 DCE (Bosch, Germany).
8. UV light to analyze TLCs, e.g., a NU-4-KL UV lamp (Konrad Benda—Herolab).
9. Rotary evaporator, e.g., a Laborota 4000 efficient (Heidolph) with a VACUUBRAND CVC 3000 pump (VWR).
10. Electroporation apparatus, e.g., a Micro Pulser (Bio-Rad).
11. Sonication bath, e.g., an ultrasonic cleaner (VWR).
12. Table top centrifuge, e.g., an Eppendorf 5418 centrifuge (Eppendorf).
13. SDS-PAGE apparatus, e.g., a Mini-Protean electrophoresis system (Bio-Rad).
14. Ultrasonic homogenizer, e.g., a Sonoplus HD 2070 homogenizer equipped with a SGH 213G booster horn and a titanium flat tip (TT13) (Bandelin).
15. Spectrophotometer, e.g., a V-650 spectrophotometer (JASCO) or a NanoDrop 2000/2000c (Thermo Scientific).

2.2 Equipment

1. Reaction tubes of 1.5, 2, 15, and 50 mL.
2. Baffled flasks for culturing.
3. Round bottom flasks.
4. Erlenmeyer flasks.
5. Graduated mixing cylinder.
6. Separating funnel.
7. Beakers.
8. Stirring bars.
9. Dialysis bags or Ultra Centrifugal Filter Unit with a cutoff of 100 kDa.
10. TLC slides.
11. Pipettes.
12. Drigalski spatula.
13. Electroporation cuvettes.

2.3 Chemicals and Enzymes

Purchase chemicals and enzymes with the highest purity grade available from your local supplier. For comparison, we list our supply sources. Chemicals for molecular biology and biochemistry are bought in "biochemistry" grade.

1. *Escherichia coli* BL21-Gold(DE3) strain (Novagen, Merck-Millipore).
2. Plasmids pETM11-$P450_{BM3}$, pET28a-$P450_{BM3}$, and pRSF-$P450_{BM3}$.
3. MicroPulser™ Electroporation Apparatus. Prepare electro-competent cells following the MicroPulser™ Electroporation Apparatus Operating Instructions and Applications Guide, which is based on the Inoue method [31]. http://www.bio-rad.com/webroot/web/pdf/lsr/literature/4006174B.pdf
4. Ultrapure water. This water, sometimes called Millipore water, has the highest possible electrical resistivity (>18 MΩ/cm) or the lowest conductivity (<0.05 μS/cm).
5. Lysogeny broth (LB) powder (Carl Roth, Germany).
6. Terrific broth (TB) powder (Carl Roth, Germany).
7. Kanamycin (AppliChem) stock solution (50 mg/mL) in water. Filtered sterile and stored at −20 °C.
8. Trace metal mix (TMM): Mix 675.75 mg (50 mM) $FeCl_3 \cdot 6H_2O$, 110.99 mg (20 mM) $CaCl_2$, 84.51 mg (10 mM) $MnSO_4 \cdot H_2O$, 143.77 mg (10 mM) $ZnSO_4 \cdot 7H_2O$, 23.79 mg (2 mM) $CoCl_2 \cdot 6H_2O$, 17.05 mg (2 mM) $CuCl_2 \cdot 2H_2O$, 23.77 mg (2 mM) $NiCl_2 \cdot 6H_2O$, 24.19 mg (2 mM) $Na_2MoO_4 \cdot 2H_2O$, and 6.18 mg (2 mM) H_3BO_3. To this, add 312.5 μL concentrated HCl and fill up with ultrapure water to 50 mL. This protocol was modified from Studier et al. [32]. An updated version of the article from 2008 can be found on the author's homepage.
9. Power mix: saturated glutamic acid (>11.1 g/L at 25 °C) and 40% glycerol (v/v) in potassium phosphate buffer (100 mM, pH 7.4). Sterilize after preparation.
10. P450 Expression Media: supplement TB medium with 1 mM $MgCl_2$, a 1:1000 dilution of the trace metal mix and a 1:100 dilution of the power mix.
11. IPTG (isopropyl ß-D-thiogalactopyranoside, MW = 238.3 g/mol) stock solution of 0.1 M in water. Sterilize by filtration using a syringe filter of 0.2 μm.
12. Potassium phosphate buffer (1 L, 100 mM, pH 7.4): Mix 13.97 g (or 802 mL of a 0.1 M solution) of K_2HPO_4 (174.18 g/mol) and 2.694 g (or 198 mL of a 0.1 M solution) of KH_2PO_4 (136.09 g/mol) in ultrapure water to achieve a pH of 7.4. Nevertheless, the pH is checked after mixing and if needed adjusted with H_3PO_4 or KOH.
13. Glucose dehydrogenase. We use either GDH-105 (Codexis), GDH-60 (Evoxx; formerly Evocatal) or the engineered GDH from Bacillus subtilis 168 (*Bs*GDH E170K/Q252L).

14. NADP-disodium salt, min. 97%, for biochemistry (Carl Roth). Prepare a 100 mM stock solution in 100 mM potassium phosphate buffer (pH 7.4). Store at −20 °C in aliquots of 1 mL.
15. Glycerol (1,2,3-propanetriol, ROTIPURAN® ≥99.5%, p.a., water free; Carl Roth, Germany): Prepare an 80% (v/v) solution in ultrapure water and sterilize by heat.
16. SDS-PAGE analysis: We either buy precasted gradient gels (4–15%) or prepare self-casted 10% gels. The following ingredients are needed for self-casting: acrylamide–bisacrylamide solution, Tris–HCl buffer, Sodium dodecyl sulfate (SDS), ammonium persulfate, tetramethylethylenediamine (TEMED or TMEDA).
17. Coomassie dye.
18. Plasmid miniprep kit (*see* **Note 2**).
19. Carbon monoxide (CO) gas.
20. Sodium dithionite ($Na_2S_2O_4$).
21. Liquid nitrogen.

3 Methods

Ensure that you fulfill all regulations of your country when importing and working with genetically modified organisms (GMO) in your lab. Most groups use *E. coli* BL21-Gold(DE3) transformants with pETM11-$P450_{BM3}$, pET28a-$P450_{BM3}$, or pRSF-$P450_{BM3}$ plasmids. Such GMOs are classified as biosafety level 1 and all the following molecular biology and biochemistry experiments need to be performed in a genetic engineering laboratory of biosafety level 1.

All steps from here on assume the use of either plasmid pETM11, pET28, or pRSF and use of *E. coli* BL21-Gold(DE3). If other combinations are used, antibiotics need to be adjusted.

3.1 *Escherichia coli* BL21-Gold(DE3) Transformation with $P450_{BM3}$ Variants

This chapter does not cover the process of cloning and mutagenesis. The reader is referred to the original research articles of the $P450_{BM3}$ variant of interest for such techniques.

In general, approach the group and obtain the desired variant as plasmid, preferably ask for a plasmid in *E. coli*, optimally already in *E. coli* BL21-Gold(DE3). When a GMO is delivered, streak it out on LB agar plates containing the appropriate antibiotics to obtain single colonies, and proceed to Subheading 3.1.2. When a plasmid has been provided, use it to transform *E.coli* BL21-Gold(DE3) electrocompetent cells as described in Subheading 3.1.1.

3.1.1 Creation of Expression Strains via Electrotransformation

1. Prechill an electroporation cuvette (gap width: 0.2 cm) and a 1.5 mL safe-lock reaction tube on ice. Use the ice to thaw the electrocompetent cells, too.

2. Place 10 ng (1–3 μL, in ultrapure water) of plasmid DNA into a prechilled 1.5 mL safe-lock reaction tube followed by 50 μL of electrocompetent cells (*see* **Note 3**).
3. Incubate for 2–5 min on ice.
4. Set the MicroPulser to Ec2 settings (2.5 kV), according to the recommendations of the manufacturer.
5. Transfer the mixture of cells and DNA into the precooled electroporation cuvette; ensure no air bubbles are formed. Remove them by taping the cuvette once or twice on the bench.
6. Pulse immediately.
7. Immediately afterward, add 1 mL of LB media (without any additives), place the mixture into a culture tube and incubate at 37 °C, 200 rpm for 60 min.
8. Spread 200 μL of the sample with a Drigalski spatula on LB plates containing 50 μg/mL of kanamycin (*see* **Note 4**) and incubate overnight at 37 °C in a cell incubator.
9. On the following day, small colonies should have appeared on the plates.
10. Proceed with Subheading 3.1.2.
11. The plates can be stored several weeks in a fridge as a backup. Wrap Parafilm around the gap to prevent drying out.

3.1.2 Plasmid Preparation and Sequencing

1. Inoculate a single colony from an agar plate into 4 mL of sterile LB media containing 50 μg/mL kanamycin and incubate at 37 °C and 200 rpm (*see* **Note 5**) overnight.
2. Use 4 μL of this culture for Subheading 3.1.3 and 0.75 mL for Subheading 3.1.5.
3. Harvest cells of the remaining culture in a 2 mL safe-lock reaction tube by centrifugation at 11,337 × *g* in a microfuge.
4. Discard the medium and tap the tube rim on a piece of paper towel to remove the last drop.
5. Obtain the plasmid from this cell pellet according to a Plasmid Miniprep Kit manual (*see* **Note 2**). Elute the sample in the last step with ultrapure water.
6. Measure the dsDNA concentration and purity by UV, using a spectrophotometer, applying 1 μL of the obtained sample (*see* **Note 6**).
7. To ensure that the DNA sequence has not been compromised upon transport and handling, submit a sample from this batch for sequencing. Store the remaining plasmid sample at −20 °C.

3.1.3 P450$_{BM3}$ Test Expression

The second experiment to ensure that the organism and plasmid DNA was not compromised upon transport is to conduct a small test expression. Ideally, use the same culture or cells from the same colony you used in Subheading 3.1.2.

1. Inoculate 4 mL $P450_{BM3}$ expression medium supplemented with 50 μg/mL of kanamycin with 4 μL (1:100) from the overnight culture of Subheading 3.1.2 and inoculate at 37 °C and 200 rpm until an OD_{600} of 0.6–0.8 is reached.
2. Take a sample corresponding to an OD_{600} of 3, pellet the cells by centrifugation at 11337 × *g* and save it for SDS-PAGE analysis as T_0 point (T_0 = prior to induction), by storing it on ice or freezing it [33]. If the measured 100 μL sample has an OD_{600} of 0.6, you would need to take 500 μL from the main culture to have a biomass density of OD_{600} = 3.
3. Induce protein expression by adding a final concentration of 0.1 mM IPTG (*see* **Note 7**).
4. Temperature is lowered to 30 °C and cultures are incubated for 7 h (*see* **Note 8**) at 200 rpm.
5. Take a sample with an optical cell density of 3 (OD_{600} = 3) as T_1 (= after induction) sample. Remember that OD_{600} measurements are nonlinear at higher densities, therefore dilute the sample 1:10 prior to OD_{600} determination. Pellet the cells at 11,337 × *g* in a 1.5 mL safe-lock reaction tube and save it for SDS-PAGE analysis.

3.1.4 SDS-PAGE Analysis

Polyacrylamide gel electrophoresis (PAGE), describes a technique to separate proteins according to their electrophoretic mobility. When performed under denaturing condition using SDS, mobility is a function of the length, conformation, and charge of the molecule. This section is kept short, since excellent step-by-step protocols are available [33, 34].

1. Resuspend the cell pellets obtained in Subheading 3.1.3 in 100 μL ultrapure water.
2. Lyse cells for 10 min on ice in a sonication bath.
3. Centrifuge the sample for 10 min at 11,337 × *g* to separate supernatant and pellet. If SDS-PAGE analysis is used to monitor expression as performed in Subheading 3.2, proceed without separation.
4. Add SDS-PAGE sample buffer to each sample (supernatant and pellet) and incubate at 98 °C for 10 min.
5. Centrifuge again at 11,337 × *g* for 10 min.
6. Set up the SDS-PAGE cell according to the manual.
7. Load 10 μL sample onto a SDS-PAGE gel (*see* **Note 9**).
8. Load T_0 supernatant and pellet, as well as T_1 supernatant and pellet samples next to each other. It accounts to four samples

when one variant is analyzed. This way, it can be analyzed if the $P450_{BM3}$ variant was expressed as soluble form.

9. Add a protein marker and run according to the manufacturer's instructions. The run is finished when the coomassie blue stain from the sample buffer reaches the end of the gel.
10. Carefully take out the gel from the glass slides and fixate the proteins to the gel by staining. We use a commercial coomassie dye stain following the manufacturer's instructions.
11. $P450_{BM3}$ band is observed at ~120 kDa.

3.1.5 Preparation of Bacterial Glycerol Stocks for Long-Term Storage of Plasmids

Bacterial glycerol stocks are important for long-term storage of plasmids. Although you can store your plasmid DNA at −20 °C, we create glycerol stocks to store the desired plasmid in the desired bacterial strain to reduce work load by avoiding retransformation and recharacterization. The addition of glycerol stabilizes the frozen bacteria, preventing damage to the cell membranes and keeping the cells alive. A glycerol stock of bacteria can be stored stably at −80 °C for many years, but the optimal concentration of long-term glycerol storage is unknown. We store bacteria in 20% glycerol.

1. Take 750 μL sample from the overnight culture of Subheading 3.1.2 and add 250 μL of 80% glycerol in a 2 mL screw top tube or cryovial (*see* **Note 10**).
2. Gently mix by inverting the tube.
3. Flash-freeze the tube in liquid nitrogen and store at −80 °C.

The stock is now stable for years, as long as it is kept at −80 °C. Subsequent freeze and thaw cycles reduce shelf life (*see* **Note 11**).

3.2 $P450_{BM3}$-Expression and Partial Purification

1. Scratch material from the glycerol stock of the target variant with a sterile toothpick or pipette tip to directly inoculate 7 mL of LB (50 μg/mL kanamycin) in a 100 mL sterile culture flask overnight at 37 °C, 200 rpm in a shaker.
2. Prepare two baffled flasks of 1 L volume with 300 mL $P450_{BM3}$ expression media containing 50 μg/mL of kanamycin each.
3. Dilute the overnight culture 1:100 into the expression media.
4. Save the rest of the overnight culture as a glycerol stock in case you need plasmid or cells for trouble shooting of a particular expression batch.
5. The culture is shaken at 37 °C until an OD_{600} of 0.6–0.8 is reached (*see* **Note 12**).
6. Prior to induction, take cell mass corresponding to an OD_{600} of 3 for SDS-PAGE analysis as T_0 point.
7. Induce protein expression by adding a final concentration of 0.1 mM IPTG (*see* **Note 7**).

8. The temperature is lowered to 30 °C and cultures are incubated for 7 h (*see* **Note 8**) at 200 rpm.
9. Cells are collected by centrifugation at 2819 × *g* for 10 min at 4 °C. *From here on, samples are kept constantly on ice.*
10. Cells are washed by resuspending (*see* **Note 13**) them once in 20 mL chilled 100 mM potassium phosphate buffer (pH 7.4). The sample is transferred to a 50 mL reaction tube and pelleted again at 2819 × *g* for 10 min at 4 °C.
11. The pellet can be stored at −20 °C until further usage.
12. Thaw the pellet in a chilled water bath (*see* **Note 14**).
13. Resuspend the cell pellet in potassium phosphate buffer (100 mM, pH 7.4). Keep the suspension volume as low as possible (~5–10 mL) to obtain a maximal concentrated sample (*see* **Note 15**).
14. Make sure all cell material is in solution and the walls and lid of the tube are clean. If not, spin down the tube at 2819 × *g* for 10 s, using the fast-spin mode of the centrifuge.
15. To obtain cleared cell lysate, cells are now broken with an ultrasonic homogenizer. The 50 mL reaction tube containing the sample is placed in a water–ice bath.
16. The probe of the sonicator is inserted halfway into the sample. Avoid touching the wall of the plastic tube with the sonication probe. We pulse 5 × 30 s, with 30 s beak intervals at 40% power (5 × 10 cycles) (*see* **Note 16**).
17. Ultrasonic homogenization leads to local heat development. Ensure that the sample is kept cool at all times. Carefully swirl the sample in the tube in the 30 s break to ensure better cooling.
18. After homogenization, the sample is transferred to a superspeed centrifugation tube and the supernatant is recovered by centrifugation (30 min, 1118 × *g*, 4 °C).
19. The pellet normally has a slimy consistency, therefore avoid excessive movements of the centrifugation tube which could lead to partially resuspension of the soft pellet (*see* **Note 17**).
20. Carefully separate supernatant and pellet using a pipette.
21. This solution, the so called cleared cell lysate, is now passed through a 0.45 μm PSE syringe filter (*see* **Note 18**), aliquoted in 1.5 mL tubes (1000 μL each) and flash-frozen in liquid N_2 prior to storage at −80 °C.
22. The solution will be red to brown, depending on the concentration, since $P450_{BM3}$ variants are colorful enzymes due to the absorption of the heme complex.

Cell lysates of most $P450_{BM3}$ variants are at least stable for 2–3 years under these conditions. Determine protein concentration after thawing of an aliquot, prior to use.

3.2.1 $P450_{BM3}$ Concentration Determination

The $P450_{BM3}$ protein concentration of the lysate is determined via light absorption of the reduced and CO bound iron–heme complex at 450 nm following a modified protocol reported by Guengerich et al. [35].

Record a full spectrum of your first CO and $Na_2S_2O_4$ containing sample to determine the exact maxima on your device. In our case, we find 449 and 423 nm instead of 450 and 420 nm, respectively.

1. After thawing a sample, dilute it 1:50 (*see* **Note 19**) in a final volume of 3 mL potassium phosphate buffer (100 mM, pH 7.4) and distribute the sample into three cuvettes.
2. The first cuvette serves as a blank which is measured before and after each measurement, if there is a difference, the difference value is used for correction in the final calculation.
3. The second cuvette is bubbled with CO (60 bubbles for 1 min) (*see* **Note 20**).
4. To the second (contains sample and CO) cuvette add roughly ~1 mg $Na_2S_2O_4$ by using a leveled spoon spatula of 2 mm diameter. The spoon is leveled at the wall of a plastic 1.5 mL reaction tube that contains the dry $Na_2S_2O_4$ (*see* **Note 21**).
5. The sample is mixed by inversion of the cuvette, using a tiny piece of Parafilm as lid, and measured immediately at the blanked photometer.
6. The absorption of the second cuvette is measured until the maximum at 449 nm is constant, which takes 30 s to 5 min, depending on the $P450_{BM3}$ variant. Sometimes the absorption decreases again over time. In such cases we take the time point with the highest value.
7. Repeat **steps 4–6** with the third cuvette (contains sample only).
8. The mean values of three measurements at the absorption at 449, 490, and 423 nm are used for calculation of the concentration.
9. The concentration is calculated from the difference of cuvette 2 and 3 at the absorption of 490 and 449 nm, corrected by the offset value of cuvette 1 using a table calculation program (e.g. Excel) as shown in Fig. 2.

3.3 $P450_{BM3}$-Based Testosterone Hydroxylation Using Cell Lysate and a NADPH-Recycling System: Biocatalysis on Milligram Scale

For the biotransformation the obtained cleared cell lysate is used.

1. Thaw the sample in a water bath set to room temperature (*see* **Note 22**).
2. After thawing, determine the $P450_{BM3}$ enzyme concentration as described in Subheading 3.2.1.
3. Calculate all ingredients as listed in Table 1.
4. All compounds should be mixed in a graduated mixing cylinder made of glass (*see* **Note 23**).

BM3	dilution factor	if Blank shows offset, add the values here for baseline correction			A449		A490		A423	
		A449	A490	A423	wo CO	w CO	wo CO	w CO	wo CO	w CO
variant 1	40	0.0009	0.001	0.00085	-0.0008	0.1422	-0.0189	0.0138	-0.0078	0.0069
variant 1	40	0.0009	0.001	0.00085	-0.0011	0.1412	-0.0192	0.0119	-0.0081	0.004
variant 1	40	0.0009	0.001	0.00085	-0.0017	0.1363	-0.0198	0.0085	-0.0087	0.0018

$P450_{diluted}$	$P450_{sample}$	$P423_{diluted}$	$P423_{sample}$	active concentration	std. deviation	inactive concentration	std. deviation
μM	μM	μM	μM	μM	μM	μM	μM
1.2	48.5	0.3	12.0	48.6	0.3	12.0	0.1
1.2	48.9	0.3	11.8				
1.2	48.3	0.3	12.0				

'=N3*B3

=((K3-J3)-(I3-H3)-(L3*(-0.041))-(E3-D3))/0.106

=L3*B3

=((G3-F3)-(I3-H3)-((C3-D3)))/0.091

Fig. 2 Excel sheet showing used equation to calculate $P450_{BM3}$ active and inactive concentrations

Table 1
Biocatalysis scheme for 2.8 mg testosterone transformation at 10 mL scale and 0.0025 mol% catalyst loading, with an exemplarily $P450_{BM3}$ concentration (*see* Note 24)

Reagent	Equivalents	Molar mass	Working concentration	Stocks	Mixing scheme
Testosterone	1 eq	288.43 g/mol	1 mM	100 mM in acetonitrile	100 μL
D-Glucose monohydrate	250 eq	198.17 g/mol	250 mM	–	0.495 g
Glycerol	–	–	5% v/v	80% in water	0.625 mL
$P450_{BM3}$ variant	0.0025 eq	–	2.5 μM	30 μM in assay buffer	833 μL
$NADP^+$	0.5 eq	–	0.5 mM	0.1 M in assay buffer	50 μL
GDH	–	–	1 U/mL	100 U/mL in assay buffer	1 mL
Assay buffer	–	–	100 mM potassium phosphate buffer, pH 7.4		Fill up to 10 mL

5. Add glucose and glycerol to the mixing cylinder and dissolve in a part of the assay buffer (~5 mL buffer).
6. Mix by inverting until homogeneous solution is achieved.
7. Add GDH, $P450_{BM3}$, and $NADP^{+}$.
8. Mix carefully again. Prevent foam formation while mixing samples containing enzymes.
9. Fill up to 10 mL with buffer and pour the solution into the 100 mL reaction flask.
10. Start the reaction by adding the steroid from the stock solution (*see* **Note 25**). Swirl the sample solution in the flask with one hand and add the steroid quickly to the solution with the pipette. The reddish solution may become turbid, since steroids have a very low solubility in water (*see* **Note 26**). Hydroxylation will increase solubility of the product.
11. Close the flask with a cotton plug or metal cap allowing oxygen influx and place it at 30 °C and 130 rpm into the shaker.
12. You will notice that the enzyme is not stable under assay conditions very long and a coiled-like plug will form rapidly, often within the first 30 min (*see* **Note 27**).
13. Draw 100 μL samples every 15 min to monitor the reaction (*see* **Note 28**). Often, full conversion can be observed after 1–3 h under given conditions, if not, higher $P450_{BM3}$ concentrations or addition of a second batch of $P450_{BM3}$ variant can drive the reaction to completion [36]. Once the reaction time is determined, this step could be shortened or skipped.
14. When full or constant conversion is observed, stop the reaction by extraction with addition of ethyl acetate as described in the following.

3.3.1 Liquid–Liquid Extraction (Separation) of Steroids

Caution, the organic solvents in the following steps will contain a high concentration of steroids. Avoid coming in contact with the solution. Wear two layers of one-time-use nitrile gloves or thicker gloves suitable for ethyl acetate. Read the material safety data sheet to make yourself familiar with the risk of exposure to high potency active pharmaceutical ingredients. CAUTION, your organic solvent solutions will contain a dangerously high steroid concentration (0.5 mg/mL).

Perform the next steps in a fume hood.

1. Transfer the reaction mixture to a 50 mL reaction tube and add 5 mL ethyl acetate (*see* **Note 29**).
2. Close tightly, shake vigorously (*see* **Note 30**).
3. Separate the phases by centrifugation at 1707 × *g* and 4–15 °C. Samples need to be slightly cooled to avoid overpressure due to the low boiling point of ethyl acetate.

4. Carefully remove the tube from the centrifuge and avoid excessive movement of the tube. You may observe a solid layer at the interphase.
5. Open tube carefully and use a glass pipette to transfer the organic layer to a 100 mL Erlenmeyer flask. It is more important to avoid the water phase than to transfer the whole organic phase, since the procedure is done three times to ensure complete extraction.
6. In total, extract three times with 5 mL ethyl acetate each. You now have an Erlenmeyer flask containing the organic solvent fraction.
7. Dry the organic solvent sample by adding dry $MgSO_4$ under swirling until the salt has a fluffy appearance. If it still forms clumps, it is still wet (*see* **Note 31**).
8. Filter the salt off by passing the mixture through a funnel containing a cotton or glass-wool plug. Remember to wash the plug, funnel, and salt with ~5 mL clean ethyl acetate.
9. Collect the filtered solvent in a round bottom flask and concentrate the sample to 1–2 mL using a rotary evaporator.
10. Perform silica column chromatography as explained here [37] (*see* **Note 32**). The elution system is ethyl acetate–methanol (5:1). Use HPLC grade solvents or distilled solvents for column chromatography.
11. Characterize the purified compound by ^{1}H and ^{13}C NMR (use deuterated methanol as NMR solvent), mass (positive APCI mode is recommended for steroids) and when possible by HPLC (*see* **Note 33**).
12. Store the obtained steroids as solids.

4 Notes

1. JOVE and YouTube are good sources for video tutorial showing common laboratory techniques. http://www.jove.com/ and www.youtube.com. A brilliant step-by-step tutorial on "How to perform column chromatography and TLC (thin-layer chromatography)" can be found at Chemistry Views. http://www.chemistryviews.org/details/education/2101817/Tips_and_Tricks_for_the_Lab_Column_Choices.htmL.
2. There are several sellers of Plasmid Miniprep Kits. We have had a good experience with Zymo Research, Qiagen, Sigma-Aldrich, Biozym, and Peqlab.
3. Make sure the DNA sample is placed at the bottom of the tube and does not remain at or in the pipette tip. Addition of the bigger volume to the smaller ensures proper mixing. Mixing

cannot be achieved with the 1–10 μL pipette tip. In the worst case, the thin diameter of the tip will destroy some cells through shearing forces.

4. Check the antibiotic resistance of your obtained plasmid, e.g., pRSF and pETM11 encode for kanamycin resistance.
5. We used a Multitron shaker with 25 mm shaking diameter from HT-Infors, Einsbach, Germany. When another shaker is used, the rpm values given here need to be adjusted depending on the shaking diameter of your own shaker.
6. Any UV–Vis spectrophotometer is useable to determine dsDNA concentration. Blank the photometer with ultrapure water in a cuvette suitable for UV wavelength and record a whole spectrum of the sample. For calculations use the following two equations:
   (a) Concentration (μg/mL) = (A260 reading – A320 reading) × dilution factor × 50 μg/mL
   (b) DNA purity (A260/A280) = (A260 reading – A320 reading) ÷ (A280 reading – A320 reading)
7. For the pETM11 plasmid, expression works best with 0.05 mM IPTG concentration and induction at an OD_{600} of 0.8–1.0, whereas expression with the pRSF plasmid is best at 0.1 mM IPTG and induction at an OD_{600} of 0.6–0.8. Often, formation of inclusion bodies can be observed in addition to sufficient soluble $P450_{BM3}$ production.
8. When overnight expression is desired, the temperature has to be lowered to 25 °C for pRSF and pETM11 plasmids.
9. Design an unsymmetrical gel loading pattern for your samples on the SDS-Page or you will not be able to distinguish the samples after the staining process. Gels can be bought or self-made. In case you buy a gel, invest in a 4–15% gradient gel for analysis to reach a better resolution.
10. Snap top tubes are not recommended as they can open unexpectedly at −80 °C.
11. We recommend preparing two copies as glycerol stocks, one as a backup and one as the working stock.
12. Monitor the process by taking samples of 1 mL culture and measure the OD at 600 nm in a UV–Vis photometer, blank with the culture media. We often take the first sample after 1.5 h. The closer you get to the desired OD, the shorter the intervals need to be, since ideally *E. coli* doubles in cell density every 20 min.
13. We use a simple, small silicon cake scrapper or jar spatula, commonly used in household baking, to quickly resuspend the pellet. Vortexing is an option, too, but often takes more time.

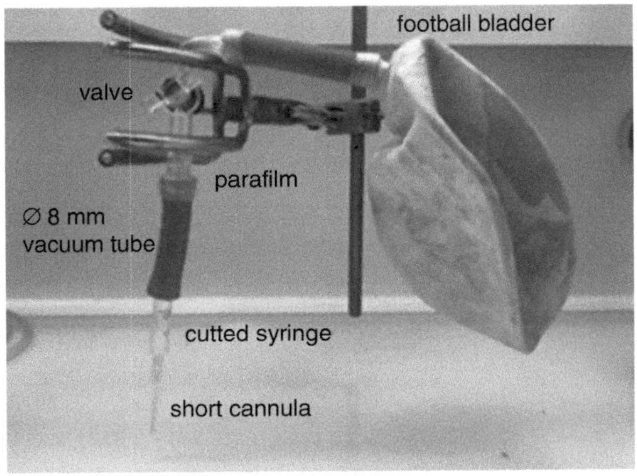

Fig. 3 Football bladder-based CO bubbling setup

14. The tube needs to be in contact with water and not with air. Air is an insulator and will increase the time until the pellet is completely thawed.
15. Due to the volume of the pellet, the final volume of the suspension will be higher than 10 mL.
16. Do not touch the reaction tube during sonication. Wear ear protection against the ultrasound to avoid ear damage. Ensure that no excessive foam is formed. Adjust probe position in sample to avoid foam formation.
17. Some $P450_{BM3}$ variants produce indigo blue insoluble pigments during fermentation. This step will separate the pigments from the lysate.
18. Avoid excessive force applied by the syringe; be patient, as the solution might still be very viscous. Either you have small particles from the pellet picked up (need to centrifuge again) or another sonication is needed to shear DNA and RNA molecules. If all this does not help, dilute the sample a little bit.
19. The more light red the solution is, the lower the concentration of the $P450_{BM3}$ variant. In such cases, use dilution factor of 1:20 or 1:40.
20. We use a football bladder-based setup in a fume hood filled with CO gas from a pressure bottle (*see* Fig. 3). Working with a carbon monoxide (CO) pressure bottle needs proper training and precaution. The bottle needs to be located in a ventilated area or fume hood. Make yourself familiar with the material data sheet prior to the use of CO gas.
21. Sodium dithionite degrades rapidly in solution; therefore, no stock solution can be prepared. Store the powder dry, cool and in the dark.

22. It is important to let the sample reach room temperature (20–25 °C), especially for short reaction times, to ensure reproducibility. $P450_{BM3}$ undergoes conformational changes at 4 °C toward less active states. Do not freeze the sample again, most variants loose activity after repeated freeze–thaw cycles.
23. Steroids are poorly soluble in water and starting material might be lost when working with plastic equipment in the absence of an organic solvent, due to hydrophobic interaction.
24. The concentration of the $P450_{BM3}$ variant is given exemplarily here; it has to be determined as described in Subheading 3.2.1 before reaction. The reaction can easily be scaled up. Our biggest reaction was 15.2 mg testosterone, dissolved in 500 μL acetonitrile, with mutant KSA-4 (R47Y/T49F/F87A/K224E/I314V), which completed in 1.5 h.
25. Do not use DMSO as a stock solvent; it inhibits the enzyme by binding to the iron of the heme complex [38]. Also avoid using DMF, since it will be hard to remove after extraction and disturbs HPLC analysis. Acetonitrile is a good solvent for most steroids.
26. The low solubility is further decreased by buffer salts, glucose and glycerol additives and increased by addition of cosolvents. In our case, beta-cyclodextrins did not improve the overall reaction. Under realistic biotransformation conditions, solubilities of around 125 μM testosterone and 22 μM progesterone were determined [39]. Using concentrations above the solubility level lead to microcrystalline dispersions of the substrates when dissolved in an organic solvent prior to fast addition [40]. The positive effect of such a technique is a minimization of mass transfer limitations. The low solubility has advantages, since it keeps the substrate concentration low and constant at the beginning of the reaction.
27. As long as the solution stays red, there is some active enzyme left. When the solution turns yellow and milky, all enzyme is degraded and the reductase has released both flavin cofactors into solution, causing the yellow color. This can often be observed after several hours or in overnight reactions [36].
28. Monitoring every 15 min can either be done by TLC analysis [37] or by HPLC analysis and becomes redundant when the reaction time was determined. Use ethyl acetate–methanol (5:1) or pure ethyl acetate for TLC. Prepare HPLC samples by mixing 100 μL of sample with 100 μL of acetonitrile, followed by debris removal via centrifugation at 11,337 × g for 10 min.
    (a) Mutant KSA-14 (R47F/T49F/V78 L/A82M/F87A/K224E/V314I) was used at 1.84 μM final concentration. The reaction with KSA-5 (V78 L/A82F/F87A) and KSA-14 is completed within 30 min. KSA-1(F87A/A330W) completes within 2.5 h.

29. Solvents should be purified by distillation on a rotary evaporator when bought at technical grade. A small separating funnel could be used instead of centrifugation. Unfortunately often three phases are observed and are almost impossible to separate by gravity alone. Scale up the organic solvent volumes when you scale up the reaction volume.
30. Do not open the tube between shaking and centrifugation. If needed, spin down the solvent mix briefly, before opening. Avoid at any means that the lid contains sample. Wipe the lid with a piece of toilet paper if this happens, otherwise the centrifugal force will push this part of the sample into the centrifuge and contaminate everything with a toxic sample!
31. If unsure how the procedure is done, check out this video: https://www.youtube.com/watch?v=vcwfhDhLiQU.
32. Alternatively, the sample can be purified on a preparative, reverse phase (C18) HPLC using an acetonitrile–water gradient.
33. A standard reverse-phase C18 column is suitable. We use a reverse-phase "250 Eclipse XDB" C18 column of 250 mm (1.8 μm size particle, 4.6 inner diameter) together with a corresponding precolumn bought from Agilent (Waldbronn, Germany) as stationary phase at 40 °C. The mobile phase consists of a mixture of high-purity water, HPLC-grade acetonitrile (CH_3CN), and methanol (MeOH). A program with a flow rate of 1 mL/min of total 8 min based on a CH_3CN–MeOH–H_2O mixture was used: 0 → 3 min (15:15:70), 3 → 5 min (20:20:60), 5 → 6 min (30:30:40), 6 → 7 min (15:15:70). This protocol allows the separation of almost all oxidation products of testosterone. Alternatively, and with lower resolution of peaks, an acetonitrile–water isocratic mobile phase of 45:55 could be used for separating testosterone and hydroxylated testosterone isomers.

References

1. Furuya T, Shibata D, Kino K (2009) Phylogenetic analysis of *Bacillus P450* monooxygenases and evaluation of their activity towards steroids. Steroids 74(12):906–912. doi:10.1016/j.steroids.2009.06.005
2. Agematu H, Matsumoto N, Fujii Y, Kabumoto H, Doi S, Machida K, Ishikawa J, Arisawa A (2006) Hydroxylation of testosterone by bacterial cytochromes P450 using the *Escherichia coli* expression system. Biosci Biotechnol Biochem 70(1):307–311. doi:10.1271/bbb.70.307
3. Bernhardt R, Urlacher VB (2014) Cytochromes P450 as promising catalysts for biotechnological application: chances and limitations. Appl Microbiol Biotechnol 98(14):6185–6203. doi:10.1007/s00253-014-5767-7
4. Bracco P, Janssen DB, Schallmey A (2013) Selective steroid oxyfunctionalisation by CYP154C5, a bacterial cytochrome P450. Microb Cell Factories 12:95. doi:10.1186/1475-2859-12-95
5. Bureik M, Bernhardt R (2007) In: Schmid R, Urlacher VB (eds) Modern biooxidation. Enzymes, reactions and applications. Wiley-VCH, Weinheim, Chichester, pp 155–176. doi:10.1002/9783527611522.ch6
6. Fernandes P, Cruz A, Angelova B, Pinheiro H, Cabral J (2003) Microbial conversion of steroid compounds. Enzyme Microb Technol

32(6):688–705. doi:10.1016/S0141-0229(03)00029-2

7. Girvan HM, Munro AW (2016) Applications of microbial cytochrome P450 enzymes in biotechnology and synthetic biology. Curr Opin Chem Biol 31:136–145. doi:10.1016/j.cbpa.2016.02.018
8. O'Reilly E, Kohler V, Flitsch SL, Turner NJ (2011) Cytochromes P450 as useful biocatalysts: addressing the limitations. Chem Commun (Camb) 47(9):2490–2501. doi:10.1039/c0cc03165h.
9. Narhi LO, Fulco AJ (1986) Characterization of a catalytically self-sufficient 119,000-dalton cytochrome P-450 monooxygenase induced by barbiturates in *Bacillus megaterium*. J Biol Chem 261(16):7160–7169
10. Hollmann F, Arends IWCE, Buehler K, Schallmey A, Bühler B (2011) Enzyme-mediated oxidations for the chemist. Green Chem 13(2):226–265. doi:10.1039/C0GC00595A
11. Roiban G-D, Reetz MT (2015) Expanding the toolbox of organic chemists: directed evolution of P450 monooxygenases as catalysts in regio- and stereoselective oxidative hydroxylation. Chem Commun (Camb) 51(12):2208–2224. doi:10.1039/c4cc09218j.
12. Schmid R, Urlacher VB (eds) (2007) Modern biooxidation: enzymes, reactions and applications. Wiley-VCH, Weinheim, Chichester
13. Wong L-L (2011) P450(BM3) on steroids: the Swiss Army knife P450 enzyme just gets better. Chembiochem 12(17):2537–2539. doi:10.1002/cbic.201100606
14. Kumar S (2010) Engineering cytochrome P450 biocatalysts for biotechnology, medicine and bioremediation. Expert Opin Drug Metab Toxicol 6(2):115–131. doi:10.1517/17425250903431040
15. Whitehouse CJC, Bell SG, Wong L-L (2012) P450(BM3) (CYP102A1): connecting the dots. Chem Soc Rev 41(3):1218–1260. doi:10.1039/c1cs15192d
16. Pflug S, Richter SM, Urlacher VB (2007) Development of a fed-batch process for the production of the cytochrome P450 monooxygenase CYP102A1 from *Bacillus megaterium* in *E. coli*. J Biotechnol 129(3):481–488. doi:10.1016/j.jbiotec.2007.01.013
17. van Vugt-Lussenburg BMA, Damsten MC, Maasdijk DM, Vermeulen NPE, Commandeur JNM (2006) Heterotropic and homotropic cooperativity by a drug-metabolising mutant of cytochrome P450 BM3. Biochem Biophys Res Commun 346(3):810–818. doi:10.1016/j.bbrc.2006.05.179
18. Kille S (2010) Flavoproteins in directed evolution: iterative CASTing to evolve YqjM and P450 BM3. Dissertation, Ruhr-Universität Bochum
19. Kille S, Zilly FE, Acevedo JP, Reetz MT (2011) Regio- and stereoselectivity of P450-catalysed hydroxylation of steroids controlled by laboratory evolution. Nat Chem 3(9):738–743. doi:10.1038/nchem.1113
20. Hoebenreich S, Zilly FE, Acevedo-Rocha CG, Zilly M, Reetz MT (2015) Speeding up directed evolution: combining the advantages of solid-phase combinatorial gene synthesis with statistically guided reduction of screening effort. ACS Synth Biol 4(3):317–331. doi:10.1021/sb5002399
21. Ren X, Yorke JA, Taylor E, Zhang T, Zhou W, Wong LL (2015) Drug oxidation by cytochrome P450BM3. Chemistry 21(42):15039–15047. doi:10.1002/chem.201502020.
22. Vottero E, Rea V, Lastdrager J, Honing M, Vermeulen NPE, Commandeur JNM (2011) Role of residue 87 in substrate selectivity and regioselectivity of drug-metabolizing cytochrome P450 CYP102A1 M11. J Biol Inorg Chem 16(6):899–912. doi:10.1007/s00775-011-0789-4
23. Venkataraman H, Beer SBAD, van Bergen LAH, van Essen N, Geerke DP, Vermeulen NPE, Commandeur JNM (2012) A single active site mutation inverts stereoselectivity of 16-hydroxylation of testosterone catalyzed by engineered cytochrome P450 BM3. Chembiochem 13(4):520–523. doi:10.1002/cbic.201100750
24. Rea V, Kolkman AJ, Vottero E, Stronks EJ, Ampt KAM, Honing M, Vermeulen NPE, Wijmenga SS, Commandeur JNM (2012) Active site substitution A82W improves the regioselectivity of steroid hydroxylation by cytochrome P450 BM3 mutants as rationalized by spin relaxation nuclear magnetic resonance studies. Biochemistry 51(3):750–760. doi:10.1021/bi201433h
25. Lewis JC, Mantovani SM, Fu Y, Snow CD, Komor RS, Wong C-H, Arnold FH (2010) Combinatorial alanine substitution enables rapid optimization of cytochrome P450BM3 for selective hydroxylation of large substrates. Chembiochem 11(18):2502–2505. doi:10.1002/cbic.201000565.
26. Cha GS, Ryu SH, Ahn T, Yun C-H (2014) Regioselective hydroxylation of 17beta-estradiol by mutants of CYP102A1 from *Bacillus megaterium*. Biotechnol Lett 36(12):2501–2506. doi:10.1007/s10529-014-1628-9

27. Venkataraman H, Te Poele EM, Rosloniec KZ, Vermeulen N, Commandeur JNM, van der Geize R, Dijkhuizen L (2015) Biosynthesis of a steroid metabolite by an engineered *Rhodococcus erythropolis* strain expressing a mutant cytochrome P450 BM3 enzyme. Appl Microbiol Biotechnol 99(11):4713–4721. doi:10.1007/s00253-014-6281-7
28. Reinen J, Postma G, Tump C, Bloemberg T, Engel J, Vermeulen NPE, Commandeur JNM, Honing M (2016) Application of a cocktail approach to screen cytochrome P450 BM3 libraries for metabolic activity and diversity. Anal Bioanal Chem 408(5):1425–1443. doi:10.1007/s00216-015-9241-x
29. Reinen J, Vredenburg G, Klaering K, Vermeulen NP, Commandeur JN, Honing M, Vos JC (2015) Selective whole-cell biosynthesis of the designer drug metabolites 15- or 16-betahydroxynorethisterone by engineered Cytochrome P450 BM3 mutants. J Mol Catal B Enzym 121:64–74. doi:10.1016/j.molcatb.2015.08.003
30. Zhao H, van der Donk WA (2003) Regeneration of cofactors for use in biocatalysis. Curr Opin Biotechnol 14(6):583–589. doi:10.1016/j.copbio.2003.09.007
31. Inoue H, Nojima H, Okayama H (1990) High efficiency transformation of *Escherichia coli* with plasmids. Gene 96(1):23–28. doi:10.1016/0378-1119(90)90336-P
32. Studier FW (2005) Protein production by auto-induction in high-density shaking cultures. Protein Expr Purif 41(1):207–234. doi:10.1016/j.pep.2005.01.016
33. Burgess RR (2009) In: R. R. Burgess, M. P. Deutscher (Eds.), Guide to protein purification, methods in enzymology, v. 463, Elsevier/Academic Press, Amsterdam, Boston, 813–820, DOI: 10.1016/S0076-6879(09)63044-5.
34. Kurien BT, Scofield RH (eds) (2012) Protein electrophoresis: methods and protocols. Humana Press, Totowa, NJ
35. Guengerich FP, Martin MV, Sohl CD, Cheng Q (2009) Measurement of cytochrome P450 and NADPH-cytochrome P450 reductase. Nat Protoc 4(9):1245–1251. doi:10.1038/nprot.2009.121
36. Brummund J, Muller M, Schmitges T, Kaluzna I, Mink D, Hilterhaus L, Liese A (2016) Process development for oxidations of hydrophobic compounds applying cytochrome P450 monooxygenases in-vitro. J Biotechnol. doi:10.1016/j.jbiotec.2016.07.002
37. Millar S (2012) Tips and tricks for the lab. ChemViews. doi:10.1002/chemv.201200074
38. Kuper J, Wong TS, Roccatano D, Wilmanns M, Schwaneberg U (2007) Understanding a mechanism of organic cosolvent inactivation in heme monooxygenase P450 BM-3. J Am Chem Soc 129(18):5786–5787. doi:10.1021/ja067036x
39. Zehentgruber D (2010) Biotransformationen mit Cytochrom P450 Monooxygenasen. Dissertation, Universität Düsseldorf, 2009; Forschungszentrum, Zentralbibliothek, Jülich
40. Schmitz G, Franke D, Stevens S, Takors R, Weuster-Botz D, Wandrey C (2000) Regioselective oxidation of terfenadine with *Cunninghamella blakesleeana*. J Mol Catal B Enzym 10(1–3):313–324. doi:10.1016/S1381-1177(00)00132-6

Chapter 18

Obtaining of 11α-Hydroxyandrost-4-ene-3,17-dione from Natural Sterols

Dmitry Dovbnya, Sergey Khomutov, Vyacheslav Kollerov, and Marina V. Donova

Abstract

Two-step one-pot microbial transformation enables obtaining of valuable steroids that are difficult to produce chemically. Here we describe a method for obtaining 11α-hydroxyandrost-4-ene-3,17-dione (11α-HAD) from cheap and available natural sterols (phytosterols or cholesterol).

11α-HAD is a primary adrenal steroid in mammals and also a key precursor in the syntheses of halogenated corticoids. Conventional routes for its obtaining are based on chemical synthesis, or microbial hydroxylation of androst-4-ene-3,17-dione (AD). AD in turn is produced primarily with microbial biotransformation of natural sterols by some actinobacteria.

Consequent bioconversions of sterols using two microbial strains in one bioreactor vessel without separation and purification of AD provides high yield of 11α-HAD. At the first fermentation step, phytosterol is converted to AD with *Mycobacterium neoaurum* NRRL 3805B, or relative strains, to yield about 70% (mol/mol). At the second step, AD is almost fully (98%) hydroxylated at the position 11α with *Aspergillus ochraceus* VKM F-830, or other suitable organisms, in the same bioreactor. At the average, 30% (w/w) of the high-purity crystalline 11α-HAD can be obtained.

The method can be exploited for production of 11α-HAD for practical use.

Key words Phytosterol, Cholesterol, Steroid, 11α-Hydroxyandrost-4-ene-3,17-dione, Bioconversion

1 Introduction

11α-Hydroxyandrost-4-ene-3,17-dione (11α-HAD) and its structural analogs are key precursors in the syntheses of steroid pharmaceuticals, such as halogenated corticoids [1–3]. Conventional production routes are based on chemical synthesis, or microbial hydroxylation of androst-4-ene-3,17-dione (AD) [4]. AD in turn is produced primarily by microbial biotransformation of natural sterols by some Actinobacteria, such as *Mycobacterium neoaurum* NRRL 3805B, VKM Ac-1815D, or their relative strains [5, 6]. Here we show a method for obtaining 11α-HAD in a two-step one-pot bioconversion from cheap and easily accessible natural

José-Luis Barredo, Ignacio Herráiz (eds.), *Microbial Steroids: Methods and Protocols*, Methods in Molecular Biology vol. 1645, DOI 10.1007/978-1-4939-7183-1_18, © Springer Science+Business Media LLC 2017

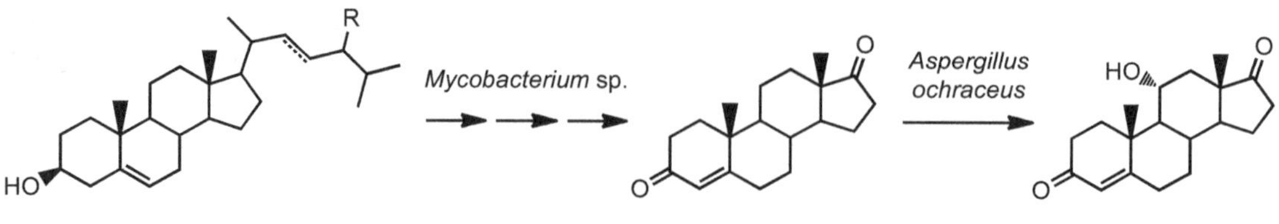

Fig. 1 Bioconversions of natural sterols to 11α-HAD. Cholesterol: R = H. Phytosterol: R = C_2H_5—β-sitosterol, CH_3—campesterol, R = C_2H_5, Δ^{22}—stigmasterol, R = CH_3, Δ^{22}—brassicasterol

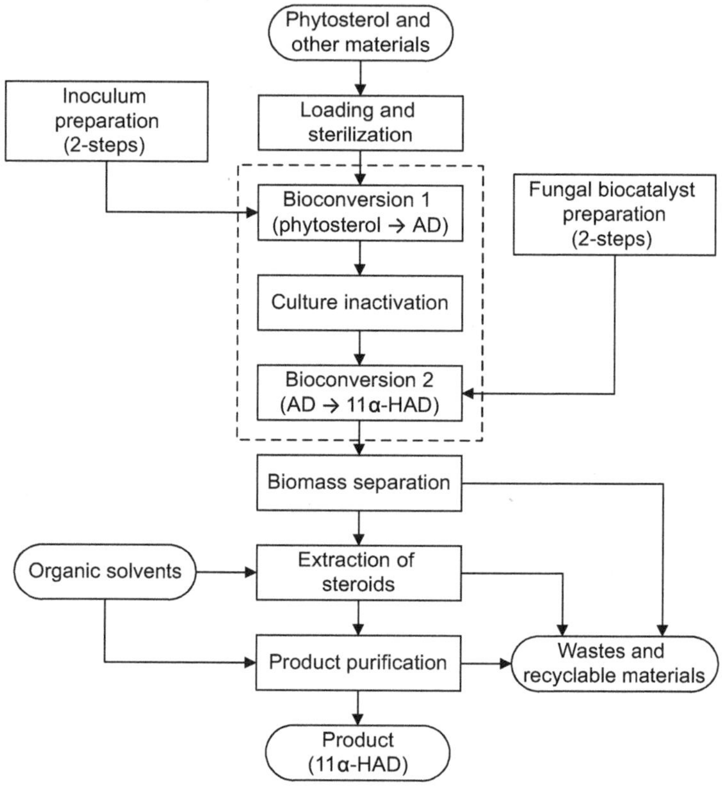

Fig. 2 Flow diagram of the method for obtaining of 11α-HAD. The stages *boxed with dashed line* are performed in the same bioreactor vessel (flask or fermenter) without unloading

sterols (phytosterol or cholesterol). The method is based on the specific biochemical activities of two microbial strains (Fig. 1) applied consecutively in one reaction vessel that excludes separation and purification of the intermediate product (Fig. 2). On a first stage sterols are converted into AD by a growing culture of *Mycobacterium neoaurum* NRRL 3805B with about 70% molar yield. AD is accumulated in the bioconversion broth in a fully soluble form owing to inclusive complexation with statistically methylated β-cyclodextrin [6]. Then AD is regiospecifically and stereospecifically hydroxylated by the mycelium of the fungus *Aspergillus ochraceus* VKM F-830 [7] (or other suitable 11-hydroxylating fungal strain) to 11α-HAD with about 98% molar yield. The target product is extracted from an aqueous broth with water-immiscible organic solvent, separated by stepwise crystallization and purified to desired quality (95–99%) by recrystallization with polar organic solvents.

2 Materials

2.1 Cultivation and Bioconversion

1. Biocatalyst for sterol conversion to AD: *Mycobacterium* sp. NRRL 3805B [5]. Relative strains with similar biochemical activity may be used instead: *Mycobacterium neoaurum* VKM Ac-1815D [6], *Mycobacterium neoaurum* ZADF-4 and others [8, 9].
2. Biocatalyst for AD conversion to 11α-HAD: *Aspergillus ochraceus* VKM F-830 [7] or analogous *A. ochraceus* strains, or possibly other strains with high 11α-hydroxylation activity toward AD [10–12].
3. KOH 10% solution in distilled water.
4. H_2SO_4 10% solution in distilled water.
5. Nutrient medium MYCA: 10 g/L glycerol, 1.5 g/L $(NH_4)_2HPO_4$, 0.55 g/L K_2HPO_4•$3H_2O$, 0.5 g/L KH_2PO_4, 0.2 g/L $MgSO_4$•$7H_2O$, 0.005 g/L $FeSO_4$•$7H_2O$, 0.002 g/L $ZnSO_4$•$7H_2O$, and 20 g/L agar. Adjust to pH 7.1 with 10% KOH if necessary (*see* **Note 1**).
6. Nutrient medium MYCB: the same as MYCA without agar (*see* **Note 1**).
7. Malt extract agar (Oxoid, UK). Adjust to pH 5.5 with 10% KOH or 10% H_2SO_4 if necessary.
8. Nutrient medium ASP1: 25 g/L corn steep solids (Sigma-Aldrich, USA), 1.2 g/L yeast extract (type D), and 20 g/L dextrose. Adjust to pH 6.5 with 10% KOH.
9. Nutrient medium ASP2: 10 g/L peptone soybean (HiMedia, India), 50 g/L sucrose, 5 g/L KH_2PO_4, and 5 g/L $MgSO_4$•$7H_2O$. Adjust to pH 6.5 with 10% KOH or 10% H_2SO_4.
10. Bioconversion medium BC: 1 g/L KH_2PO_4, 4.5 g/L K_2HPO_4•$3H_2O$, 3 g/L $(NH_4)_2SO_4$, 5 g/L glycerol, 68 g/L statistically methylated β-cyclodextrin (Cavasol W7 M, Wacker Chemie AG, Germany), 1.5 g/L silicone antifoam (for fermenters only; 30% aqueous emulsion; Sigma-Aldrich, USA), bioconversion substrate (BS). Where BS is 12 g/L of high grade commercial phytosterol containing ≥95% of convertible sterols (Sigma-Aldrich, USA; Arboris, USA) or 11.2 g/L of cholesterol with purity not less than 95% (Serva Electrophoresis GmbH, Germany).

11. KOH 10% sterile solution in distilled water (for fermenters only).
12. H_2SO_4 10% sterile solution in distilled water (for fermenters only).
13. Sterile distilled water.
14. Shake flask or fermenter.
15. Orbital shaker.
16. Benchtop centrifuge.
17. Autoclave.
18. Extraction funnel or reactor.
19. Vacuum rotary evaporator.
20. Glass beaker.
21. Porosity 3 fritted glass filtration funnel.
22. Porcelain mortar.
23. Vacuum drying oven.
24. Cyclomixer.
25. pH meter.
26. HPLC system.

2.2 Analytical Procedures

1. HPLC system minimum requirements: 1-channel pump, column oven, UV absorbance detector with 10 mm optical cell, injector, computer, and integration software (Infinity 1260, Agilent, USA or equivalent).
2. HPLC column symmetry C_{18}, 5 μm, 250 × 4.6 mm (Waters, USA or equivalent).
3. Guard column symmetry C_{18}, 20 × 3.9 mm (Waters, USA or equivalent).
4. Conditioned 50% ACN: mix equal volumes of ACN (HPLC grade) and deionized water (ASTM Type I), warm to an ambient temperature and degas by vacuum filtration through 0.45 μm pore size nylon filtration disk (47 mm).
5. Conditioned HPLC mobile phase: mix 52 volumes of ACN (HPLC grade), 48 volumes of deionized water (ASTM Type I) and 0.01 volumes of acetic acid (HPLC grade), warm to an ambient temperature and degas by vacuum filtration as described above.
6. Standard solution of AD: weight a precisely known amount (about 10 mg) of analytical standard of androst-4-en-3,17-dione (Steraloids, USA) using analytical balance and dissolve it in the conditioned 50% ACN in a 100-mL class A volumetric flask. Calculate the accurate concentration of AD (about 0.1 mg/mL).

7. Standard solution of 11α-HAD: prepare the solution of analytical standard of 11α-hydroxyandrost-4-ene-3,17-dione (Steraloids, USA) and calculate its accurate concentration as described above for AD (about 0.1 mg/mL).

3 Methods

3.1 Choosing the Bioconversion Scale

Choose the proper scale and equipment (shake flasks or fermenter) for the bioconversion stages depending from the desired amount and purity of the 11α-HAD product (*see* **Note 2**). For operations in volumes of more than 1–2 L the use of fermenter is preferable. Necessary amounts of supplementary media and seed cultures are presented as percentage from the desired scale on the first bioconversion stage.

3.2 Maintenance of Stock Cultures

Subculture *Mycobacterium* sp. on agar slants with MYCA at 30 °C for 3–5 days (*see* **Notes 3** and **4**). Subculture *A. ochraceus* on large agar slants (18 × 200 mm test tubes) with malt extract agar at 30 °C for 7–10 days (*see* **Notes 4** and **5**). 1–2 slants are sufficient for obtaining up to 5 L of bioconversion broth.

3.3 Preparation of Mycobacterium sp. Inoculum

1. Prepare sterile (120 °C, 30 min) MYCB dispensed by 75 mL to 1-L shake flasks. Total volume of the medium should consist at least 11% of the desired bioconversion scale.
2. Prepare the first generation seed culture: use one slant to inoculate 75 mL of the MYCB in one 1-L shake flask; incubate on orbital shaker (200 rpm) at 30 °C for 40–48 h (sufficient for obtaining up to 5 L of bioconversion broth) (*see* **Note 5**).
3. Prepare the second generation liquid seed culture: inoculate MYCB fresh medium with 10% (v/v) of the first generation culture incubate on orbital shaker (200 rpm) at 30 °C for 24–26 h (1 flask is sufficient for obtaining up to 1 L of bioconversion broth). Mean to prepare over 10% (v/v) of second generation inoculum from the desirable bioconversion scale (*see* **Note 6**).

3.4 First Stage of Bioconversion (Sterols to AD)

1. For the bioconversion in shake flasks prepare necessary amount of BC dispensed by 100 mL on 1-L shake flasks. Add the necessary amount of BS to each flask individually. Autoclave BC at 110 °C for 30 min. Shake well and sonicate the sterile medium on ultrasonic bath (~1 min) before inoculation.
2. For the bioconversion in a lab-scale fermenter assemble the apparatus with temperature, agitation, pH, and dissolved oxygen (DO) controls, pitched turbine impellers. Prepare the BC medium excluding BS, antifoam, and sulfates. Mean a working volume of the batch on the bioconversion first stage of about

50% of the total volume of the vessel. Load BS and antifoam directly to the fermenter and then add the rest BC excluding the sulfates; sterilize the medium in the fermenter (120 °C, 30–45 min).

Separately prepare the solution of sulfates in 0.2% (v/v) H_2SO_4 (in about 3–5% volume of the desired batch scale), pour to a feeding bottle, autoclave (110 °C, 30 min) and add to the sterilized fermenter. Prepare sterile 10% KOH solution in a feeding bottle (about ~2% volume of the desired batch scale), join to the fermenter and use for pH control of the broth. Equilibrate the temperature and DO in the fermenter before inoculation.

3. Inoculate the BC with 10% (v/v) of fresh second generation *Mycobacterium* sp.inoculum.
4. Incubate the shake flasks aerobically on an orbital shaker at 30 °C, 200 rpm.

   In the fermenter run the batch at 30 °C, pH 6.8–6.9, DO 35–40%, air flow 0.5–0.14 vvm (volume of air per volume of culture broth per minute).
5. Continue the biotransformation process for about 72 h in the shake flasks or 50–60 h in the fermenter until stop of AD accumulation. Follow the titer of AD by HPLC in samples of bioconversion broth taken at 24 h intervals (shake flasks) or 10–12-h intervals (fermenter) (*see* Subheading 3.11). *See* Fig. 3 for the reference bioconversion dynamics in fermenter.

3.5 Inactivation of the Bioconversion Broth

After stabilization of AD titer at a maximum level (normally 5.7 ± 0.2 g/L or 68 ± 1% of molar yield), inactivate the mycobacterial culture by pasteurization:

1. In the shake flasks: remove the flasks from the shaker, incubate on hot water bath (about 90 °C) for 20 min, and cool down to 25–30 °C.
2. In the fermenter: decrease air flow to ~0.01vvm. Fast (≤30 min) heat the broth inside the fermenter vessel to 70 °C, incubate at that temperature for 20 min, and cool down to 30 °C. Increase air flow and saturate the broth with dissolved oxygen. Adjust pH of the broth to 6.0 using the sterile solution of H_2SO_4 (or KOH). The broth is ready for the stage 2 of bioconversion.

3.6 Preparation of Aspergillus mycelium

1. Prepare sterile ASP1 (110 °C, 30 min) and ASP2 (120 °C, 30 min) dispensed by 75 mL to 1-L shake flasks.
2. Inoculate one ASP1 shake flask with *A. ochraceus* spores and aerial mycelium fragments from one slant. Incubate on orbital shaker (200 rpm) at 30 °C for 24–26 h for the first generation liquid culture (*see* **Notes 5** and 7).

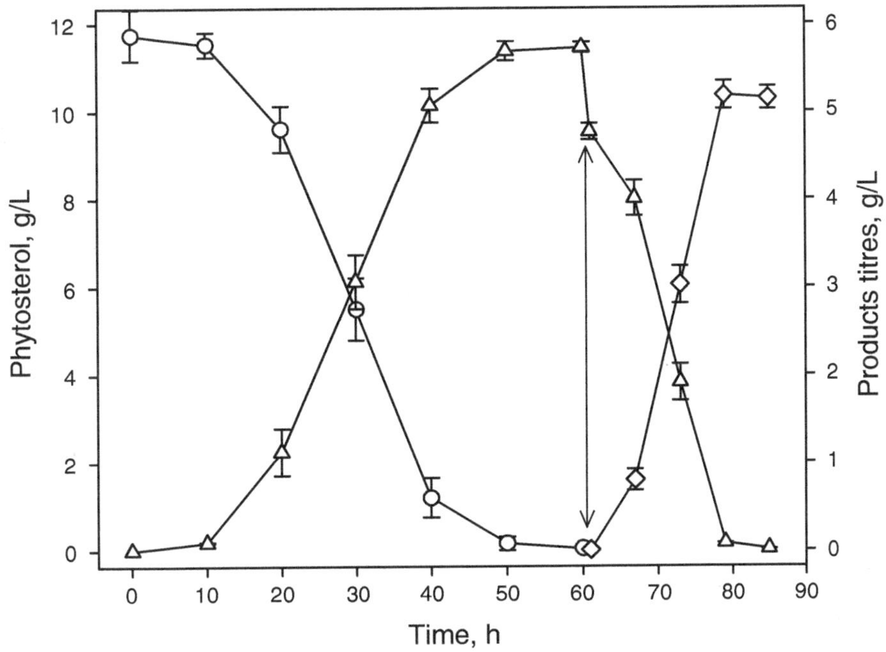

Fig. 3 Reference dynamics for the two-stage bioconversion of phytosterol in a laboratory-scale fermenter with working volume 5–6 L: ○—phytosterol, Δ—AD, ◇—11α-HAD; *arrow* indicates the point of broth inactivation and addition of the fungal biocatalyst

3. Inoculate ASP2 with 5% (v/v) of the first generation culture. Mean to prepare not less than 20% (v/v) of the second generation liquid culture from the scale on the first bioconversion stage. Incubate the flasks on orbital shaker (200 rpm) at 30 °C for 42–48 h to obtain the active mature mycelium of *A. ochraceus* in the second generation liquid culture (*see* **Note 8**).

3.7 Second Stage of Bioconversion (AD to 11α-HAD)

1. For the bioconversion in shake flasks load 20 mL of the freshly obtained *A. ochraceus* second generation liquid culture to each flask with the inactivated first stage bioconversion broth. Incubate the flask on orbital shaker (200 rpm) at 30 °C for 24–26 h.
2. For lab-scale fermenter: load 20% (v/v) of the freshly obtained *A. ochraceus* second generation liquid culture to the fermenter with inactivated first stage bioconversion broth prepared as per 3.5 (e.g., add 0.3 L of the culture for 1.5 L of the first stage bioconversion broth).
3. Run the second stage batch for 12–18 h at 30 °C, pH 5.9–6.1, and DO > 10% (preferably 20%). Increase air flow from about 0.15 vvm at the beginning to about 0.4 vvm at a maximum activity period. Apply the necessary agitation to keep and adjust the required DO. Continuously adjust pH with sterile 10%

KOH and sterile 10% H_2SO_4. Follow the titer of 11α-HAD by HPLC in samples of bioconversion broth taken in 6-h intervals (shake flasks) or 3–6-h intervals (fermenter) (*see* Subheading 3.11). *See* Fig. 3 for the reference bioconversion dynamics in fermenter.

4. After 11α-HAD titer reaches maximum level (normally 4.9 ± 0.2 g/L or 67 ± 1% of molar yield from phytosterol or cholesterol), inactivate the mycelium by pasteurization as described in Subheading 3.5.
5. After cooling down the broth to 25–30 °C, totally gather it for biomass separation. Rinse the vessels with small amount of distilled water if necessary.

3.8 Separation of Biomass

1. Distribute all the broth to a proper scale centrifuge bottles and spin on a laboratory centrifuge at 4000–6000 × *g* and 20–30 °C for 1 h.
2. Decant immediately the supernatant.
3. Kill the biomass in an autoclave (120 °C, 30–45 min).

3.9 Extraction of 11α-HAD

1. Load the supernatant of bioconversion broth to a proper scale extraction funnel (shake flasks) or extraction reactor (shake flasks or fermenter).
2. Add 0.55 volumes of ethyl acetate (research grade) and shake or agitate vigorously for 30 min to 1 h (preferably) for organic solvent extraction.
3. Keep the mixture without agitation until obtain a good separation of the phases.
4. Collect the upper (organic) phase I. Place the aqueous phase back to the reactor (funnel) (*see***Note 9**).
5. Repeat the extraction two times more as described, combining the organic phases I–III. Discard the aqueous phase.

3.10 Purification of 11α-HAD

1. Concentrate the organic phases on a vacuum rotary evaporator at water bath temperature 50–60 °C till about 0.6–0.65 × volume of the first stage bioconversion broth. Do not cool down the concentrate below 25 °C. If any precipitate appeared, warm the concentrate on a water bath until it is fully dissolved.
2. Pour the concentrate to an extraction funnel (or extraction reactor), add warm (25–30 °C) distilled water (0.1 × volume of the first stage bioconversion broth) and wash the concentrate under vigorous shaking or agitation for 30 min.
3. Keep the mixture without agitation until obtain a good separation of the phases. Collect the upper (organic) phase IV. Discard the aqueous phase.

4. Repeat washing once more as in the operations 2–3 using warm distilled water (0.05 × volume of the first stage bioconversion broth). Collect the organic phase V.
5. Concentrate the organic phase V on a vacuum rotary evaporator at water bath temperature 50–60 °C till about 0.08 × volume of the first stage bioconversion broth.
6. Pour the hot concentrate to a glass beaker immediately, cool down to room temperature, cover and place in a refrigerator for crystallization overnight (≥8 h) at 4 °C.
7. Suspend the formed 11α-HAD crystals with glass rod and separate them by vacuum filtration on porosity 3 fritted glass filtration funnel. Discard the filtrate (mother solution I).
8. Keep the crystals on the filter under fume hood for drying for 2–3 h. Gather the crystals with spatula and slightly grind in a porcelain mortar. Then dry in a vacuum drying oven at 60 °C till constant weight (about 2 h).
9. Check purity of the preparation I by HPLC. Normally in should be ≥90%.
10. If necessary, continue purification by recrystallization from ethyl acetate as follows.
11. Totally dissolve the preparation I in hot (near the boiling point) ethyl acetate taking 40 mL of the solvent for 1 g of the substance. Repeat operations as described from 6 to 8. Collect the filtrate (mother solution II).
12. Check the purity of the preparation II by HPLC. Normally it should be not less than 95%.
13. If necessary continue purification by recrystallization from ethyl acetate as described in **step 11**. Combine the mother solutions II and III (*see***Note 10**).
14. Check the purity of the preparation III by HPLC. Normally it should be not less than 99%. If necessary continue purification by recrystallization from ethyl acetate as described for the operation 11.

3.11 HPLC Analysis of Steroids

1. Run and equilibrate the HPLC system in the following conditions: column temperature 50 °C, mobile phase flow rate 1 mL/min; detection wavelength 240 nm (preferably) or 254 nm, acquisition times are 10 min for the standards and 30 min for broth samples.
2. Consecutively inject 5, 10 and 20 μL of the AD and 11α-HAD standard solutions, register stable peak signals (*see***Note 11**) and built linear calibration curves considering the precise concentration of compounds in the standard solutions.
3. Prepare bioconversion broth samples as follows: shake the sample on cyclomixer, take precise aliquot immediately and

dilute 1:50 with the conditioned 50% ACN. Spin the dilution at $\geq 1000 \times g$ for 15 min in a tightly closed centrifuge tube. Take out and keep the tube in vertical position.

4. Apply the supernatant for injection to the HPLC system (20 μL). Register and integrate the chromatogram. Calculate AD and 11α-HAD titers using the calibration curves obtained and the sample dilution factor.
5. For purity estimation of the crystalline product make the test solution of the obtained preparation in the same way as described for the standard solutions (see Subheading 2.2) and apply it for the injection (20 μL). Prepare the solution in duplicate or triplicate for better results. Calculate 11α-HAD content in the powder using the calibration curves obtained and the precise concentrations of the substance in the test solution.

4 Notes

1. For MYCA, MYCB and BC, dissolve the sulfates separately from other compounds in water and then mix.
2. Depending from the desired purity of the substance the method allows to obtain about 4.8 g (>90% purity), 3.3 g (>95% purity), or 2.6 g (>99% purity) of 11α-HAD from 1 L of bioconversion broth.
3. For inoculation of first generation liquid culture of *Mycobacterium* sp. growth from one agar slant should be totally suspended in 3–4 mL of the sterile medium and transferred to one shake flask.
4. *Mycobacterium* sp. slants can be stored at +4 °C within 2 weeks. *A. ochraceus* slants can be stored at +4 °C within 1 month. The first generation liquid cultures can be stored at +4 °C for 24 h (the second cultivation should be prolonged by ~2–4 h in this case).
5. For inoculation of agar slants and first generation liquid culture of *A. ochraceus* suspend the spores and aerial mycelium fragments in 4–5 mL of sterile water by scraping the growth with a hard loop.
6. OD_{600} of the second generation *Mycobacterium* sp. culture should be 8–10 units; wet biomass 20–25 g/L.
7. Cultivation of the *A. ochraceus* in shake flasks should be synchronized with first stage of the bioconversion. The mature mycelium of *A. ochraceus* should be applied for 11α-hydroxylation of AD at the second stage of bioconversion immediately after termination of the cultivation process to preserve the activity. By-turn the inactivated bioconversion broth after completion of first stage can be stored in the bioreactor

for few hours at 30 °C or up to 1 day if being chilled to 10–15 °C.

8. The content of *A. ochraceus* mycelium in the second generation liquid culture should be normally 120 ± 20 g/L (wet) or about 14 g/L (dried to a constant weight at 70 °C) that corresponds to early stationary phase.
9. Do all manipulations with organic solvents under fume hood and comply with fire safety rules.
10. Additional amount of the substance (>95% purity) can be obtained if necessary from the mother solutions II–III as follows. Vacuum evaporate the mother solutions to ~1/6 of the initial volume, get crystals as described in Subheading 3.10, **steps 6–8**, recrystallize as per **step 10** and check purity of the preparation by HPLC.
11. Reference retention times (Rt) at HPLC are 8.51 min for AD and 3.91 min for 11α-HAD.

Acknowledgment

This work was supported by Russian Science Foundation grant (14-24-00169) and ERA-IB (Project: MySterI).

References

1. Huy LD, Diep NT, Nhung LTK (2012) Alternative road for synthesis of 16α,17α-epoxy-pregn-4,9(11)-dien-21-ol-3,20-dione from 9α-hydroxy-androstenedione. Pharm Chem J 46:356–359
2. Tong WY, Dong X (2009) Microbial biotransformation: recent developments on steroid drugs. Recent Pat Biotechnol 3:141–153
3. Kang G, Wang Y, Liu J, Wu J, Zhao M, Li G, Li N, Peng L, Zhang X, Li L, Maira N, Peng S (2012) Development of three-component conjugates: to get nano-globes with porous surfaces, high in vivo anti-osteoporosis activity and minimal side effects. J Mater Chem 22:21740–21748
4. Donova MV, Egorova OV (2012) Microbial steroid transformations: current state and prospects. Appl Microbiol Biotechnol 94:1423–1447
5. Marsheck WJ, Kraychy S, Muir RD (1972) Microbial degradation of sterols. Appl Microbiol 23:72–77
6. Donova MV, Dovbnya DV, Kalinichenko AN, Arinbasarova AY, Morozova ZV, Vagabova LM, Koshcheyenko KA (1997) The method of androst-4-en-3,17-dione production. RU Patent 20779258
7. Dovbnya D, Kollerov V, Khomutov S, Malov D, Donova M (2014) A two-step one-pot bioprocess for production of 11α-hydroxyandrost-4-ene-3,17-dione from phytosterol. New Biotechnol 315:S119–S120
8. Liu C, Zhang X, Rao Z-M et al (2015) Mutation breeding of high 4-androstene-3,17-dione-producing *Mycobacterium neoaurum* ZADF-4 by atmospheric and room temperature plasma treatment. J Zhejiang Univ Sci B 16:286–295
9. Liu Y, Chen G, Ge F, Li W, Zeng L, Cao W (2010) Efficient biotransformation of cholesterol to androsta-1,4-diene-3,17-dione by a newly isolated actinomycete *Gordonia neofelifaecis*. World J Microb Biot 27:759–765
10. Bolten SL (2007) *Aspergillus ochraceus* 11 alpha hydroxylase and oxidoreductase. US Patent 7,033,807
11. Hosseinabadi T, Vahidi H, Nickavar B, Kobarfard F (2014) Fungal transformation of androsta-1,4-diene-3,17-dione by *Aspergillus brasiliensis*. Daru 22:71
12. Xiong Z, Wei Q, Chen H, Chen S, Xu W, Qiu G, Liang S, Hu X (2006) Microbial transformation of androst-4-ene-3,17-dione by *Beauveria bassiana*. Steroids 71:979–983

Chapter 19

Steroid 11-Alpha-Hydroxylation by the Fungi *Aspergillus nidulans* and *Aspergillus ochraceus*

Lidia Ortega-de los Ríos, José M. Luengo, and José M. Fernández-Cañón

Abstract

Steroids are a group of natural compounds derived from the cyclopentane-perhydro-phenantrene nucleus that have a great interest for the pharmaceutical industries as a consequence of their physiological effects. Among their functions are anti-inflammatory, immunosuppressive, or contraceptive activities. Nowadays, microbial transformation of steroid precursors is winning relevance opposite to the chemical synthesis, since it allows for decreasing time, expenses, and environmental pollution. Pharmaceutical industry tends to use cholesterol and phytosterols as starting materials due to their low cost. *Aspergillus ochraceus* and *Aspergillus nidulans*, a fungus whose biochemistry and genetics are well known, have been chosen because of their capacity of 11-α-hydroxylation over some steroids which confers on them their anti-inflammatory properties. We have cloned the genes encoding the 11-α-hydroxylase enzymatic activities with the aim to introduce them in other microorganisms, such as *Mycobacterium smegmatis*, used in the industry to split the side chain of phytosterols, and thus creating recombinant microorganisms able to generate useful steroids from cheap precursors in just one-step fermentation.

Key words Steroids, *Aspergillus*, Fermentation, CYP450, Hydroxylation, Biotransformation

1 Introduction

Steroids carry a distinguished commercial value because they are employed in medicine as anti-inflammatories or to fight a big number of allergic reactions, especially in the skin, with corticoids. They are also used as contraceptives, anabolic steroids, in hormone therapies or as immunosuppressant agents. Furthermore, they are also essential to face some autoimmune diseases or some types of cancer [1]. Therefore, steroid production has a prominent place in the pharmaceutical sector. Nowadays, their synthesis is based in the combination of microbiological and chemical techniques. Traditionally, sapogenins, phytosterols, and cholesterols have been steroid precursors in the pharmaceutical industry although these days the tendency is changing only to the last two substrates due to

José-Luis Barredo, Ignacio Herráiz (eds.), *Microbial Steroids: Methods and Protocols*, Methods in Molecular Biology vol. 1645, DOI 10.1007/978-1-4939-7183-1_19, © Springer Science+Business Media LLC 2017

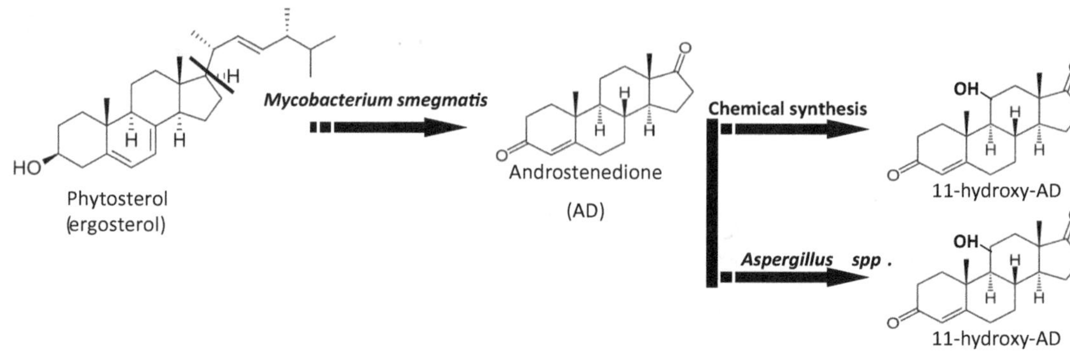

Fig. 1 Schematic representation of the process followed in pharmaceutical industry for the transformation of cholesterol in 11-hydroxy-AD

their low cost and transformation facility to steroid intermediaries by certain microorganisms [2].

The synthesis of steroids starts by the biotransformation of phytosterols (obtained from soy) or cholesterol (originated from pig fat) [3], into molecules such as progesterone or androstenedione (AD) after splitting their C17 side chains. This transformation is, generally, carried out by *Mycobacterium smegmatis* [4, 5]. Subsequently, it is necessary the introduction of certain chemical modifications to lead the synthesis of compounds of commercial interest. One of the most important modifications consists of the introduction of a hydroxyl group in C-11 over the sterane nucleus since this modification gives the molecule its anti-inflammatory properties. Hydroxylation may be accomplished by chemical synthesis, a process more time-consuming and with higher costs [6], or by a second biotransformation performed by a member of the genus *Aspergillus* spp. in which the presence of a P450 cytochrome enzyme with this activity has been widely reported [7]. However, the need of a second fermentation implies the purification of intermediary products, further addition of compounds and nutrients as well as increasing time and cost to obtain the final product. For this reason, the cloning of the gene encoding 11-α-hydroxylase and its expression in *M. smegmatis* would give rise to a microorganism able to develop both transformations in a one-step fermentation. A schematic representation of the procedure followed by the pharmaceutical companies for the synthesis of steroids is shown in Fig. 1.

We have cloned the gene encoding the 11-α-hydroxylase of *A. nidulans* following the method of suppression subtractive hybridization (SSH) described by Diatchenko et al. in 1996 [8]. Following this protocol, we have also been able to clone some genes encoding other CYP450 proteins [9, 10]. The screening of the candidate genes was made by RT-PCR using cDNAs coming from noninduced and induced mycelium (cultured in media containing progesterone, AD, or testosterone). Definitive testing was possible thanks to the creation of a knockout strain where we interrupted the candidate gene with the *pyrG* gene from *A. fumigatus* according to a technique of fusion PCR described by Szewczyk et al. in 2006 [11].

Once we cloned the gene encoding 11-α-hydroxylase of *A. nidulans*, we proceed to clone the homologous gene of *A. ochraceus*, the industrial producer, based on the identity to the *A. nidulans* gene. As the genome of *A. ochraceus* has not been described yet in the data bases, degenerated primers were built with the corresponding sequence to the gene encoding the 11-α-hydroxylase in *A. nidulans, A. flavus, A. niger*, and *A. oryzae*. In this way, we found the gene responsible for the 11-α-hydroxylation of *A. ochraceus*, which finally showed 62% of identity with the homologous gene of *A. nidulans*.

2 Materials

2.1 Growth and Sporulation Media

1. Minimal Medium Liquid (MML): 10 mM KH_2PO_4, 20 mM $(NH_4)_2SO_4$, 1 mM $MgSO_4 \cdot 7(H_2O)$, 50 mM glucose, and Hutner's trace elements 1 mL/L. pH 7 is adjusted with KOH and sterilized 120 °C, 15 min.
2. Minimal Medium Solid (MMS): MML and 2% agar (w/v).
3. Hutner's trace elements: 2.2 g $ZnSO_4 \cdot 7(H_2O)$, 1.1 g H_3BO_3, 0.5 g $MnCl_2 \cdot 4(H_2O)$, 0.5 g $FeSO_4 \cdot 7(H_2O)$, 0.16 g $CoCl_2 \cdot 6(H_2O)$, 0.16 g $CuS0_4 \cdot 5(H_2O)$, 0.11 g (NH_4) $Mo_7O_{24} \cdot 4(H_2O)$, and 5 g EDTA per 100 mL of distilled water.
4. Complete medium: Per liter of distilled water: 1.75 g malt extract, 2.75 g dextrin, 2.35 g glycerol, 0.78 g gelatin peptone, and 15 g bacteriological agar (Extract malt agar, Pronadisa).
5. Fermentation medium: MML with 100 mM KH_2PO_4 (instead of 10 mM KH_2PO_4).

2.2 HPLC

1. HPLC column Kromasil 100 C18 5 μm 25 × 0.46 cm.
2. Guard column: Nucleosil C18 5 μm 1 × 0.46 cm.
3. HPLC Alliance 2690, photodiode detector array 996, Waters.
4. HPLC software Empower3, Waters.
5. Mobile phase: acetonitrile–water (60:40).

2.3 Suppression Subtractive Hybridization (SSH) and Fusion PCR

1. TRIzol (TRI; Sigma).
2. Magnetic bead enrichment kit (Mag-Bind mRNA Kit; VWR Omega).
3. Double-Stranded cDNA Synthesis Kit (Invitrogen).
4. NucleoSpin gel and PCR cleanup Kit (Macherey-Nagel).

5. SSH buffer: 50 mM Tris–HCl pH 8.3, 0.5 M NaCl, 20 mM EDTA, and 10% wt/vol PEG 8000.
6. NucleoSpin Plasmid, plasmid Miniprep kit (Macherey-Nagel).
7. Phire Hot Start II DNA Polymerase (Fisher).
8. Oligonucleotide primers.
9. Maxima Minus H First Strand cDNA Synthesis Kit (Fisher).
10. *E. coli* XL1-Blue Electroporation-Competent Cells, *recA1 endA1 gyrA96 thi-1 hsdR17 supE44 relA1 lac* [F^-*proAB lacI*^{q} ZΔ*M15* Tn10(Tet^r)].
11. Ampicillin sodium salt 150 μg/mL.
12. X-Gal stock concentration: 2% in dimethylformamide.
13. Luria–Bertani (LB) broth: 10 g tryptone, 5 g yeast extract, and 5 g NaCl per litre of water, pH adjusted to 7.0 with 2 M NaOH.
14. LB solid medium: Add 1.5% of agar (w/v) to the LB broth.
15. SOC medium: 2% tryptone, 0.5% yeast extract, 10 mM NaCl, 2.5 mM KCl, 10 mM $MgCl_2{\cdot}(H_2O)_6$, 10 mM $MgSO_4{\cdot}(H_2O)_7$, and 20 mM glucose.

2.4 Fungal Transformation

1. Protoplast buffer (TP): 1.2 M $MgSO_4$ and 10 mM NaH_2PO_4, pH 5.8.
2. Lytic enzyme: Glucanex (Sigma), 100 mg/g mycelium.
3. Overlaying solution: 0.6 M sorbitol and 10 mM Tris–HCl, pH 7.5.
4. ST: 1 M sorbitol and 10 mM Tris–HCl pH 7.5.
5. STC: ST and 10 mM $CaCl_2$.
6. PEG: 60% polyethyleneglycol 6000, 10 mM Tris–HCl pH 7.5, and 10 mM $CaCl_2$.
7. Regeneration media (MR): MMS and 1 M sucrose.

2.5 Laboratory Equipment

1. Petri plates.
2. Erlenmeyer flasks.
3. Incubator Shaker (Innova 44, New Brunswick Scientific or equivalent).
4. Incubator.
5. Centrifugal evaporator Vacuum Concentrator (Jouan RC 1010 or equivalent).
6. 13 mm, PTFE membrane 0.22 μm (Jet Biofil)
7. Microscope.
8. Microscope Slides Lames Porte-Object.
9. Microscope cover glass.

10. Bench centrifuge 5415D (Eppendorf or equivalent).
11. Allegra X-22R Centrifuge (Beckman Coulter or equivalent).
12. Laminar flow cabinet.
13. Thermal cycler.
14. Ultrasounds.
15. Digital Sonifier.
16. UV–Vis spectrophotometer.
17. MicroPulser electroporation (Bio-Rad or equivalent).
18. pH meter.
19. Electronic precision balance.
20. High-Current Power Supply.

2.6 Chemicals

1. Ethanol.
2. Chloroform.
3. Phenol liquefied–Tris–HCl, pH 8.
4. Phenol–chloroform–isoamyl alcohol, pH 5.2.
5. Ethyl acetate
6. Acetonitrile.
7. 2-Propanol.
8. Androstenedione (AD) (Gadea Biopharma).
9. Testosterone (Sigma).
10. Progesterone (Sigma).
11. 11-OH-androstenedione (Gadea Biopharma).
12. 11-OH-progesterone (Gadea Biopharma).
13. EGD (Gadea Biopharma).
14. Canrenone (Gadea Biopharma).

3 Methods

3.1 Growth of Aspergillus

1. Seed the strains of *A. nidulans* in MML or MMS containing 1% glucose as a carbon source. When necessary, additional nutrients must be added (*see***Note 1**).
2. Grow the strains by incubation at 37 °C during 24 h (*see* **Note 2**).

3.2 Sporulation of Aspergillus

1. Seed the strains of *A. nidulans* in complete media.
2. Grow the strains by incubation at 37 °C during 72 h (*see* **Note 2**).
3. Collect the spores by scraping the surface of the plates and resuspending them in 0.1% Tween 80.

3.3 Fermentation

1. Inoculate about 10^6 spores/mL of *A. nidulans* or *A. ochraceus* in fermentation medium containing 1% glucose.
2. Incubate in orbital incubator the cultures at 37 °C and 250 rpm (*A. nidulans*) or 27 °C and 250 rpm (*A. ochraceus*).
3. After 24 h of growth, add the steroid compounds (1 g/L of medium).
4. Take samples (5 mL) at day 0 (20 min after steroid addition), day 1 and day 2.
5. Preserve the samples frozen (−80 °C) until their processing.

3.4 HPLC

1. Extract the samples with ethyl acetate (1:1 vol.).
2. Evaporate the solvent at 40 °C and 275 × *g* in the centrifugal evaporator.
3. Dissolve the resulting pellets in absolute ethanol.
4. Filtrate through PTFE membrane 0.22 μm filter before to introduce in the HPLC vials and analyze by HPLC.
5. To analyze the samples, use acetonitrile–water (60:40) as a mobile phase with a flow of 1.5 mL/min and read them at 240 nm.

3.5 Suppression Subtractive Hybridization (SSH)

In order to identify the DNA sequence codifying the 11-alpha-hydroxylase of *A. nidulans*, a subtraction procedure was followed [8]. This technique consists of four steps:

1. Enzyme induction: searching for the conditions that allowed and do not allow the induction of the enzyme.
2. RNA and mRNA isolation and cDNA synthesis.
3. Suppression subtractive hybridization (SSH): selection of cDNAs that only appear under induced conditions.
4. Cloning and sequencing: cloning of the subtractive sequences in a cloning vector and their subsequent sequencing to compare with the database.

3.5.1 Induction

1. Incubate two cultures with 100 mL of fermentation medium and seed with spores of the wild-type strain of *A. nidulans*.
2. After 12 h of growth, add 150 mg of androstenedione (AD) to the induced RNA (+) culture, and nothing to the noninduced RNA (−) culture.
3. After another 12 h of incubation, collect the mycelium of both flasks by filtration.

3.5.2 RNA and mRNA Isolation and cDNA Synthesis

1. Collect total RNA using TRIzol, according to the manufacturer's protocol.
2. Precipitate total RNA with 3 M sodium acetate and resuspend in DEPC water.

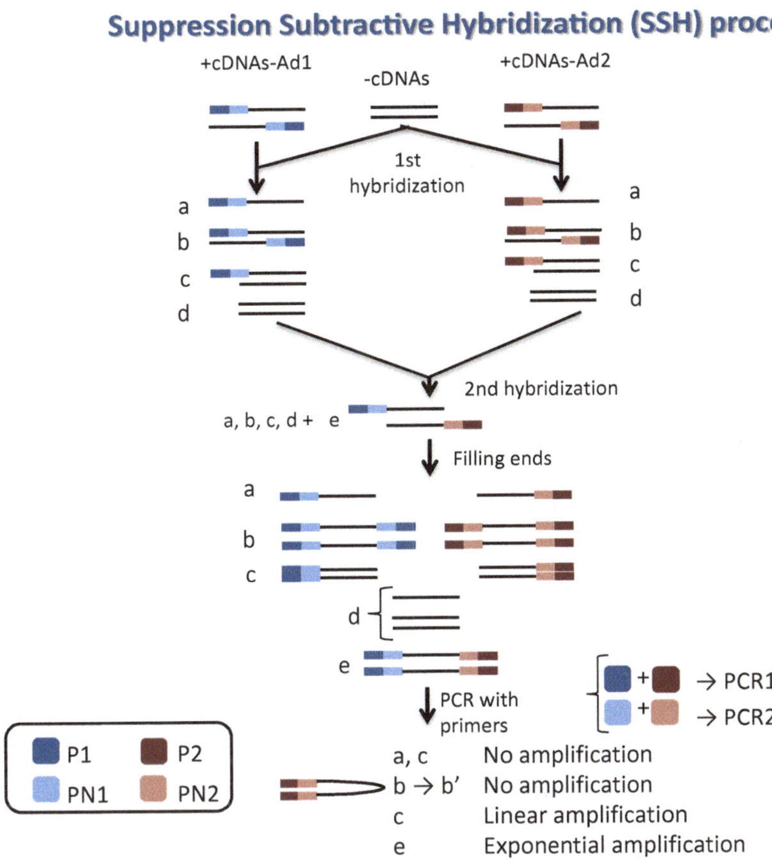

Fig. 2 Schematic representation of the hybridizations occurred in the suppression subtractive hybridization [8]

3. Purify mRNA with the Magnetic Bead Enrichment kit, starting with 110 μg of total RNA of each type.
4. Synthesize the double-stranded cDNA with 2 μg from each mRNA using the Double-Stranded cDNA Synthesis kit.

3.5.3 Suppression Subtractive Hybridization

SSH is based on the hybridization of complementary induced cDNAs and noninduced cDNAs. In this way, the sequences that are present in induced and noninduced population of cDNAs will be removed, and only those that appear as a consequence of the presence of AD (inducer) will be preserved (they will not be hybridized and will remain as a single stranded cDNA) [8].

Selection of induced cDNAs (+cDNAs), nonhybridized, is possible thanks to the joining of specific adaptors which allow for an exponential amplification of nonhybridized +cDNAs (Fig. 2). After first hybridization (with excess of noninduced cDNA) single +cDNAs with adaptor 1 and adaptor 2, separately, are mixed. After the second hybridization, there will be a-type strands complementary between them that are ligated to adaptors 1 and 2, molecules type *e*. Thanks to the presence of both adaptors, these type of strands will have an exponential amplification when using the specific primers; b-type molecules are not amplified due to internal complementary annealing (Fig. 2).

Adaptor Production and Ligation to Inductive cDNA

1. Design specific primers that will form adaptor 1, on the one hand, and adaptor 2, on the other hand.
2. Ligate the adaptors, separately, to the +cDNA using T4 ligase.
3. Incubate at 19 °C during 15 h and purified the reaction with the PCR cleanup kit.

Subtraction

The subtraction phase is composed of two parts: first hybridization and second hybridization.

1. For the first hybridization, ligate noninduced cDNA with induced cDNA to adaptor 1 (ad1).
2. Additionally, ligate noninduced cDNA with induced cDNA to adaptor 2 (ad2).
3. Precipitate separately with 3 M sodium acetate and resuspend in 5 μL of SSH buffer.
4. Denaturalize both samples at 98 °C for 5 min.
5. Hybridize both samples at 68 °C for 24 h.
6. For the second hybridization, add to each tube 5 μL of hybridization buffer.
7. Add noninduced cDNA (*see* **Note 3**).
8. Hybridize both tubes together at 68 °C for 30 h (*see* **Note 4**).
9. Add 10 μL of TE and 10 μL of chloroform.
10. Centrifuge at 10,000 × *g* for 3 min to remove the mineral oil.
11. Amplify the subtractive +cDNA by PCR according to the following conditions (*see* **Note 5**):
    (a) PCR 1 with primers (to 0.2 μM) corresponding to the exterior part of each adaptor, P1 and P2, where initial extension is required to fill ends, as Fig. 2 shows. Conditions: 75°5′, (98°5″, 68°10″, 72°30″) × 30, 72°2′.
    (b) PCR 2 with primers corresponding to interior part of adaptors, PN1 and PN2, respectively. PCR1 product is diluted ten times and we use 1 μL to carry out PCR2. In this reaction we change primers by PN1 and PN2. Conditions: 98°2′, (98°5″, 72°35″) × 20, 72°2′.

3.5.4 Cloning and Sequencing

The cloning procedure to identify those cDNAs that only appear as a consequence of presence of AD is as follows:

1. Purify PCR2 product with PCR cleanup kit.
2. Ligate to pBS opened with *Eco*RV.

1. cDNA non-induced for A gene
2. cDNA non-induced for *GPD* gene
3. cDNA induced with AD for *A* gene
4. cDNA induced with AD for *GPD* gene
5. cDNA induced with progesterone for *A* gene
6. cDNA induced with progesterone for *GPD* gene
7. cDNA induced with testosterone for *A* gene
8. cDNA induced with testosterone for *GPD* gene

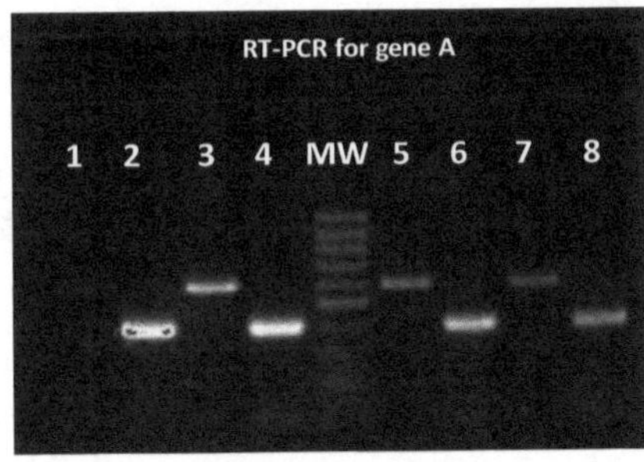

Fig. 3 Agarose gel of RT-PCR fragments of *A* gene resulted from cultures noninduced (*1* and *2*) and induced (*3–8*) with different steroids

3. Transform *E. coli* and select in ampicillin LB plates with X-Gal.
4. Test transformant clones by lysis.
5. Sequence the clones (*see* **Note 6**).

3.6 Verification of the Suitability of the Candidate Genes Found

To ensure which one of the candidate genes was responsible for the 11-alpha-hydroxilation, two tests were accomplished. The first one consisted of an initial sifting where four candidate genes were chosen and an RT-PCR was run to see which one was amplified under inducing conditions. Nonetheless, the definitive test that let us know if the candidate gene was the right one was the fulfillment of a knockout strain which lacked the 11-alpha hydroxylase function.

3.6.1 RT-PCR

1. Design specific primers for each candidate gene.
2. Make RT-PCR with four types of RNA: Noninduced RNA, AD-induced RNA, progesterone-induced RNA, and testosterone-induced RNA.
3. Do the synthesis of single-stranded cDNA using Maxima Minus H First Strand cDNA Synthesis Kit: (1) Start with 2.2 μg RNA with oligodT. (2) Final cDNA volume of 20 μL. (3) Use 2 μL of cDNA to fulfil a 25 μL PCR.
4. Develop a PCR with Taq Polymerase and 0.4 μM primers in the following conditions: 95°2′, (95°20″, 55°30″, 72°45″) × 33, 72°5′. Besides the specific primers designed for each candidate gene, use specific primers for the glyceraldehyde 3-phosphate dehydrogenase (GPD) gene in which, as it is constitutive, it should be amplification with all kind of RNAs.

Figure 3 shows an agarose gel from the RT-PCR for *A* gene (encoding the 11-alpha-hydroxylase from *A. nidulans*), where there is no amplification with noninduced cDNA while there is with the three induced types of cDNAs. However, as expected, GPD gene is amplified in all cases. The same procedure was made for the rest candidate genes but in those cases there was the same amplification in the induced and noninduced cDNAs was observed.

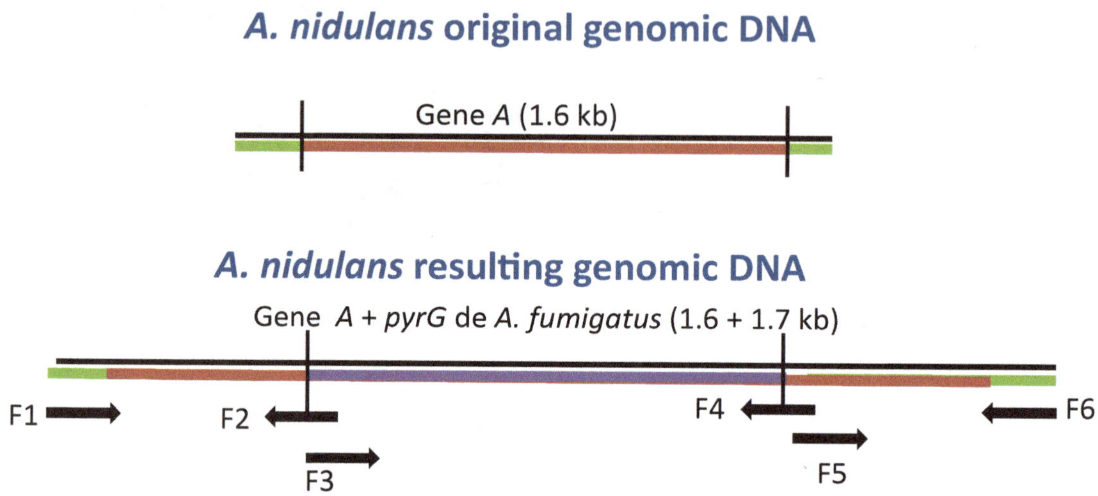

Fig. 4 Schematic representation of the fusion PCR DNA construction that allows the collection of knockout strain affecting the *A* gene

3.6.2 Fusion PCR

Fusion PCR was employed to get a DNA fragment to make a knockout strain with a nonfunctional 11-alpha-hydroxylase activity. The knockout strain was made following a protocol described by Szewczyk et al. [11]. Using this technique, identification of the flanked regions to the *A* gene (*see* **Note 7**) is needed to create a linear construction in which, besides these flanked regions, *A* gene is disrupted with a selection marker gene. This way, by homologous recombination, this linear construction will interrupt the *A* gene in the genome, thus generating a knockout strain. In our case, the selection marker chosen is the *pyrG* of *A. fumigatus.* We selected the *A. fumigatus* gene instead of *A. nidulans* to avoid homologous recombination. Figure 4 shows the construction that we used from the genomic DNA of *A. nidulans.*

To carry out the knockout, a strain deficient in the nonhomologous binding region gene, denominated *nkuA,* was used (*A. nidulans nkuA*Δ). This strain needs the presence of pyrimidines (*pyrG89*) together with pyridoxine (*pyroA4*) to grow and it also has glufosinate resistance which is the gene used to make the *nkuA* knockout.

Fusion PCR procedure

1. Amplification of flanked regions:
   (a) Amplify the P1 region with F1 and F2 primers. F2 primer includes the beginning sequence of the *pyrG* gene.
   (b) Amplify the P2 region with F5 and F6 primers. F5 primer includes the final sequence of the *pyrG* gene.
   (c) Amplify the *pyrG* gene (P3) from *A. fumigatus* with F3 and F4 primers.

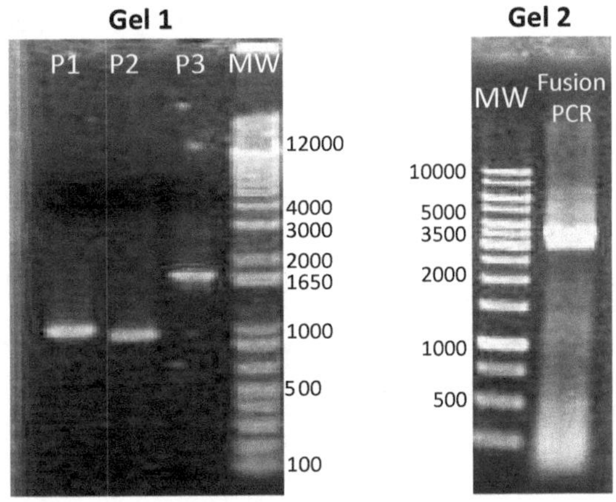

Fig. 5 Agarose gel electrophoresis of the fragments P1, P2, and P3 (gel 1) and fusion PCR product resulting from P1 + P2 + P3 (gel 2)

(d) Use Phire Hot Start II DNA Polymerase and primers to 0.3 μM. Conditions: 98°2′, (98°5″, 62.5°10″, 72°35″) × 35, 72°2′.

(e) Purify each fragment with a 1% agarose gel (Fig. 5) using NucleoSpin gel and PCR cleanup Kit.

2. Fusion PCR: 0.7 ng of P1, 0.7 ng of P2, and 1.3 ng of P3 (since it is almost double in length compared to P1 and P2).

   (a) Use Phire Hot Start II DNA Polymerase and primers to 0.4 μM. Conditions: 98°2′, (98°5″, 62.5°10″, 72°1.15″) × 35, 72°2′.

   (b) 1% agarose gel (Fig. 5) with 25 μL of fusion PCR product that has a length of nearly 5 kb (we expected a 5.3 kb band).

   (c) Purify fusion fragment with NucleoSpin gel and PCR cleanup.

   (d) Transform the *nkuA*Δ *A. nidulans* strain using this construction.

3.7 Fungal Transformation

Fungal transformation is made according to the protocol described by Tilburn in 1983 and Yelton [12, 13]. It consists of a fusion between protoplasts and plasmid or DNA fragment. Protoplasts are collected from mycelium by incubation with a lytic enzyme, glucuronidase.

3.7.1 Procedure

1. Grow mycelium (25–30 °C) (*see***Note 8**) and collect by filtration.
2. Wash mycelium with distilled water.
3. Resuspend mycelium TP at a concetration of 20 mL/g together with 100 mg/g of lytic enzyme and 1 mg/mL TP of BSA.

4. Incubate at 30 °C for about 90–120 min to let the enzyme activity.
5. Monitor protoplast liberation by microscopic observation.
6. Isolate protoplasts with a gradient of the overlaying solution after centrifugation in a swinging bucket rotor at 3900 × *g* for 25 min at 4 °C.
7. Preserve the interphase, which harbors protoplasts.
8. Purify protoplasts by washing them with ST solution three times.
9. Resuspend in 200 μL of STC and addition of 1–2 μg of DNA.
10. Before adding DNA, a little amount of the mixture is used to seed control plates: (1) MR plate containing the requirements of the *nkuA*Δ strain (pyrimidine and pyridoxine) to check the viability of the protoplasts. (2) Complete media plate to check that protoplast purification has been effective and that there was no residual mycelium.
11. Facilitate fusion between DNA and protoplasts adding 50 μL of PEG. Incubate this mixture for 15 min on ice (*see***Note 9**).
12. Add 1 mL of PEG and incubate for 15 min at room temperature.
13. Centrifuge at 10,000 × *g*, for 5 min, to recover protoplasts and resuspend in 200 μL of STC.
14. Seed three MR plates with the requirements of the resulting strain (pyridoxine, since pyrimidine requirements has been restored thanks to the presence of the *A. fumigatuspyrG* gene).
15. Transformant colonies should appear after 3–5 days of incubation.

3.7.2 Verification of Transformants

Growth in Specific Medium

1. To check if transformants are pyrimidine(+) and pyridoxine(−), seed plates with MMS, 1% glucose, and pyridoxine.
2. Check if they have glufosinate resistance in the presence of acetamide, as nitrogen source and ethanol, as carbon source.
3. Compare with the wild-type strain of *A. nidulans* to check that the wild-type strain is not resistant.

By PCR

To test the presence of the whole construction in the four transformants (called 1, 2, 4, and 10) do a PCR analysis.

1. Obtain DNA by sonication and phenol extraction of mycelium obtained from liquid cultures.
2. Design specific primers in the ±700 bp regions of the protein.
3. Make a PCR using those specific primers.
4. Conditions: 98°2′, (98°5″, 70°10″, 72°1′10″) × 33, 72°2′.
5. Make a 1% agarose gel to check the size of the fragment (*see*-**Note 10**).

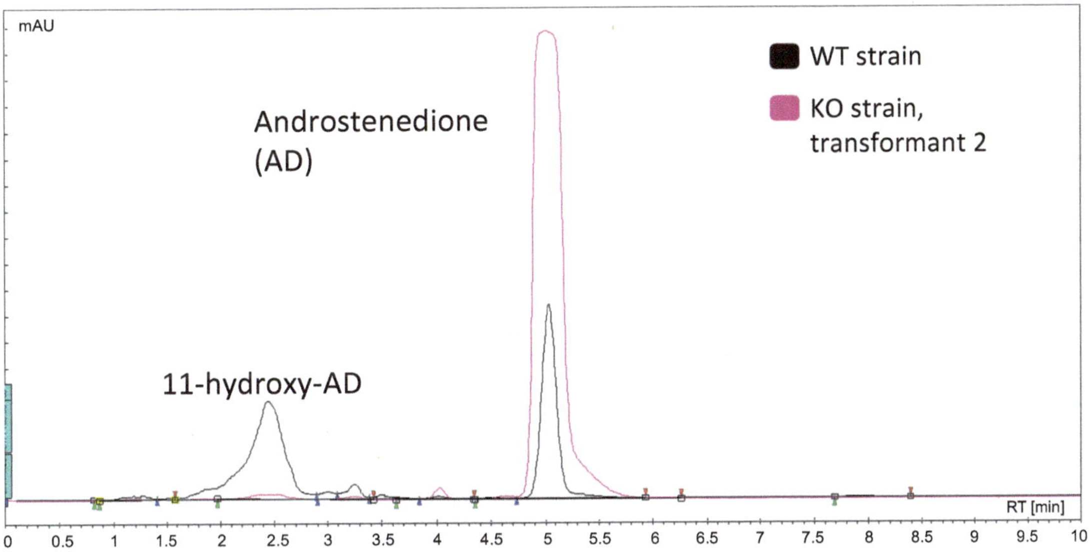

Fig. 6 Comparison of HPLC chromatograms of 11-alpha-AD production with WT strain (*black*) and KO strain (transformant 2, *pink*) after 1-day growth in the presence of AD

6. Sequence the band to confirm the *pyrG* gene inserted in the coding region of *A* gene.

By Fermentation

A fermentation using transformants 1, 2, and 10 from the KO and the WT strains is done to prove that there was no 11-alpha-hydroxylation of AD.

1. Incubate four different flasks with 100 mL of MML and 1% glucose and 0.5 mL of each kind of spores. For the flasks containing the KO strain, add pyridoxine as supplement and biotin for the flask containing the WT strain.
2. Incubate flasks at 37 °C for 24 h in agitation.
3. Then, add AD (1 g/L).
4. Take samples at day 0, 1, and 2.
5. Process the samples and analyze by HPLC as previously described.

In the Fig. 6 the chromatograms corresponding to products isolated from clone 2 of the KO strain (pink) and WT strain (black) after 1-day incubation in the presence of AD are shown. It is evident that there was no 11-hydroxy-AD production in the KO strain while most of the AD was converted to 11-hydroxy-AD in the WT one. To sum up, we conclude that the identification of the 11-alpha-hydroxylation gene has been correct and that this gene is the only one responsible for this activity.

KO strain was tested with other steroid compounds together with WT strain of *A. nidulans* and with *A. ochraceus*. In the

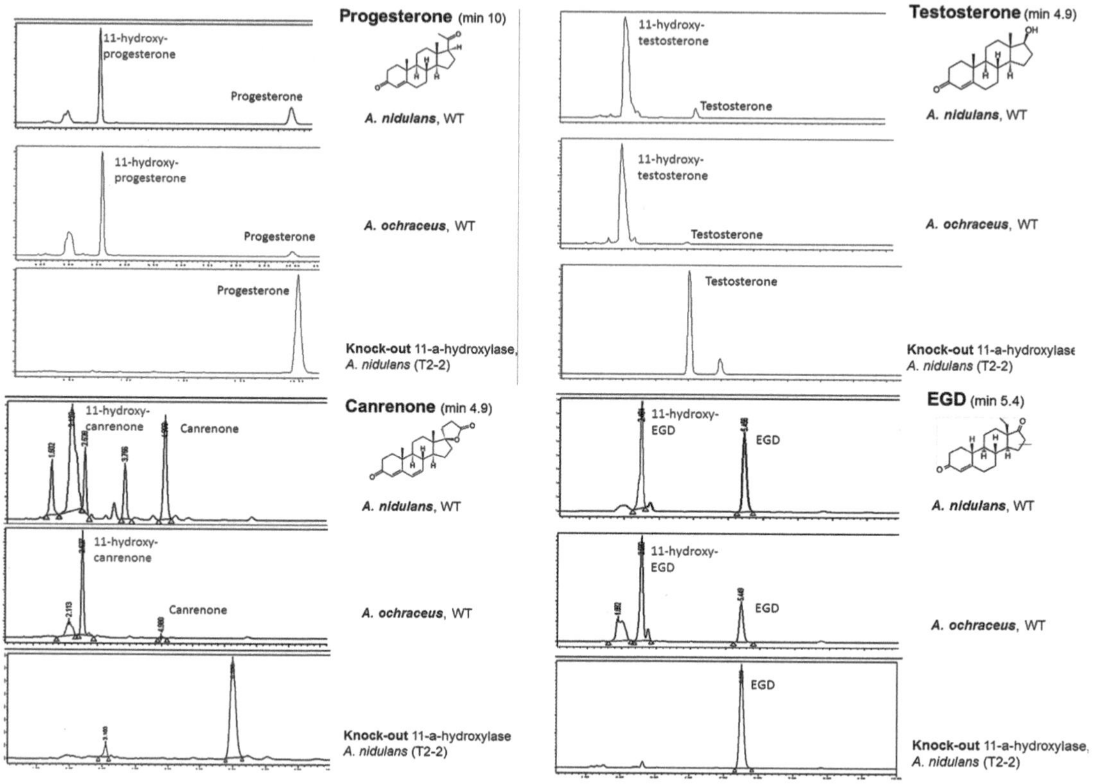

Fig. 7 Chromatograms corresponding to fermentations with different steroids (progesterone, testosterone, canrenone, and EGD) using the WT strains of *A. nidulans* and *A. ochraceus* and the KO strain of *A. nidulans*

chromatograms (Fig. 7) there is 11-hydroxylase activity with both WT strains, while there is not any 11-alpha-hydroxylation when using our KO strain. Thanks to these chromatograms too, *A. ochraceus* has been observed to have a higher 11-alpha-hydroxylation activity than *A. nidulans* and, for this reason, it is used as the industrial producer.

Moreover, we have accomplished different fermentations with other fungi such as *Rhizopus oryzae*, where 11-alpha-hydroxylation is described [14], or *Curvularia lunata*, in which 11-beta-hydroxylase activity is described [15, 16]. Therefore, in Fig. 8 we compare hydroxylation of androstenedione made by *A. ochraceus*, *R. oryzae*, and *C. lunata*. Almost total biotransformation of AD was observed in *A. ochraceus*, while in *R. oryzae* and *C. lunata* it did not take place in a big proportion. In these two last cases the amount of AD added to the culture to see the activity was higher and for that reason the peaks of the chromatograms are not as good as they should be.

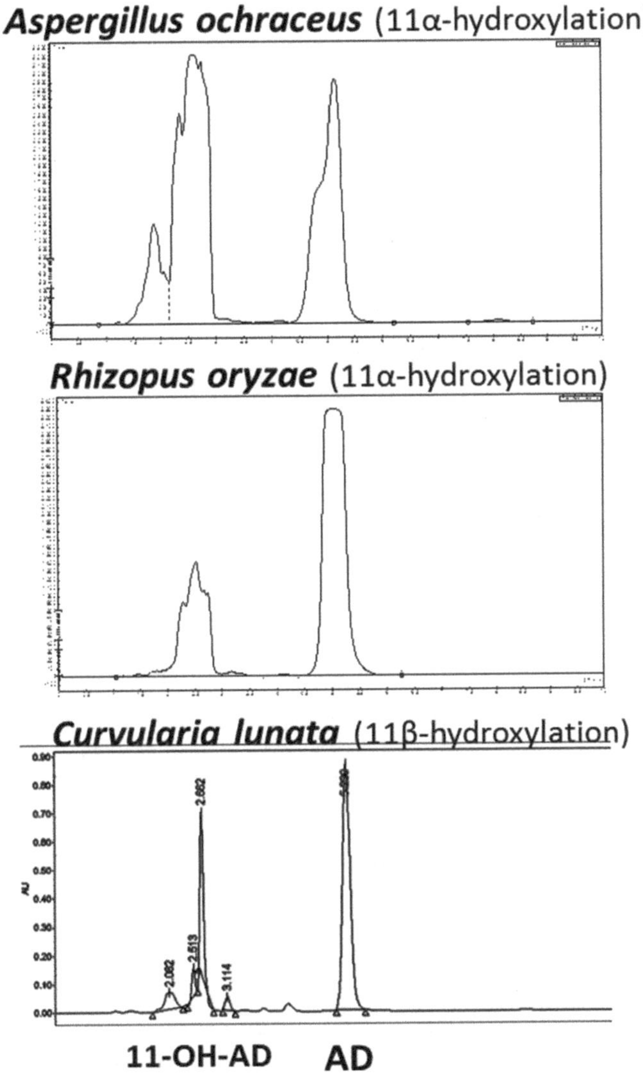

Fig. 8 Comparison of the chromatograms of the 11-α/β hydroxylation of AD made by *Aspergillus ochraceus*, *Rhizopus oryzae*, and *Curvularia lunata*

4 Notes

1. We add pyrimidines and pyridoxine as nutrients when growing the *nkuA*Δ strain because they are used as selection markers. The KO strain, resulting from the *nkuA*Δ strain, only needs the addition of pyridoxine as nutrient because the synthesis of pyrimidines, thanks to the presence of the *pyrG* gene, has been restored.

2. *Aspergillusochraceus* is grown in the same medium, but it is incubated at 27 °C.

3. As for SSH, to assure the success of this technique, it is required to expose induced cDNA to an excess of noninduced cDNA to make sure that all +cDNA present as single strands is not due to the lack of –cDNA to hybridize.
4. In order to avoid evaporation of the reaction of hybridization in the SSH process, it is necessary to cover the reaction with mineral oil.
5. Phire Polymerase leaves blunt ends, so its later cloning in plasmid pBlueScript SK+ (pBS) opened with *Eco*RV is possible.
6. The sequencing process of 100 clones was done by the CIB Sequencing Service (Madrid). Four potential candidate genes resulted.
7. Due to patent requirements the sequence of *A* gene (encoding 11-alpha-hydroxylase), is not given here. It will be provided in a further publication.
8. For an optimal protoplast collection, an excessive growth of the mycelium should be avoided.
9. To avoid protoplasts rupture, prepare fresh PEG in the same day of the transformation.
10. The knockout strain was expected to produce a DNA fragment 1.7 kb bigger than the wild-type fragment because it includes the *pyrG* from *A. fumigatus.*

Acknowledgment

We gratefully acknowledge the support of the Ministerio de Economía y Competitividad (Madrid, España, grant BIO2015-66960-C3-R; grant RTC-2014-2249-1), and the European Social Fund, Programa Operación of Castilla y León, and Junta de Castilla y León, through Consejería de Educación (grant LE114U13 and a grant to Lidia Ortega). We also thank Dr. Miguel A. Peñalba for his expertise advice and the gift of some strains and plasmids used in this work.

References

1. Wang F-Q, Yao K, Wei D-Z (2011) From soybean phytosterols to steroid hormones. In: El-Shemy H (ed) Soybean and health. InTech, Croatia
2. Malaviya A, Gomes J (2008) Androstenedione production by biotransformation of phytosterols. Bioresour Technol 99:6725–6737
3. Pérez C, Falero A, Duc HL, Balcinde Y, Hung BR (2006) A very efficient bioconversion of soybean phytosterols mixtures to androstanes by mycobacteria. J Ind Microbiol Biotechnol 33:719–723
4. Lee CY, Liu WH (1992) Production of androsta-1,4-diene-3,17-dione from cholesterol using immobilized growing cells of *Mycobacterium* sp. NRRL B-3683 adsorbed on solid carriers. Appl Microbiol Biotechnol 36:598–603

5. Smith M, Zahnley J, Pfeifer D, Goff D (1993) Growth and cholesterol oxidation by *Mycobacterium* species in Tween 80 medium. Appl Environ Microbiol 59:1425–1429
6. Bensasson CS, Hanson JR, Le Huerou Y (1999) The microbiological hydroxylation of 3 alpha,5-cycloandrostanes by *Cephalosporium aphidicola*. Phytochemistry 52:1279–1282
7. Fernandes P, Cruz A, Angelova B, Pinheiro HM, Cabral JMS (2003) Microbial conversion of steroid compounds: recent developments. Enzym Microb Technol 32:688–705
8. Diatchenko L, Lau YF, Campbell AP, Chenchik A, Moqadam F, Huang B, Lukyanov S (1996) Suppression subtractive hybridization: a method for generating differentially regulated or tissue-specific cDNA probes and libraries. Proc Natl Acad Sci U S A 93(12): 6025–6030
9. Mingot JM, Peñalva MA, Fernández-Cañón JM (1999) Disruption of phacA, an *Aspergillus nidulans* gene encoding a novel cytochrome P450 monooxygenase catalyzing phenylacetate 2-hydroxylation, results in penicillin overproduction. J Biol Chem 274:14545–14550
10. Ferrer-Sevillano F, Fernández-Cañón JM (2007) Novel phacB-encoded cytochrome P450 monooxygenase from *Aspergillus nidulans* with 3-hydroxyphenylacetate 6-hydroxylase and 3,4-dihydroxyphenylacetate 6-hydroxylase activities. Eukaryot Cell 6:514–520
11. Szewczyk E, Nayak T, Oakley CE, Edgerton H, Xiong Y, Taheri-Talesh N, Osmani SA, Oakley BR, Oakley B (2006) Fusion PCR and gene targeting in *Aspergillus nidulans*. Nat Protoc 1(6):3111–3120. doi:10.1038/nprot.2006.405
12. Tilburn J, Scazzocchio C, Taylor GG, Zabicky-Zissman JH, Lockington RA, Davies RW (1983) Transformation by integration in *Aspergillus nidulans*. Gene 26:205–221
13. Yelton MM, Hamer JE, Timberlake WE (1984) Transformation of *Aspergillus nidulans* by using a trpC plasmid. Proc Natl Acad Sci U S A 81:1470–1474
14. Petrič S, Hakki T, Bernhardt R, Zigon D, Crešnar B (2010) Discovery of a steroid 11α-hydroxylase from *Rhizopus oryzae* and its biotechnological application. J Biotechnol 150:428–437
15. Huang SH, Xu SW, Fa YH (1989) 11 beta-hydroxylation of 16 alpha-methyl-Reichstein's compound S 21-acetate by *Curvularia lunata*. Wei Sheng Wu Xue Bao 29:68–71
16. Suzuki K, Sanga K, Chikaoka Y, Itagaki E (1993) Purification and properties of cytochrome P-450 (P-450lun) catalyzing steroid 11 beta-hydroxylation in *Curvularia lunata*. Biochim Biophys Acta 1203:215–223

Chapter 20

Biotransformation of DHEA into 7α,15α-diOH-DHEA

Hui Li, Jin Sun, and Zhenghong Xu

Abstract

3β,7α,15α-Trihydroxy-5-androsten-17-one (7α,15α-diOH-DHEA) is a key intermediate of the novel oral contraceptive "Yasmin" (widely used in birth control pills and postmenopausal hormone replacement therapy pills; the active ingredient is drospirenone). It can be synthesized from dehydroepiandrosterone (DHEA) by microbial dihydroxylation at the C7 and C15 positions. Here we describe the method of bioconversion from DHEA into 7α,15α-diOH-DHEA by *Colletotrichum lini*. Using 6 g/L DHEA as a substrate, the DHEA conversion and the 7α,15α-diOH-DHEA molar yield were 72.6% and 51.2%, respectively.

Key words Biotransformation, *Colletotrichum lini*, DHEA, Dihydroxylation, 7α,15α-diOH-DHEA

1 Introduction

3β,7α,15α-Trihydroxy-5-androsten-17-one (7α,15α-diOH-DHEA) is an important intermediate of pharmacologically significant steroids, such as drospirenone. As an active ingredient of the novel oral contraceptive "Yasmin," it is effective, well tolerated and has positive effects on body weight, skin, and premenstrual symptoms [1]. 7α,15α-diOH-DHEA synthesized from DHEA by microbial dihydroxylation at the position of C7 and C15 attracts lots of attention.

Dihydroxylation of DHEA by microorganisms offers several advantages over conventional chemical reactions, including high stereoselectivity, minimal side reactions, mild reaction conditions, and it is environment-friendly [2–4]. Because of such obvious advantages and the huge market demand for safer and ecological chemical compounds, the bioconversion from DHEA to 7α,15α-diOH-DHEA performed by microorganisms has become more and more popular [5, 6]. According to literature, several fungi are able to transform DHEA into 7α,15α-diOH-DHEA, such as *Colletotrichum lini*, *Gibberella intermedia*, *Acremonium felinum* VKM F-1300, *Gibberella zeae* VKM F-2598, *Fusarium oxysporum*

José-Luis Barredo, Ignacio Herráiz (eds.), *Microbial Steroids: Methods and Protocols*, Methods in Molecular Biology vol. 1645, DOI 10.1007/978-1-4939-7183-1_20, © Springer Science+Business Media LLC 2017

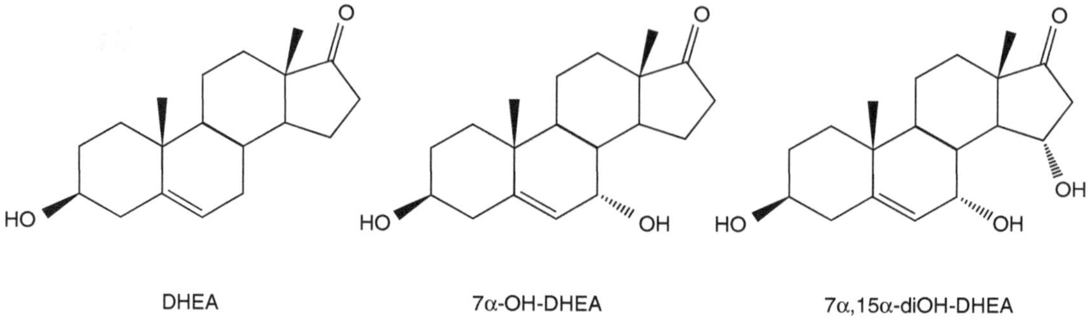

Fig. 1 The chemical formula of DHEA, 7α-OH-DHEA, and 7α,15α-diOH-DHEA

VKM F-1600, and *Nigrospora oryzae* RCM F-1939 [7, 8]. Among them, *C. lini* showed greatest potential [8]. However, due to the poor water solubility of substrate (1–100 μM), the transport of it into the cells is limited. The poor availability of substrate for whole-cell biocatalysts has been widely considered to be one of the main limiting factors in the DHEA biotransformation process, because the substrate concentration and product yield are not enough for the scale-up to an industrial application.

Previously, our group has reported that DHEA conversion and 7α,15α-diOH-DHEA molar yield by *C. lini* ST-1 were only 72.6% and 51.2%, respectively, using 6 g/L DHEA as a starting material. A lot of efforts have been made to improve the production by increasing substrate solubility, such as adding an organic cosolvent, rapeseed oil [9], or cyclodextrin (CD) [10], which have led to notably improved results. Using 10 g/L DHEA as a substrate, forming inclusion complexes with CD, the highest productivities of hydroxylation products (7α-OH-DHEA and 7α,15α-diOH-DHEA) were 8.06 g/L with M-β-CD and 8.45 g/L with HP-β-CD. In addition, the conversion efficiency of DHEA in the presence of 2% (w/v) Tween 80 increased drastically [11]. The DHEA conversion and the 7α,15α-diOH-DHEA molar yield increased to 90.0% and 72.0%, respectively, at the DHEA concentration of 10 g/L by *C. lini* ST-1 resting cells (Fig. 1).

2 Materials

Steroid compounds, solvents, and other chemicals are of analytical grade (unless indicated otherwise) and are obtained from commercial sources. Prepare all media using deionized water (*see* **Note 1**). Prepare HPLC solvents using ultrapure water (prepared by purifying deionized water, to attain a sensitivity of 18 MΩ cm at 25 °C). Prepare and store all reagents at room temperature.

2.1 Cultivation and Transformation

1. *Colletotrichum lini* ST-1. Deposit under CGMCC 6051 in the China General Microbiological Culture Collection Center (Beijing, China).
2. Slant medium: 30 g/L glucose, 0.01 g/L $FeSO_4$, 3 g/L $NaNO_3$, 1 g/L K_2HPO_4, 0.5 g/L KCl, 0.5 g/L $MgSO_4{\bullet}7H_2O$, and 20 g/L agar at natural pH.
3. Seed medium: 15 g/L glucose, 15 g/L yeast extract, 3 g/L corn steep liquor, and 10 g/L bean cake powder at natural pH.
4. Transformation medium: 15 g/L glucose, 15 g/L yeast extract, and 3 g/L corn steep liquor at natural pH.
5. DHEA with a purity of 99% (Xianju Pharmaceutical Company, Zhejiang, China).
6. 7α,15α-diOH-DHEA with a purity of 98.4% (Xianju Pharmaceutical Company, Zhejiang, China).
7. 7α-OH-DHEA with a purity of 98% (isolated, purified, and identified by our laboratory).
8. Erlenmeyer flasks (250 and 500 mL).
9. Rotary shaker.

2.2 Sample Preparation and HPLC Analysis

1. Centrifuge tubes.
2. Plastic container.
3. Ethyl acetate.
4. Thermo-shaker.
5. Acetonitrile (HPLC grade).
6. Ultrapure water.
7. Volumetric flask (100 mL).
8. Vacuum filter.
9. 0.22 μm microfiltration membrane (organic and inorganic).
10. High performance liquid chromatography equipment (HPLC).
11. C18 column (5.0 μm, 4.6 mm × 250 mm) (Agilent).
12. Vacuum oven.

3 Methods

3.1 Cultivation of the Microorganism

1. Seed *C. lini* ST-1 on slant medium from vials stored at −80 °C (*see* **Note 2**).
2. Transfer the mycelium into 500 mL Erlenmeyer flask, which contains 100 mL seed medium (*see* **Note 3**).
3. Then culture it on a rotary shaker (220 rpm) at 30 °C for 2–3 days.

4. Inoculate the culture with 10% volume into fresh transformation medium (250 mL flask with 30 mL medium).
5. Cultivate for further 24 h in transformation medium to reach the stationary phase.

3.2 Transformation of DHEA

1. Add 0.18 g of fine powder of DHEA into the culture (*see* **Note 4**).
2. Carry out the transformation reaction under the same conditions as described previously (*see* Subheading 3.1, **step 3**) for further 60–72 h (*see* **Note 5**).

3.3 Sample Preparation and Pretreatment

1. Withdraw 1 mL of the fermentation broth at regular intervals of about 6 h, and use the samples for the analysis of DHEA transformation products (*see* **Note 6**).
2. Add 1 mL ethyl acetate and shock the mixture on a thermo-shaker for 30 min (1000 rpm, 25 °C). Remove the ethyl acetate layer into a plastic container.
3. Repeat **step 2** for at least four times (*see* **Note 7**).
4. Dry the extract under vacuum at 80 °C.
5. Subsequently, redissolve the residue in 8 mL acetonitrile for HPLC analysis.

3.4 HPLC for Product Analysis

1. Prepare acetonitrile and ultrapure water as the mobile phase for HPLC. Then filter through a 0.22 μm microfiltration membrane (*see* **Note 8**) by vacuum filter and ultrasonically exhaust for 20 min at least.
2. Prepare the standard by weighing approximately 50 mg of standard material into a 100 mL volumetric flask and dissolving it in acetonitrile.
3. Filter the samples and standards used for HPLC through a 0.22 μm organic microfiltration membrane.
4. Run the analysis on the Agilent C18 column at the wavelength of 206 nm, column temperature of 30 °C, with acetonitrile and ultrapure water (7:3) as the mobile phase at a flow rate of 0.5 mL/min.
5. Observe three major peaks in the analysis results by HPLC. The retention time of the peaks is about 6.5, 7.2, and 11.5 min (Fig. 2; *see* **Notes 9** and **10**).

3.5 Calculation Formulas

1. Calculate the DHEA conversion according to the following formula:

$$\text{DHEA conversion}\,(\%) = \frac{C_S - C_R}{C_S} \times 100\%$$

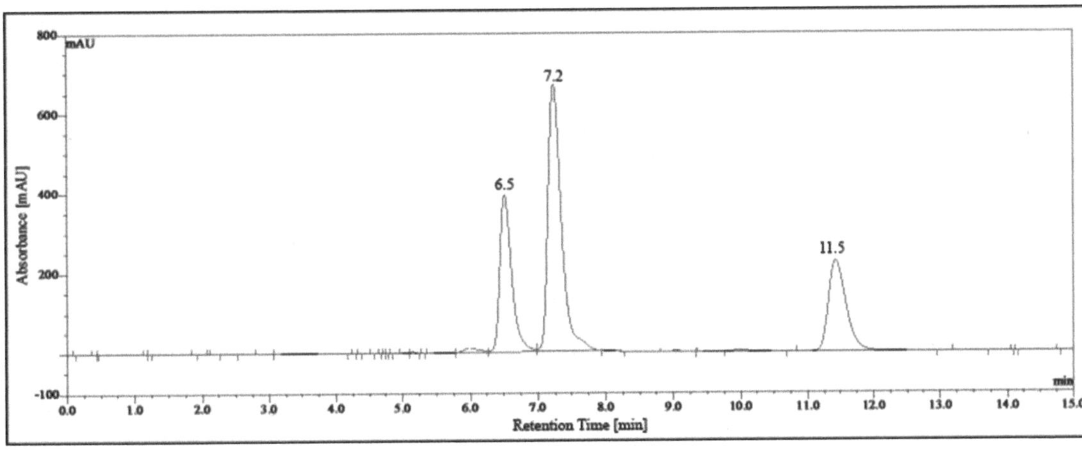

Fig. 2 HPLC chromatogram of a sample. The compounds at 6.5, 7.2, and 11.5 min are 7α,15α-diOH-DHEA, 7α-OH-DHEA, and DHEA, respectively

2. Calculate the molar yield of 7α,15α-diOH-DHEA according to the following formula (*see* **Note 11**):

$$\text{Molar yield of } 7\alpha,15\alpha-\text{diOH}-\text{DHEA}\,(\%) = \frac{C_\text{P} \times M_\text{S}}{C_\text{S} \times M_\text{P}} \times 100\%$$

3. Calculate the DHEA concentration according to the following formula:

$$C_\text{R}\,(\text{g}/\text{L}) = \frac{A_\text{S}}{A_1} \times C_1 \times D$$

4. Calculate the 7α-OH-DHEA concentration according to the following formula:

$$C_{BP}\,(\text{g}/\text{L}) = \frac{A_{BP}}{A_2} \times C_2 \times D$$

5. Calculate the 7α,15α-diOH-DHEA concentration according to the following formula:

$$C_\text{P}\,(\text{g}/\text{L}) = \frac{A_\text{P}}{A_3} \times C_3 \times D$$

(C_S: the initial concentration of DHEA; C_R, C_BP, and C_P: the concentration of the samples of DHEA, 7α-OH-DHEA, and 7α,15α-diOH-DHEA, respectively; M_S: the molecular weight of DHEA; M_P: the molecular weight of 7α,15α-diOH-DHEA; D: the dilution ratio (In this protocol the dilution ratio is 8); A_S, A_BP, and A_P: the peak area of the samples of DHEA, 7α-OH-DHEA, and 7α,15α-diOH-DHEA, respectively; A_1, A_2, and A_3: the peak area of the standards of DHEA, 7α-OH-DHEA, and 7α,15α-diOH-DHEA, respectively; C_1, C_2, and C_3: the concentration of the standards of DHEA, 7α-OH-DHEA, and 7α,15α-diOH-DHEA, respectively).

4 Notes

1. The medium should be sterilized at 121 °C for 20 min right after preparation, and the sterilized medium has to be used within a week.
2. The microorganism should be activated frequently to avoid strain degeneration. The strain is preserved with 20% glycerol at −80 °C in an ultralow temperature freezer. The strain on slant medium can be stored at 4 °C for 1 month.
3. To prevent microbial contamination, much attention has to be paid to all the operations, especially inoculation, which should be done in an aseptic, dustless, and clean working environment.
4. DHEA is ground into a fine powder before being added into the fermentation broth as a substrate, otherwise conversion efficiency will be affected.
5. The biotransformation period should be controlled strictly. If the biotransformation time is longer than what it should be, the 7α,15α-diOH-DHEA can get additionally hydroxylated and even fully catabolized into carbon dioxide and water by the fungus.
6. Before sampling from the fermentation broth for analysis of transformation products, the fermentation broth should be mixed thoroughly aiming at reducing measurement errors.
7. Repeat extraction at least for four times, otherwise the efficiency will obviously decrease.
8. Filter acetonitrile through an organic membrane and ultrapure water through an inorganic membrane.
9. The retention time is based on the polarity of steroids. The steroids at 6.5, 7.2, and 11.5 min are 7α,15α-diOH-DHEA, 7α-OH-DHEA, and DHEA, respectively.
10. A large proportion of the 7α-OH-DHEA by-product is observed during the conversion, which indicated low catalytic specificity. To obtain the higher substrate conversion and 7α,15α-diOH-DHEA yield, the concentration of DHEA can be reduced. For example, with 4 g/L DHEA, the DHEA conversion and the molar yield of 7α,15α-diOH-DHEA increased to 93.6% and 80.2%, respectively.

11. The substrate concentration and product yield are relatively low on account of the low steroid aqueous solubility. Several strategies for improving substrate solubility can be adopted to enhance product yield, such as liquid polymer, ionic liquids, cyclodextrin inclusion (CD), cloud-point systems, organic–aqueous biphasic systems, water-miscible organic cosolvents, resting cells biotransformation, and combinations of the abovementioned strategies [12–15].

Acknowledgment

This work is financially supported by the National High-Tech R&D Program (No. 2011AA02A211).

References

1. Mansour D (2002) Experiences with Yasmin (R): the acceptability of a novel oral contraceptive and its effect on well-being. Eur J Contracept Reprod Health Care 7:35–41
2. Petzoldt K, Laurent H, Wiechert R (1983) A novel synthetic route to the aldosterone-antagonist spirorenone. Angew Chem Int Ed 22:406–407
3. Sabrieh G, Maryam M, Zohreh H (2014) Biotransformation of testosterone and testosterone heptanoate by four filamentous fungi. Steroids 92:7–12
4. Torres S, Castro GR (2004) Non-aqueous biocatalysis in homogeneous solvent systems. Food Technol Biotechnol 42:271–277
5. Boaventura MAD, Lopes RFAP, Takahashi JA (2004) Microorganisms as tools in modern chemistry: the biotransformation of 3-indolylacetonitrile and tryptamine by fungi. Braz J Microbiol 35:345–347
6. Donova MV, Egorova OV (2012) Microbial steroid transformations: current state and prospects. Appl Microbiol Biotechnol 94:1423–1447
7. Lobastova TG, Gulevskaya SA, Sukhodolskaya GV, Donova MV (2009) Dihydroxylation of dehydroepiandrosterone in positions 7α and 15α by mycelial fungi. Appl Biochem Microbiol 45:617–622
8. Li H, Fu ZZ, Zhang XM, Li H, Shi JS, Xu ZH (2014) The efficient production of 3β,7α,15α-trihydroxy-5-androsten-17-one from dehydroepiandrosterone by *Gibberella intermedia*. Appl Biochem Biotechnol 174:2960–2971
9. Li H, Fu ZZ, Li H (2014) Enhanced biotransformation of dehydroepiandrosterone to 3β,7α,15α-trihydroxy-5-androsten-17-one with *Gibberella intermedia* CA3-1 by natural oils addition. J Ind Microbiol Biotechnol 41:1497–1504
10. Wu Y, Li H, Xu ZH (2014) Enhancement of steroid hydroxylation yield from dehydroepiandrosterone by cyclodextrin complexation technique. Steroids 84:70–77
11. Li H, Xu ZH, Yin SQ (2016) Enhanced 3β,7α,15α-trihydroxy-5-androsten-17-one production from dehydroepiandrosterone by *Colletotrichum lini* ST-1 resting cells with Tween-80. Appl Biochem Biotechnol 178:91–100
12. Avramova T, Spassova D, Mutafov S, Momchilova S, Boyadjieva L, Damyanova B, Angelova B (2010) Effect of Tween 80 on 9α-steroid hydroxylating activity and ultrastructural characteristics of *Rhodococcus* sp. cells. World J Microbiol Biotechnol 26:1009–1014
13. Shen Y, Liang J, Li H (2015) Hydroxypropyl-β-cyclodextrin-mediated alterations in cell permeability, lipid and protein profiles of steroid-transforming *Arthrobacter simplex*. Appl Microbiol Biotechnol 99:387–397
14. Mao SH, Hu XJ, Hua B (2012) 15α-Hydroxylation of a steroid (13-ethyl-gon-4-en-3,17-dione) by *Penicillium raistrickii* in an ionic liquid/aqueous biphasic system. Biotechnol Lett 34:2113–2117
15. Kim Y, Han J, Lee SS (2007) Steroid 9-hydroxylation during testosterone degradation by resting *Rhodococcus equi* cells. Arch Pharm Res 340:209–214

Chapter 21

Stabilization of Enzymes by Using Thermophiles

Ana Luisa Ribeiro, Mercedes Sánchez, Aurelio Hidalgo, and José Berenguer

Abstract

Manufactured steroid compounds have many applications in the pharmaceutical industry. Due to the chemical complexity and chirality of steroids, there is an increasing demand for enzyme-based bioconversion processes to replace those based on chemical synthesis. In this context, thermostability of the involved enzymes is a highly desirable property as both the increased half-life of the enzyme and the enhanced solubility of substrates and products will improve the yield of the reactions. Metagenomic libraries from thermal environments are potential sources of thermostable enzymes of prokaryotic origin, but the number of expected hits could be quite low for enzymes handling substrates such as steroids, rarely found in prokaryotes. An alternative to metagenome screening is the selection of thermostable variants of well-known steroid-processing enzymes. Here we review and detail a protocol for such selection, where error-prone PCR (epPCR) is used to introduce random mutations into a gene to create a variants library for further selection of thermostable variants in the thermophile *Thermus thermophilus*. The method involves the use of folding interference vectors where the proper folding of the enzyme of interest at high temperature is linked to the folding of a reporter encoding a selectable property such as thermostable resistance to kanamycin, leading to a life-or-death selection of variants of reinforced folding independently of the activity of the enzyme.

Key words *Thermus thermophilus*, Enzyme variants library, Thermostabilization, Folding-interfering reporter

1 Introduction

Bioactive steroids are biologically and commercially relevant compounds. Manufactured steroid compounds have a vast therapeutic applicability as anti-inflammatory, diuretic, contraceptive agents, breast and prostate cancer and anti-obesity agents, and others [1]. Chemical synthesis of steroids is often difficult, or not advantageous due to process complexity and low yields. Recently, bioconversion processes involving engineered microorganisms and their enzymes have emerged as efficient and cost-effective alternatives to chemical production and modification of steroid compounds [2].

José-Luis Barredo, Ignacio Herráiz (eds.), *Microbial Steroids: Methods and Protocols*, Methods in Molecular Biology vol. 1645, DOI 10.1007/978-1-4939-7183-1_21, © Springer Science+Business Media LLC 2017

Although some of these processes are well established, increasing the efficiency of the processes as well as identifying new potentially useful bioconversions, are two goals in steroid drugs production [3].

Thermozymes—thermostable enzymes that function optimally between 60 and 125 °C—have attracted a great industrial and biotechnological interest because they are better suited for harsh industrial processes. In addition to sustained catalytic activity at high temperatures, they are resistant to organic solvents, detergents, low and high pH, and other denaturants, and have a lowered susceptibility to proteolysis [4–7]. The low solubility of sterols and steroids in water [8] entails a need for higher temperatures during modification reactions. Therefore, thermostable variants of the steroid bioconverting enzymes used in the industry or new enzymes from thermophilic organisms capable of bioconverting steroids would be of great value.

Thermozymes can be of natural origin or obtained by selection of thermostable variants of normal enzymes. Natural thermozymes are produced by thermophiles, microorganisms that are adapted to grow at temperatures above 50 °C. Access to steroid-modifying enzymes from these microorganisms requires either culture and further selection of a producing strain, or the selection of the appropriate enzyme from a genomic or metagenomic library from a thermophile or a thermophilic environment. As few thermophiles are easily grown under laboratory conditions, metagenomics has evolved as a powerful cultivation-independent approach, which can be applied to gain access to new sterol bioconverting enzymes from difficult to grow thermophilic microorganisms. Over the past decade, metagenomic libraries derived from extreme habitats have been constructed and novel enzymes with potential use in industrial applications have been found [9, 10] (Table 1). Two screening methodologies can be used on metagenomic material: the function-driven screening of expression libraries, and the sequence-driven analysis based on gene homology. The first approach is the most commonly used for the identification of industrial target proteins and requires the uncovering of positive clones expressing a desired trait by either direct or indirect detection. Direct detection is achieved by the addition of dyes or chromophore-containing chemicals into the growth medium [18, 19]; the indirect approach consists on the use of host strains able to grow under selective conditions only when expressing the targeted gene product in an active form [20].

Despite several successes, metagenomic strategies typically have low rates of target identification, in most cases due to the inability of the host to transcribe and or translate or fold properly the gene(s) and protein(s) of interest [21]. As an alternative, mesophilic enzymes can be engineered to become more thermostable. Thermal stability is a consequence of the concurrent combination of multiple factors [22–24]. When mesophilic and thermophilic

Table 1
Methods to obtain thermostable proteins

Approach	Method	Technique	Advantages	Disadvantages	References
Search for existing enzymes	Metagenomic library	Cloning, sequencing, functional expression	Potential to discover proteins that would not have been recognized by their sequences; large molecular diversity	Hits do not always meet the required demands in terms of activity, selectivity and stability	[11]
Protein engineering: Directed evolution	Random mutagenesis	Error-prone PCR (epPCR)	High mutation rates; nearly even mutational spectrum; simplicity	Limited amino acid substitutions; accumulated deleterious mutations; polymerases bias	[12, 13]
		Mutator strains	Easy to use	Low mutation rates; uneven mutational spectrum; limited amino acid substitutions	[12]
		Physical–chemical mutagenesis	Dose-dependent mutation rates, simplicity	Low efficiency; hazardous	[12]
		Saturation utagenesis	Simplicity; access to all amino acid repertoire; focus on specific genetic locus	Gene sequence required; limited diversity; surplus of inactive clones within simultaneous saturation libraries	[13]
	Homologous recombination	DNA shuffling StEP RACHITT	Robust; flexible	Rely on high sequence homology; low crossover rate	[13]
		Incremental truncation methods	Eliminate recombination bias; structural knowledge not necessary	Limited to two parents; complex, laborious	

(continued)

Table 1
(continued)

Approach	Method	Technique	Advantages	Disadvantages	References
Rational design	Computational tools	Sequence–function relationship prediction	Prediction of functionally relevant residues; capable of using large volume data	Affected by incorrect annotations in databases	[14, 15]
		Structure-based molecular modeling	Sequence identities as low as 25% are usually enough to predict reliable structure models	Difficulties in accurately predicting stabilizing mutations without affecting activity or selectivity	[14]
	Site-directed mutagenesis	PCR-based site-directed mutagenesis	Allows direct modification of specific residues; simplicity	The information gathered relies entirely on the selection of mutants produced	
		Gene truncations and deletions	Valuable in mapping specific regions of functionally; may improve solubility	Requires structural and sequence knowledge	[16]
De novo design	Computational tools	Fixed backbone methods Flexible backbone methods	De novo design can be extended to 3° and 4° structure	Requires advanced sampling techniques	[17]

counterpart of an enzyme sharing >30% of sequence identity are compared, their three-dimensional structures are easily superimposable, sharing also similar architecture of their respective active centres, thus allowing the study of thermophilic adaptations [7, 24]. In many cases, thermophilic adaptation comes along with a higher packing efficiency through hydrophobic interactions and an increased density of salt bridges or charge-assisted hydrogen bonds. Often, these amino acid changes contribute to an increase in mechanical rigidity of the structure concomitant with a decrease in flexibility at low temperatures [25–27]. Most of these changes are located on surface areas, at the ends of conserved domains and in loops. A predominant accumulation of nonpolar alanine and positively charged arginine is frequently observed in thermostable proteins of *Thermus thermophilus* [22]. However, it is important to stress that the general trend is that the local distribution of rigid or functionally important flexible regions in a thermophilic protein is similar to that of a mesophilic homolog, provided that the proteins are compared at their respective optimal temperatures [28, 29].

However, many studies have produced conflicting results on factors and evolutionary mechanisms that lead to increased protein thermostability. They may differ from taxon to taxon, organism to organism, and even from protein to protein within the same organism [30], making it very difficult to assemble a common strategy for enzyme thermostabilization by protein engineering.

The first approach adopted to thermostabilize relevant enzymes was rational design (Table 1), which uses molecular modeling and structural and mechanistic information derived from the comparisons commented above to predict the amino acid(s) substitution required to produce the desired changes in the protein. Actually, the difficulties of making reliable predictions imposed a change of path to the so-called "directed evolution" strategies (Table 1). This approach relies on the generation of thousands or millions of protein variants by random mutagenesis ("variants library") and further screening or selection for those that have the desired properties. In recent years, the line that separated rational design and directed evolution became blurred because most researchers prefer to use combinations of both practices to improve proteins [31].

De novo design (Table 1) is the latest methodology adopted by protein engineers to thermostabilize enzymes and it relies solely on creating any functional enzyme by specifying their amino acid sequence on synthetic genes. This is not impossible, since it is known that the primary sequence of a protein determines its structure [32]. However, despite impressive advances in computational enzyme design, the actual improvements of enzyme performance obtained by the de novo design systems are rather poor [33].

Whichever approach chosen, the main obstacle remains the screening of large mutant libraries or metagenomic libraries for variants with the desired functionality at high temperature. This

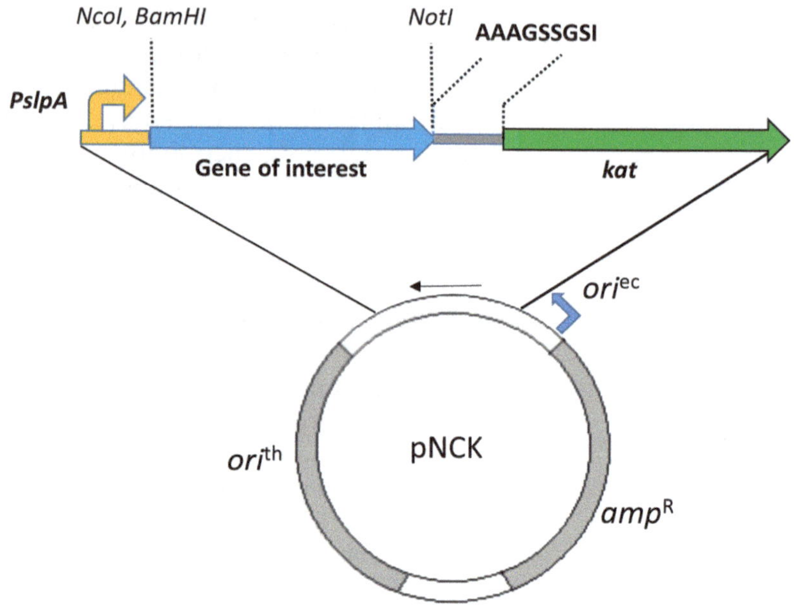

Fig. 1 Schematic structure of plasmid pNCK. The gene of interest is fused to the gene encoding a thermostable kanamycin nucleotidyl transferase variant (*kat*) through a 9-amino acid linker (AAAGSSGSI). The promoter of the *slpA* gene (*PslpA*) is represented as *yellow bar*, upstream of the gene of interest. The plasmid contains the ampicillin resistance gene (amp^R) that codes for the enzyme *β-lactamase*. The *Thermus* spp. origin of replication (ori^{th}), *E. coli* origin of replication (ori^{ec}), and a simplified cloning site for the target genes are showed

can be very expensive and time consuming in the absence of an appropriate high-throughput screening or a growth-dependent selection assay. The folding-interference approach using a thermophile as a host is an activity independent high-throughput screening technology that selects for enzyme variants with reinforced folding at high temperatures in a life-or-death manner, decreasing the screening efforts needed by orders of magnitude [34]. In this strategy, the non-thermostable enzyme variants are fused to a thermostable protein reporter, and the resulting protein fusion is expressed in a thermophile. This way, the non-thermostable variants will not fold properly at high temperature, leading to protein aggregation and to the "interference" with the folding of the reporter. On the contrary, those variants more thermostable will fold properly at high temperature, allowing the proper folding and the activity of the reporter. As the selection method is independent of the activity of the enzyme under selection, a secondary screening will be needed among those "thermostable" variants to eliminate the enzymes without activity. The most commonly used folding interference vector is pNCK (*see* Fig. 1), a bifunctional *Escherichia coli–Thermus* spp. vector in which the protein of interest is expressed from the $slpA_p$ promoter fused to a thermostable

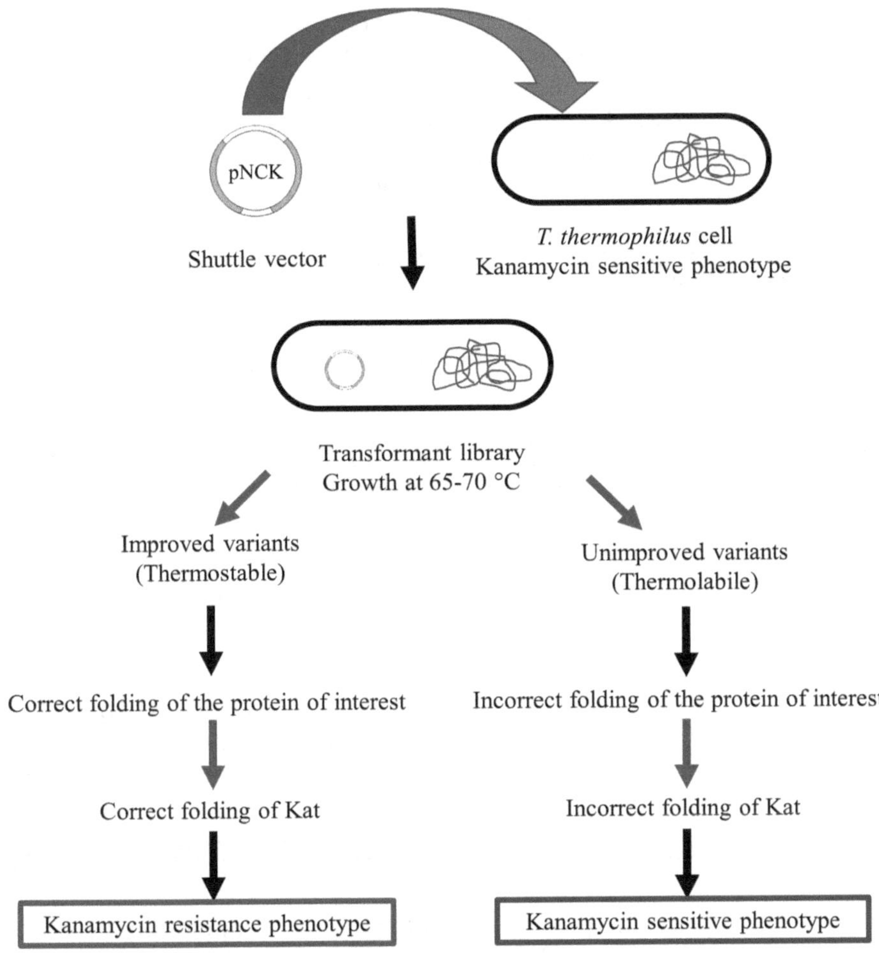

Fig. 2 Selection of thermostable proteins by folding interference. A fusion between a library of mutants of the protein of interest and the thermostable Kat protein is expressed in *T. thermophilus*. When the protein of interest is thermostable, it will fold correctly at high temperatures and as result Kat can fold properly exhibiting a kanamycin resistance phenotype. Instead, when the protein of interest is thermolabile, its wrong folding interferes with Kat folding and a kanamycin sensitive phenotype is obtained

kanamycin nucleotidyltransferase, conferring resistance to kanamycin in a wide temperature range. With this promoter, a life-or-death strategy is easily followed, allowing the selection of thermostable variants from large libraries upon transformation in high transformation-efficiency mutants of *T. thermophilus* HB27 (*see* Fig. 2).

However, different conditions should be taken into account before deciding to use folding interference selection methods. The main requirement is the limitation in the size and structural complexity of the enzyme to thermostabilize, as multidomain proteins do not easily transmit the unfolding information to the reporter through an intermediate protein domain. Also, limitations due to factors such as different codon usage with respect to the thermophile host, the putative toxicity of the protein in vivo, or the transformation efficiency, could be limiting factors to implement this technology [33].

A prime decision to be made for the implementation of this method is the appropriate selection of the screening host. To the best of our knowledge, *Thermus thermophilus* has been the only thermophile for which this strategy has been used successfully for the selection of thermostable proteins and enzyme variants [34, 35]. Here we detail the optimized protocol for the folding-interfering selection of thermostable enzyme variants in this thermophile.

2 Materials

2.1 Bacterial Growth Media

1. *Thermus* Broth (TB): 8 g tryptone, 4 g yeast extract, and 3 g NaCl per litre of mineral water (*see* **Note 1**), pH adjusted to 7.5 with 2 M NaOH.
2. Luria–Bertani (LB) broth: 10 g tryptone, 5 g yeast extract, and 5 g NaCl per litre of water, pH adjusted to 7.0 with 2 M NaOH.
3. SOC liquid medium: 2% (w/v) tryptone, 0.5% (w/v) yeast extract, 10 mM NaCl, 2.5 mM KCl, 10 mM $MgCl_2$, and 20 mM glucose. Prepare a solution containing the first four reagents, sterilize at 121 °C, and then add sterile $MgCl_2$ and glucose.
4. SOC solid medium: Add 1.5% of agar to the SOC liquid medium.
5. If required, after autoclaving and upon cooling down, supplement media with 30 μg/mL kanamycin for LB plates or at the concentration required for selection as defined in Subheading 3.3.

2.2 Bacterial Strains

1. *Escherichia coli* DH5α: *fhuA2 lac(del)U169 phoA glnV44 Φ80′ lacZ(del)M15 gyrA96 recA1 relA1 endA1 thi-1 hsdR17*.
2. *Thermus thermophilus* HB27 (ATCC BAA-163) derivative $HB27^{EC}$ showing high transformation efficiency due to the absence of the Argonaute protein [36] (*see* **Note 2**).

2.3 Biological and Chemical Materials

1. DNA primers for PCR and epPCR amplification (*see* **Note 3**).
2. DNA polymerase.
3. DNA template encoding for the enzyme.
4. PCR and epPCR reaction solution: DNA polymerase reaction buffer, dNTP mix, DNA polymerase, $MgCl_2$, $MnCl_2$ (*see* **Note 4**).
5. pNCK vector (*see* Fig. 1, *see* **Notes 5** and **6**).
6. Agarose gels, stained with 0.5 μg/mL ethidium bromide.
7. 10× TBE electrophoresis buffer: 1 M Tris base, 1 M boric acid, and 20 mM EDTA, pH 8.0.

8. DNA loading buffer and DNA ladder.
9. Gel and PCR purification kit: Wizard SV Gel and PCR Clean-Up System (Promega).
10. Plasmid extraction and purification kit: GeneJet Plasmid MiniPrep kit (Thermo Scientific).
11. T4 DNA ligase (Promega).
12. FastDigest *NcoI* or *BamHI* and *NotI* (Thermo Scientific).

2.4 Equipment

1. Thermocycler (i.e., Bio-Rad C1000 Touch thermocycler).
2. Agarose gel electrophoresis apparatus.
3. High precision static incubator (±0.1 °C at 60–80 °C; i.e., UN55, Memmert).
4. Thermomixer (i.e., Comfort, Eppendorf).
5. Orbital incubator (i.e., MaxQ 420 HP, Thermo Scientific).
6. NanoDrop (i.e., ND-1000, Thermo Scientific).
7. Thermo button for the recording of actual selection temperature (i.e., Thermotrack).

3 Methods

3.1 epPCR

1. With a PCR reaction amplify the gene coding for the protein you want to thermostabilize. Include in the primers the restriction sites appropriate to get the fusion (*NcoI* and *BamHI* or *NotI*).
2. Purify the PCR products with Wizard SV Gel and PCR Clean-Up System and quantify the DNA concentration with a NanoDrop.
3. Calculate the amount of DNA that is required for the desired mutation rate (*see* **Note 7**).
4. Prepare the epPCR reagents for ten reactions (*see* **Notes 8** and **9**; Table 2).
5. Mix the sample and divide it in ten thin-walled, 0.2 mL tubes (*see* **Note 10**).
6. Place the tubes in the thermocycler. Run the epPCR program (Table 3). Select the conditions according to the G + C content and melting temperature of the primers.
7. Verify the quality of the epPCR reaction by agarose gel electrophoresis.
8. Purify the reaction products with a PCR purification kit.
9. Mix the products of the epPCR reactions, digest the mix with the pair of restriction enzymes specified in the forward and reverse

Table 2
Reaction mix for epPCR

Component	Final concentration
Nuclease-free water	Up to 100 μL
10× *Taq* buffer	1×
dNTPs	0.2 mM dATP/dGTP; 1 mM dCTP/dTTP
$MnCl_2$	0.1–0.5 mM (*see* **Note 10**)
Forward primer	0.5 μM
Reverse primer	0.5 μM
Template DNA	5 ng/100 μL
NZY*Taq* DNA polymerase	0.05 U/μL

Table 3
Example of thermocycler program for the randomization by epPCR

Cycle	Step	Temperature (°C)	Time
1	Denaturation	95	3 min
2–31	Denaturation	95	30 s
	Annealing	65	30 s
	Elongation	72	75 s
32	Elongation	72	10 min

primers (*NcoI/BamHI/NotI*) using Thermo Scientific FastDigest enzymes, according to the manufacturer's instructions.

10. Digest pNCK with the aforementioned enzymes and load the digestion product directly onto an agarose electrophoresis gel.
11. Purify both the excised band from the gel and the digested products using Wizard SV Gel and PCR Clean-Up System or any other similar kit (*see* **Note 11**).
12. Ligate pNCK and the PCR products with T4 DNA ligase. A typical reaction setup contains 50 ng of pNCK, 1:3 (plasmid–insert) molar ratio of PCR product, 0.01–0.1 U/μL of ligase, and 1× ligase buffer. Adjust the final volume to 10 μL with Milli-Q water and incubate overnight at 16 °C.
13. Transform 50 μL of chemically competent *E. coli* DH5α with 5 μL of the ligation. After a 30 min incubation on ice, put the tubes on a thermoblock at 42 °C for 90 s and subsequently

incubate again on ice for 5 min. Add 200 μL of SOC medium to the transformed cells and let them at 37 °C and 180 rpm for 3 h. Plate the transformation on LB agar plates with ampicillin (*see* **Note 12**). Other ready to use commercial hypercompetent *E. coli* cells can be used also (i.e., One-shot Top10 chemically *E. coli* competent cells, Invitrogen).

3.2 Mutant Library Preparation

1. Select ten colonies of transformants and inoculate each one onto 5 mL of LB with ampicillin. Let them grow overnight at 37 °C and 180 rpm.
2. Extract the plasmid DNA with GeneJet Plasmid MiniPrep kit (Thermo Scientific).
3. Determine the concentration and quality of the plasmid preparation with a NanoDrop and sequence the extracted clones to obtain a quick overview of the mutational spectrum.
4. If the mutational spectrum and library size are satisfactory (*see* **Note 13**), pool all the colonies of the plates with LB and extract the plasmids using the MiniPrep kit. Keep records of the total number of colonies in the library.
5. Grow *T. thermophilus* in TB medium at 65 °C and 180 rpm until it reaches exponential phase (OD_{600} 0.3–0.4).
6. Transform *T. thermophilus* HB27 by natural competence or electroporation with the variants library and the wild type fusions (*see* **Notes 14** and **15**) by adding 0.2 μg of the library to 0.5 mL of cell culture in a 5 mL sterile tube (*see* **Note 16**).
7. Incubate for 4 h at 60 °C under strong aeration.
8. Spread the cell suspension directly onto TB agar selection plates as described below.

3.3 Selection of Thermostable Mutants

The selection of thermostable mutants is carried out using both temperature of selection and kanamycin concentration in the plates as selective pressure (*see* **Note 17**). However, prior to selection the appropriate conditions have to be determined (**steps 1–4**).

1. Prepare TB agar selective plates containing kanamycin at 20, 40, 60, 80, and 100 μg/mL of kanamycin in triplicate (*see* **Note 18**).
2. Spread 500 μL of the suspension of *T. thermophilus* cells transformed with the fusion to the wild-type protein (*see* Subheading 3.2).
3. Each replicate will be incubated at 60, 65, and 70 °C for 48 h (*see* **Note 19**).
4. Identify the lowest kanamycin concentration and incubation temperature at which the cells transformed with the fusion to the wild type enzyme do not grow. These are the selective conditions for the primary screening of the variants library.

5. Spread 500 μL of the suspension of *T. thermophilus* transformed with the variants library on TB agar plates supplemented with the selected kanamycin concentration.
6. Incubate the plates at the selective temperature determined in **step 4** for 48 h. Register the temperature inside the incubator with a Thermo-track button to control putative temperature changes along the incubation period.
7. Select several colonies for successive rounds of selection at higher temperature (*see* **Notes 20** and **21**).
8. Extract the plasmidic DNA and sequence it to determine which mutations confer thermostability.
9. Combine the most frequent mutations in a single gene and produce the protein in your standard system. Analyse the activity by your standard procedures at different temperatures to confirm the increase in thermostability.

4 Notes

1. Transformation of *T. thermophilus* depends very much on yet undefined components present in mineral water rich in bicarbonate salts. "Evian" water works well, but if not available, test more than one local commercial mineral water to determine which one gives optimum transformability. Test your tap water as well.
2. *Thermus thermophilus* HB27 grows in TB broth at 60–70 °C under strong aeration and plates are incubated at the same temperature in a closed container in a water-saturated atmosphere. Colonies of *T. thermophilus* on TB medium are punctiform, show entire margin and are flat. They have yellow pigmentation due to the presence of the natural carotenoids.
3. PCR primers should be designed to amplify the gene coding for the enzyme that is going to be thermostabilized. The gene is to be cloned in the right translation frame for expression as fusion protein to the thermostable kanamycin nucleotidyltransferase (Kat), and therefore, it might be necessary to add nucleotides. The forward primer should contain the *NcoI* or *BamHI* recognition sequence whereas the reverse primer should have *NotI* restriction site. Both primers have to be noncomplementary and have similar melting temperatures. Melting temperatures can be estimated accurately using the OligoAnalyzer tool from Integrated DNA Technologies: http://www.idtdna.com/analyzer/Applications/OligoAnalyzer/ and the following website can be used to design optimal primer sequences: http://bioweb.uwlax.edu/GenWeb/Molecular/seq_anal/primer_design/primer_design.htm

4. Optimal PCR conditions should be determined for each gene. DNA polymerase without proofreading ability, such as *Taq*, should be used for epPCR.
5. Other vectors could be used but we have more experience in our laboratory with the pNCK *E. coli/T. thermophilus* shuttle vector (*see* Fig. 2). It was constructed by inserting a replication origin from *T. thermophilus* and the *slpA*$_p$-*kat* cassette conferring thermostable resistance to kanamycin [37]. The P*slpAp-kat* cassette was further modified by inserting unique cloning sites (*NcoI, BamHI*, and *NotI*) downstream the promoter followed by a 9-amino acid long linker [34].
6. The linker (AAAGSSGSI) was designed to be flexible, hydrophilic, and resistant to any protease.
7. The average rate of mutation by epPCR depends on the presence of Mn^{2+} and the number of cycles of duplication. If the initial amount of template is high, it will undergo few duplications in the epPCR. On the other hand, a low amount of template will result in a greater number of duplications, and more mutations will be introduced. The ideal number of cycles depends also on the size and sequence of the template DNA. It is recommended evaluating the PCR efficiency before performing an epPCR. In our experience, up to 20 fmol template can be used in an epPCR reaction [38, 39].
8. By running several epPCR reactions in parallel we are decreasing the bias resulting from the errors introduced during the first cycles.
9. The *Taq* DNA polymerase fidelity is further decreased by the addition of Mn^{2+} and unequal dNTP concentrations. By varying the amount of $MnCl_2$ added to the reaction, the mutation rate can be varied from ~1–5 nucleotides per 1 kb [40]. The $MnCl_2$ should be added to the reaction last to prevent dNTP precipitation.
10. It is a good idea to prepare several different error-prone libraries with varying concentrations of $MnCl_2$ (0–0.5 mM) in the PCR reaction.
11. In addition to determining the DNA concentration the NanoDrop ND-1000 (Thermo Scientific) also gives information regarding the 260/280 and 260/230 absorbance ratios. For a successful ligation you need to have a very pure DNA, therefore, the 260/280 ratio should be ≥1.8 and the 260/230 ratio ≥ 2.0.
12. Electrocompetent cells exhibit a rather reproducible transformation efficiency. However, transformation by electroporation requires expensive, specialized apparatus (the electroporator)

that may not be available in every laboratory. Normal rates of transformation are approximately 10^6 to 5.0×10^9 cfu/μg DNA for chemically competent *E. coli* cells and 10^8 to 2.0×10^{10} cfu/μg DNA for electrocompetent cells.

13. A library with an unbiased spectrum of mutations will be maximally diverse to have better chances to contain at least one improved variant. There are three key indicators of bias: (a) the ratio of transition to transversion mutations; (b) the ratio of AT → GC transitions to GC → AT transitions; and (c) the frequency of mutations at A–T base pairs, to mutations at G–C base pairs [41].
14. To reach higher levels of transformation efficiency, libraries can be electrotransformed. A *T. thermophilus* culture is grown in TB at 60 °C until OD_{600} reached 0.3–0.4. The culture is then centrifuged (10,000 × *g*, 5 min), washed twice in ½ volumes of 10% glycerol at 4 °C, and resuspended in a small volume of the same solution (1% of the volume of the starting culture). The DNA (200 ng) is added to 40 μL of this concentrated cell solution, and the mix is electroporated (12,500 V/cm, 5 ms). Cells are incubated at 60 °C for 4 h in 1 mL TB and then spread on TB plates containing increasing concentrations of kanamycin (20–100 μg/mL). Plates are then incubated at the selective temperature as defined for the wild type fusion [34].
15. *Thermus thermophilus* HB27 with this pNCK plasmid reaches 10^5 colonies/μg by natural competence and around 10^6 colonies/μg by electroporation [34].
16. Depending on the manufacturer of the 5 mL sterile tube, sometimes it is necessary to make a small hole with a needle on the lid of the tube to ensure that it does not come out due to broth evaporation at 65 °C.
17. A secondary selection by activity should be made afterward.
18. At least 30 min before the plating dry the plates in a laminar flow cabinet and preheat them in the static incubator at the desired temperature of selection.
19. The selection temperature usually varies between 60 and 72 °C. Start with the lowest temperature.
20. Increase the temperature 1–2 °C for each selection round until you do not have colonies.
21. Plates may be stored afterward for several days at room temperature or at 4 °C up to 1 month. Avoid picking colonies for subsequent assays after this time of storage.

Acknowledgments

This work has been supported by grant RTC-2014-1439-1 from the Spanish Ministry of Economy and Competence. An institutional grant from Fundación Ramón Areces to CBMSO is also acknowledged.

References

1. García JL, Uhía I, Galán B (2012) Catabolism and biotechnological applications of cholesterol degrading bacteria. Microb Biotechnol 5:679–699
2. Plessis-Rosloniec KZD (2011) Steroid transformation by *Rhodococcus* strains and bacterial cytochrome P450 enzymes. Dissertation, University of Groningen
3. Fernandes P, Cruz A, Angelova B, Pinheiro HM, Cabral JMS (2003) Microbial conversion of steroid compounds: recent developments. Enzym Microb Technol 32:688–705
4. Diane W, Stephan R, Wolfgang Z (1997) Application of thermostable enzymes for carbohydrate modification. In: Praznik W, Huber A (eds) Carbohydrates as raw materials IV: proceedings of the fourth international workshop on carbohydrates as organic raw materials. WUV-Universitatverlag, Vienna
5. Fredrich A, Antrakian G (1996) Keratin degradation by *Fervidobacterium pennavorans*, a novel thermophilicanaerobic species of the order thermotogales. Appl Environ Microbiol 62:2875–2882
6. Leuschner C, Antranikan G (1995) Heat stable enzymes from extremely thermophilicand hyperthermophilic microorganisms. World J Microbiol Biotechnol 11:95–114
7. Zeikus JG, Vieille C, Savchenko A (1998) Thermozymes: biotechnology and structure-function relationships. Extremophiles 2: 179–183
8. Goetschel R, Bar R (1992) Formation of mixed crystals in microbial conversion of sterols and steroids. Enzyme Microb Technol 14:462–469
9. Fernández-Arrojo L, Guazzaroni ME, López-Cortés N et al (2010) Metagenomic era for biocatalyst identification. Curr Opin Biotechnol 21:725–733
10. Lorenz P, Liebeton K, Niehaus F, Eck J (2002) Screening for novel enzymes for biocatalytic processes: accessing the metagenome as a resource of novel functional sequence space. Curr Opin Biotechnol 13:572–577
11. Otte KB, Hauer B (2015) Enzyme engineering in the context of novel pathways and products. Curr Opin Biotechnol 35:16–22
12. Rubin-Pitel SB, Cho CMH, Chen W, Zhao H (2007) Directed evolution tools in bioproduct and bioprocess development. In: Yang ST (ed) Bioprocessing for value-added products from renewable resources. New technologies and applications. Elsevier B.V., Oxford, pp 49–72
13. Packer MS, Liu DR (2015) Methods for the directed evolution of proteins. Nat Rev Genet 16:379–394
14. Widmann M, Pleiss J, Samland AK (2012) Computational tools for rational protein engineering of aldolases. Comput Struct Biotechnol J 2:e201209016
15. Lee D, Redfern O, Orengo C (2007) Predicting protein function from sequence and structure. Nat Rev Mol Cell Biol 8:995–1005
16. Sobti M, Mabbutt BC (2013) Rational-based protein engineering: tips and tools. Methods Mol Biol 996:233–250
17. Khoury GA, Smadbeck J, Kieslich CA et al (2014) Protein folding and de novo protein design for biotechnological applications. Trends Biotechnol 32:99–109
18. Bastien G, Arnal G, Bozonnet S et al (2013) Mining for hemicellulases in the fungus-growing termite *Pseudacanthotermes militaris* using functional metagenomics. Biotechnol Biofuels 6:78–92
19. Rashamuse K, Sanyika W, Ronneburg T et al (2012) A feruloyl esterase derived from a leachate metagenome library. BMB Rep 45:14–19
20. Chen IC, Lin WD, Hsu SK et al (2009) Isolation and characterization of a novel lysine racemase from a soil metagenomic library. Appl Environ Microbiol 75:5161–5166
21. Guazzaroni ME, Silva-Rocha R, Ward RJ (2015) Synthetic biology approaches to improve biocatalyst identification in metagenomic library screening. Microb Biotechnol 8:52–64
22. Kumwenda B, Litthauer D, Bishop OT et al (2013) Analysis of protein thermostability enhancing factors in industrially important thermus bacteria species. Evol Bioinformatics Online 9:327–342
23. Zhou XX, Wang YB, Pan YJ, Li WF (2008) Differences in amino acids composition and

coupling patterns between mesophilic and thermophilic proteins. Amino Acids 34:25–33

24. Vieille C, Zeikus GJ (1996) Thermozymes: identifying molecular determinants of protein structural and functional stability. Trends Biotechnol 14:183–190
25. Robinson-Rechavi M, Alibes A, Godzik A (2006) Contribution of electrostatic interactions, compactness and quaternary structure to protein thermostability: lessons from structural genomics of *Thermotoga maritima*. J Mol Biol 356:547–557
26. Szilágyi A, Závodszky P (2000) Structural differences between mesophilic, moderately thermophilic and extremely thermophilic protein subunits: results of a comprehensive survey. Structure 8:493–504
27. Vogt G, Argos P (1997) Protein thermal stability: hydrogen bonds or internal packing? Fold Des 2:S40–S46
28. Jaenicke R, Bohm G (1998) The stability of proteins in extreme environments. Curr Opin Struct Biol 8:738–748
29. Jaenicke R (1991) Protein stability and molecular adaptation to extreme conditions. Eur J Biochem 202:715–728
30. Trivedi S, Gehlot HS, Rao SR (2006) Protein thermostability in archea and bacteria. Genet Mol Res 5:816–827
31. Bornscheuer U, Kazlauskas RJ (2011) Survey of protein engineering strategies. Curr Protoc Protein Sci. Chapter 26:Unit 26.7
32. Rubingh DN, Grayling RA (2009) Protein engineering. In: Encyclopedia of life support systems. UNESCOEOLSS Joint Committee. EOLSS Publishers, Vol 3. p 140–165
33. Van Rossum T, Kengen SW, Van Der Oost J (2013) Reporter-based screening and selection of enzymes. FEBS J 280:2979–2996
34. Chautard H, Blas-Galindo E, Menguy T et al (2007) An activity-independent selection system of thermostable protein variants. Nat Methods 4:919–921
35. Rivera NR (2013) Termostabilización de proteínas de interés biotecnológico. Dissertation, Department of Molecular Biology. Autonomous University of Madrid, Madrid
36. Swarts DC, Makarova K, Wang Y et al (2014) The evolutionary journey of Argonaute proteins. Nat Struct Mol Biol 21:743–753
37. Lasa I, Castón JR, Fernández-Herrero LA et al (1992) Insertional mutagenesis in the extreme thermophilic eubacteria *Thermus thermophilus* HB8. Mol Microbiol 6:1555–1564
38. Cirino PC, Mayer KM, Umeno D (2003) Generating mutant libraries using error-prone PCR. Methods Mol Biol 231:3–9
39. Wilson DS, Keefe AD (2001) Random mutagenesis by PCR. Curr Protoc Mol Biol. Chapter 8: Unit 8.3
40. Zhao H, Moore JC, Volkov AA, Arnold FH (1999) Methods for optimizing industrial enzymes by directed evolution. In: Demain AL, Davies JE (eds) Manual of industrial microbiology and biotechnology, 2nd edn. ASM, Washington, DC, pp 597–604
41. Hanson-Manful P, Patrick WM (2013) Construction and analysis of randomized protein-encoding libraries using error-prone PCR. Methods Mol Biol 996:251–267

Chapter 22

Biocatalysis of Steroids with *Mycobacterium* sp. in Aqueous and Organic Media

Carla C.C.R. de Carvalho and Pedro Fernandes

Abstract

Mycobacterium sp. can convert steroids such as β-sitosterol, campesterol, and cholesterol into commercially interesting products. In aqueous systems, the biocatalysis is limited by the low solubility of the steroids in water. This may be overcome by the introduction of an organic phase acting as a substrate and/or product reservoir.

In this chapter, we describe the biocatalysis of β-sitosterol to 4-androstene-3,17-dione (AD) and 1,4-androstadiene-3,17-dione (ADD) both in aqueous and organic–aqueous systems. In the latter case, both traditional organic solvents and green solvents are proposed.

Key words Biotransformation, Bioconversion, Organic solvents, Two-phase systems, β-Sitosterol, Cholesterol, Green solvents

1 Introduction

Steroids, like many other industrially relevant substrates, are poorly water-soluble and may often present an inhibitory or toxic effect on whole cell biocatalysts [1, 2]. Among these are substrates such as β-sitosterol, campesterol, and cholesterol (Fig. 1), which may be converted by *Mycobacterium* cells. The introduction of an organic solvent which will act as a reservoir for substrates and/or products is a well established approach to circumvent this [3–5]. Nevertheless, organic solvents may influence the physiology and activity of the bacterial cells [2, 6, 7]. Recent environmental, health, and safety concerns have led to the development of sustainable processes in a "Design for the Environment" approach [8, 9]. Steroid bioconversion by *Mycobacterium* cells has been successfully carried out in green solvents such as silicone oil, polyethylene glycol, polypropylene glycol, and ionic liquids [10].

In general, the bioconversion of steroids with *Mycobacterium* cells is carried out with resting cells [11–13]. The cells are firstly produced by fermentation, and then washed with phosphate buffer

José-Luis Barredo, Ignacio Herráiz (eds.), *Microbial Steroids: Methods and Protocols*, Methods in Molecular Biology, vol. 1645, DOI 10.1007/978-1-4939-7183-1_22, © Springer Science+Business Media LLC 2017

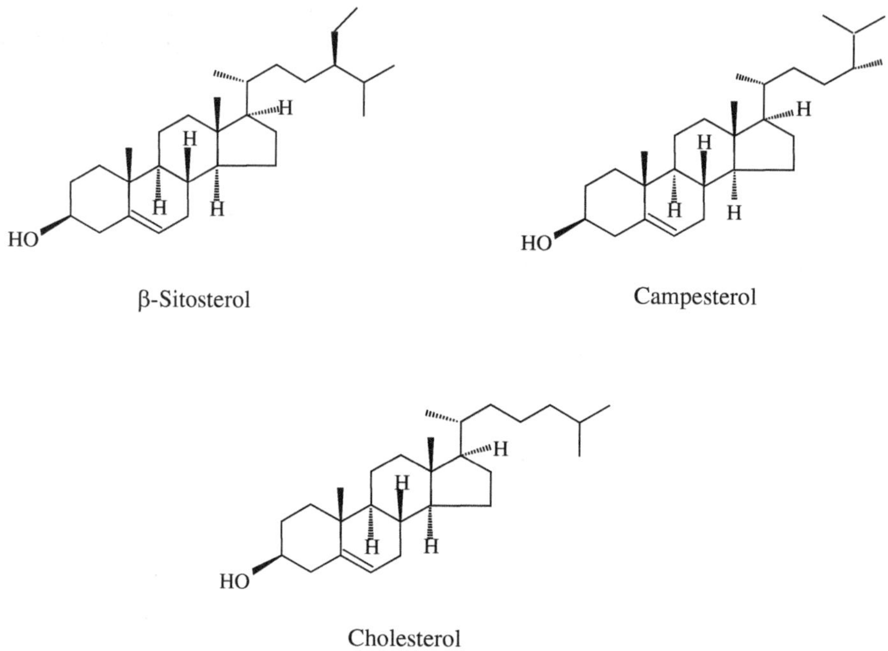

Fig. 1 Chemical structure of β-sitosterol, campesterol, and cholesterol

and recovered by filtration as a wet paste. During the bioconversions they remain in a vegetative state by lack of growth nutrients. These cells are able to remain viable for several days both in aqueous, organic-aqueous, and even single organic bioconversion systems [2, 14]. Nevertheless, other authors have proposed the production of steroids during the fermentative process [15–17]. In these cases, solubilizing agents such as polypropylene glycol or silicone are added since they are biocompatible with *Mycobacterium* sp. growth.

In this chapter, we present the side-chain cleavage of β-sitosterol (*see* **Note 1**) to 4-androstene-3,17-dione (AD) and 1,4-androstadiene-3,17-dione (ADD) as an example of a well-established steroid conversion system.

2 Materials

Prepare buffer solutions and defined growth media preferentially with ultrapure or deionized water. Store prepared buffers and culture media at 4 °C. Solvents for the biocatalytic system include phthalates such as bis(2-ethylhexyl) phthalate, diethyl phthalate, and di-isodecyl phthalate, and green solvents such as poly(methylphenylsiloxane), polyethylene glycol, and polypropylene glycol. β-Sitosterol is a suitable biotransformation substrate while progesterone may be used as an internal standard during HPLC analysis. Standards for AD and ADD, the products of the side-chain cleavage of sitosterol, should be used to determine their retention time in the HPLC analysis. Use ACS (American Chemical Society) and Reagent Grade solvents and analytical grade compounds as standards. Follow all safety handling and storage recommendations provided by the suppliers and diligently respect waste disposal regulations.

2.1 Media Preparation

1. Potato dextrose agar slants. Suspend the appropriate amount of dehydrated potato dextrose agar in 1 L of water, according to the supplier. Heat with agitation and boil for 1 min to completely dissolve the powder. Transfer 5–10 mL of molten agar to test tubes (the amount of medium will be dependent on the volume of the test tubes) and close the tubes with loose caps. Autoclave at 121 °C for 15 min. While the medium is hot, tilt the tubes over a surface so that a long slanted surface is formed during solidification of the agar. After cooling, tighten the caps.
2. Growth medium. This defined mineral medium contains the following compounds per litre: 20 g glycerol, 4 g NH_4Cl, 0.8 g Tween® 20 (*see* **Note 2**), 0.2 g $MgSO_4{\bullet}7H_2O$, 1 g β-sitosterol (*see* **Note 3**), and 100 mM of di-sodium/potassium phosphate buffer, pH 7.0 [10]. Autoclave at 121 °C for 20 min.
3. Disodium/potassium phosphate buffer, pH 7.0. To prepare 1 L of 100 mM buffer dissolve 2.65 g of $NaH_2PO_4{\bullet}H_2O$ and 14 g of K_2HPO_4 in two separate flasks, using 60–70% of the total volume of water. Add both solutions and measure the pH. If necessary, adjust the pH to 7.0 by adding the acid or base species. Adjust the volume to 1 L by adding water in a volumetric flask (*see* **Note 4**).
4. 50 mM phosphate buffer, pH 7.0. To prepare 1 L of buffer, dissolve 5.31 g of K_2HPO_4 and 2.65 g of KH_2PO_4 in two separate flasks using ca. 70% of the total volume of water. After adding both solutions, measure the pH and correct it by adding the necessary acid or base species. Adjust the volume of the mixture to 1 L by adding water in a volumetric flask (*see* **Note 4**).

2.2 Bioreactors

Bioconversion assays may be carried out either in shaken microtiter plates (*see* **Note 5**), Erlenmeyer type flasks or similar plasticware/glassware, or in magnetically/mechanically stirred reactors. The working volume should be maintained at approximately 15% of the total volume of the reactor vessel. When working with organic solvents such as phthalates, glass reactors should be used. Use temperature controlled shakers or place magnetic or mechanical stirrers inside an incubator.

2.3 Bacterial Strains

Mycobacterium strains able to convert steroids into AD and ADD include the following (*see* **Note 6**):

1. *Mycobacterium* sp. NRRL B-3805 [18].
2. *Mycobacterium* sp. NRRL MB 3683 [16, 19].
3. *Mycobacterium* sp. NRRL B-3683 [20].
4. *Mycobacterium neoaurum* VKPM strain As-1634 [17].
5. *Mycobacterium neoaurum* NwIB-yV [11].

2.4 HPLC

1. HPLC equipment.
2. Lichrospher Si-60 column, 5 μm particle size (Merck, Germany), or equivalent.
3. Mobile phase for steroid detection at 215 nm: *n*-heptane and ethanol (98:2, v/v).
4. Mobile phase for steroid detection at 254 nm: *n*-heptane and ethanol (95:5, v/v).

Store prepared samples in a freezer at −16 °C, or at lower temperature, until analysis. The mobile phase should be prepared only before use to prevent solvent evaporation.

3 Methods

3.1 Bacterial Growth

1. Grow *Mycobacterium* sp. NRRL B-3805 cells in potato dextrose agar slants, to use as an inoculum in the fermentation step, for 3 days at 30 °C.
2. Grow *Mycobacterium* cells in orbital incubators at 30 °C and 200 rpm in 1 L Erlenmeyer flasks containing 200 mL of growth medium (*see* **Note 7**).
3. Growth may be monitored by measuring for example the turbidity of the culture medium at 600 nm with a spectrophotometer or by dry weight measurements. However, extensive cell aggregation may occur with *Mycobacterium* cells, thus preventing a reliable monitoring.
4. Harvest the cells by filtration with a 0.2 μm filter (*see* **Note 8**), wash the cells with 50 mM phosphate buffer pH 7.0 and store the cells as a wet paste (approximately 120–200 mg dry cell weight/g wet paste) at −20 °C until use.

3.2 Side-Chain Cleavage of Sitosterol in Aqueous Systems

1. Add the necessary volume of 50 mM potassium phosphate buffer, pH 7.0 (*see* **Note 9**) to reach a ratio of liquid to total bioreactor volume of 1:10–1:5.
2. Weight the necessary wet biomass to achieve ca. 2 g/L of cell dry weight (approximately 10–18 g/L of wet biomass; *see* **Note 10**) and add the cells to the phosphate buffer.

3. Start the bioconversion by adding β-sitosterol at a concentration of 1–2.5 mM (*see* **Note 11**).
4. Run the conversion assay for ca. 24 h at 30 °C and 200 rpm. Agitation may be achieved by an incubator, or by magnetic or mechanical stirring.
5. Take samples along the assay and extract the dissolved steroids with an organic solvent such as *n*-heptane containing 0.2 g/L progesterone as internal standard. In this case, use one volume of internal standard solution, centrifuge at 10,000 × *g* for 2 min and collect the organic phase for HPLC analysis (*see* **Note 12**).

3.3 Side-Chain Cleavage of Sitosterol in Organic–Aqueous Systems

1. Add the necessary volume of 50 mM potassium phosphate buffer, pH 7.0 (*see* **Note 9**) to reach a ratio of liquid to total bioreactor volume of 1:10–1:5.
2. Weight the necessary wet biomass to achieve ca. 2 g/L of cell dry weight (approximately 10–18 g/L of wet biomass; *see* **Note 10**).
3. Add from 1:3 to 1:1 (v/v) organic solvent–aqueous phase (phosphate buffer 50 mM, pH 7.0). As organic phase the following phthalates have been tested and are known to allow the bioconversion of β-sitosterol: bis(2-ethylhexyl) phthalate [2, 12], diethyl phthalate and di-isodecyl phthalate [21] (*see* **Note 13**). Nevertheless, bis(2-ethylhexyl) phthalate allows the highest cellular activities [21]. Add β-sitosterol to achieve a concentration of 5 g/L, referred to the organic phase.
4. Run the conversion assay for ca. 120 h at 30 °C and 200 rpm. Agitation may be achieved by an incubator, or by magnetic or mechanical stirring.
5. Take 300 μL samples along the assay, centrifuge them at 10,000 × *g* for 1 min and dilute 80 μL of the organic phase with 720 μL of *n*-heptane containing 0.2 g/L of an internal standard such as progesterone.

3.4 Side-Chain Cleavage of Sitosterol Using Green Solvents

1. Use 50 mM potassium phosphate buffer, pH 7.0 (*see* **Note 9**) as aqueous phase.
2. Weight the necessary wet biomass to achieve ca. 2 g/L of cell dry weight (approximately 10–18 g/L of wet biomass; *see* **Note 10**).
3. Add 5 g/L of β-sitosterol in a green solvent such as a liquid polymer (e.g., poly(methylphenylsiloxane), polyethylene glycol, or polypropylene glycol) [10, 22]. Use a 1:1 green solvent–aqueous phase ratio.
4. Carry out the experiment at 35 °C and 200 rpm. Take 300 μL samples along the assay, centrifuge them at 10,000 × *g* for

1 min, and dilute 80 μL of the organic phase with 720 μL of *n*-heptane containing 0.2 g/L of an internal standard such as progesterone.

3.5 Determination of Side-Chain Cleavage Activity

1. Perform steroid and sterol quantification by HPLC with 1 mL/min isocratic elution. A column with silica particles with a pore diameter of 60 Å should be used.
2. As mobile phase use (1) *n*-heptane and ethanol (98:2, v/v) for steroid detection at 215 nm (to determine the concentration of β-sitosterol; *see* **Note 14**), or (2) *n*-heptane and ethanol (95:5, v/v) for steroid detection at 254 nm (for quantification of the products AD and ADD).
3. Inject samples containing only substrate or each of the products to determine their respective retention time in the HPLC system used.

4 Notes

1. Besides β-sitosterol, other sterols may be used as a substrate to yield the same steroid products.
2. Tween is intended at decreasing the formation of clusters by the *Mycobacterium* cells during growth. Tween 80 may also be used for the same purpose (*see* for example [23]).
3. The substrate of the bioconversion, β-sitosterol, should be added during growth to stimulate the corresponding metabolic pathway.
4. To prevent growth of microorganisms a 100× concentrated buffer solution may be prepared, which may be stored at 4 °C until being diluted for use.
5. In order to minimize evaporation, bioreactors should be closed with sealing tape, viz., AeraSeal™ or similar, while allowing for gas transfer, as defective aeration hampers sterol side-chain cleavage [18].
6. Several studies have proposed the production of mutants to separate AD production from that of ADD. In 1972, Marsheck et al. produced a mutant by UV irradiation, *Mycobacterium* sp. NRRL B-3683, capable of converting sterols into ADD [24]. Another mutant, unable to 1-dehydrogenate steroids, *Mycobacterium* sp. NRRL B-3805, allowed the production of AD from sterols. Liu et al. selected *M. neoaurum* mutants able to produce large amounts of AD using an atmospheric and room temperature plasma treatment [25]. Gulla et al. obtained a mutant, *M. fortuitum* subsp. *fortuitum* NCIM 5239, able to convert soysterols to AD by combining mitomycin C and UV treatment [15]. Huang

et al. used nitrosoguanidine mutagenesis to produce two *Mycobacterium* sp. NRRL B-3683 mutants, one able to produce AD and the other ADD from phytosterol [26].

7. The ratio volume liquid/Erlenmeyer volume should not be higher than 0.2 to prevent oxygen limitation during growth.
8. Since *Mycobacterium* sp. cells may form large cell aggregates, even when Tween 80 is used, a filter with a lower particle retention capacity may be used to facilitate the filtration process.
9. To increase *Mycobacterium* sp. activity and stability, Marques et al. proposed the addition of the following salts to the reaction medium (in g/L): 0.75 $CaCl_2$, 0.01 $CuSO_4$, 0.01 $CoCl_2{\cdot}6H_2O$, 0.03 $Fe_2(SO_4)_3$, 0.01 $MnSO_4{\cdot}H_2O$, 0.01 $ZnSO_4{\cdot}7H_2O$, and 10.0 $MgSO_4{\cdot}7H_2O$ [27].
10. The wet paste may be weighted frozen for easier handling.
11. β-Sitosterol concentration may be increased with the addition of a water-miscible cosolvent such as dimethylformamide, methanol, acetone, or 1,2-propanediol. The amount of cosolvent should be maintained below 1.5–5% (v/v) to prevent biocatalyst inhibition.
12. If a concentration of β-sitosterol above the solubility of this steroid in aqueous phase is used, suicidal assays must be performed. In this case, parallel assays are made and the entire content of one bioreactor is extracted with an organic solvent such as *n*-heptane containing an internal standard (e.g., progesterone) for analysis in each data point.
13. Wear protection gloves, clothes and glasses when handling phthalates. Use a fume hood and always follow manufacturer's safety instructions when handling organic solvents.
14. Adjustments of ±1% in ethanol concentration may be required to improve peak resolution.

Acknowledgments

C.C.C.R. de Carvalho acknowledges Fundação para a Ciência e a Tecnologia, I.P. (FCT), Portugal, for financial support under program "FCT Investigator 2013" (IF/01203/2013/CP1163/CT0002).

References

1. Schmid A, Dordick JS, Hauer B, Kiener A, Wubbolts M et al (2001) Industrial biocatalysis today and tomorrow. Nature 409:258–268
2. de Carvalho CCCR, Cruz A, Angelova B, Fernandes P, Pons MN et al (2004) Behaviour of *Mycobacterium* sp. NRRL B-3805 whole cells in aqueous, organic-aqueous and organic media studied by fluorescence microscopy. Appl Microbiol Biotechnol 64:696–701
3. León R, Fernandes P, Pinheiro HM, Cabral JMS (1998) Whole-cell biocatalysis in organic media. Enzyme Microb Technol 23:483–500

4. Schmid A, Kollmer A, Mathys GR, Witholt B (1998) Developments toward large-scale bacterial bioprocesses in the presence of bulk amounts of organic solvents. Extremophiles 2:249–256
5. de Bont JAM (1998) Solvent-tolerant bacteria in biocatalysis. Trends Biotechnol 16:493–499
6. Heipieper HJ, Neumann G, Cornelissen S, Meinhardt F (2007) Solvent-tolerant bacteria for biotransformations in two-phase fermentation systems. Appl Microbiol Biotechnol 74:961–973
7. de Carvalho CCCR (2010) Adaptation of *Rhodococcus* to organic solvents. In: Alvarez MH (ed) Biology of *Rhodococcus*. Springer, Berlin, Heidelberg, pp 109–131
8. Little AD (2001) Making EHS an integral part of process design. American Institute of Chemical Engineers, CWRT, CCPS, New York. 164p
9. Tao J, Kazlauskas RJ (2011) Biocatalysis for green chemistry and chemical process development. John Wiley & Sons, Inc., Hoboken, NJ. 496p
10. Marques MPC, Carvalho F, de Carvalho CCCR, Cabral JMS, Fernandes P (2010) Steroid bioconversion: towards green processes. Food Bioprod Process 88:12–20
11. Gao XQ, Feng JX, Wang XD, Hua Q, Wei DZ (2015) Enhanced steroid metabolites production by resting cell phytosterol bioconversion. Chem Biochem Eng Q 29:567–573
12. Cruz A, Fernandes P, Cabral JMS, Pinheiro HM (2001) Whole-cell bioconversion of β-sitosterol in aqueous–organic two-phase systems. J Mol Catal B: Enzym 11:579–585
13. Donova MV, Egorova OV (2012) Microbial steroid transformations: current state and prospects. Appl Microbiol Biotechnol 94:1423–1447
14. Angelova B, Fernandes P, Spasova D, Mutafov S, Pinheiro HM et al (2006) Scanning electron microscopy investigations on bis(2-ethylhexyl) phthalate treated *Mycobacterium* cells. Microsc Res Tech 69:613–617
15. Gulla V, Banerjee T, Patil S (2010) Bioconversion of soysterols to androstenedione by *Mycobacterium fortuitum* subsp. *fortuitum* NCIM 5239, a mutant derived from total sterol degrader strain. J Chem Technol Biotechnol 85:1135–1141
16. Kutney JP, Milanova RK, Vassilev CD, Stefanov SS, Nedelcheva NV et al (1999) Process for the microbial conversion of phytosterols to androstenedione and androstadienedione. CA 2326121 A1; WO9949075-A1
17. Rodina NV, Molchanova MA, Voishvillo NE, Andryushina VA, Stytsenko TS (2008) Conversion of phytosterols into androstenedione by *Mycobacterium neoaurum*. App Biochem Microbiol 44:48–54
18. Cruz A, Angelova B, Fernandes P, Cabral JMS, Pinheiro HM (2004) Study of key operational parameters for the side-chain cleavage of sitosterol by free mycobacterial cells in bis-(2-ethylhexyl)phthalate. Biocatal Biotransformation 22:189–194
19. Sedlaczek L, Gorminski BM, Lisowska K (1994) Effect of inhibitors of cell-envelope synthesis on beta-sitosterol side-chain degradation by *Mycobacterium* sp. NRRL-MB-3683. J Basic Microbiol 34:387–399
20. Llanes N, Hung B, Falero A, Perez C, Aguila B (1995) Glucose and lactose effect on AD and ADD bioconversion by *Mycobacterium* sp. Biotechnol Lett 17:1237–1240
21. Cruz A, Fernandes P, Cabral JMS, Pinheiro HM (2004) Solvent partitioning and whole-cell sitosterol bioconversion activity in aqueous-organic two-phase systems. Enzyme Microb Technol 34:342–353
22. Carvalho F, Marques MPC, de Carvalho CCCR, Cabral JMS, Fernandes P (2009) Sitosterol bioconversion with resting cells in liquid polymer based systems. Bioresour Technol 100:4050–4053
23. Sih CJ (1988) Process for preparing steroids. US Patent 4,755,463 A
24. Marsheck WJ, Kraychy S, Muir RD (1972) Microbial degradation of sterols. Appl Microbiol 23:72–76
25. Liu C, Zhang X, Rao ZM, Shao ML, Zhang LL et al (2015) Mutation breeding of high 4-androstene-3,17-dione-producing *Mycobacterium neoaurum* ZADF-4 by atmospheric and room temperature plasma treatment. J Zhejiang Univ Sci B 16:286–295
26. Huang C-L, Chen Y-R, Liu W-H (2006) Production of androstenones from phytosterol by mutants of *Mycobacterium* sp. Enzyme Microb Technol 39:296–300
27. Marques MPC, de Carvalho CCCR, Claudino MJC, Cabral JMS, Fernandes P (2007) On the feasibility of the microscale approach for a multi-step biotransformation: sitosterol side chain cleavage. J Chem Technol Biotechnol 82:856–863

Chapter 23

Downstream Process Synthesis for Microbial Steroids

Fabian B. Thygs and Juliane Merz

Abstract

Systematic and holistic process development, considering upstream and downstream processing simultaneously, is crucial for designing effective and economical production processes. Based on a multiphasic fermentation process, where oil is added to improve substrate solubility, a systematic approach for the purification of microbial steroids is presented. The design methodology incorporates expert knowledge in the form of heuristics to generate different downstream processing alternatives for processing the complex, multiphasic fermentation broth, and recovering the target steroid precursor androst-4-ene-3,17-dione (androstenedione; AD). The resulting alternative tree of different purification techniques is evaluated using scouting experiments to check the performance of each technique. Purification steps such as extraction, adsorption–desorption, and precipitation seem most promising and have been investigated in detail. Each process step selected is optimized and connected to a process route for recovering AD.

Key words Downstream process, Heuristics, Extraction, Adsorption, Desorption, Precipitation, Solvent

1 Introduction

The steroid compounds androst-4-ene-3,17-dione (androstenedione; AD) and 1,4-androstadiene-3,17-dione (androstadienedione; ADD) are direct metabolic precursors of testosterone or estradiol, and are of special interest for industry [1, 2]. Different substrates can be used to produce AD(D) such as cholesterol from wool grease or animal fat, sapogenins like hecogenin or diosgenin, and phytosterols such as β-sitosterol or campesterol [1]. Due to low costs and adequate conversion rates phytosterols are appropriate substrates [3]. They commonly appear in plants with a high fat content such as vegetable corn oils (soybean [4] or rape seed oil [5]), tall oils [6], and waste materials of sugarcane [7–9] or pulp/paper-industry [10, 11]. Different production approaches for the bioconversion of phytosterols to steroid precursors AD(D) are available like chemical, semisynthetic, and completely biochemical

José-Luis Barredo, Ignacio Herráiz (eds.), *Microbial Steroids: Methods and Protocols*, Methods in Molecular Biology, vol. 1645, DOI 10.1007/978-1-4939-7183-1_23, © Springer Science+Business Media LLC 2017

processes [12–14]. For this case study, the biochemical conversion of phytosterols via the fermentation of *Mycobacterium* species is investigated, as other authors have done before [1, 14–23]. As the fermentation media is commonly of aqueous nature the extremely low solubility of the substrate (≤100 μmol/L) [24] limits a successful industrial production. Hence, different fermentation strategies, as investigated by [11, 14, 25–28], are needed to increase the substrate solubility and with it the microorganism accessibility. A promising approach is the use of solubilizing agents like Tween and lecithin [29], cyclodextrins [30–34], or the addition of a second phase such as polypropylene glycol, silicone oils, or vegetable oils such as soybean- or sunflower oil [35]. The application of a second hydrophobic liquid phase to the fermentation medium is advantageous as the substrate concentration in the aqueous fermentation phase is kept constant on a high level due to the partition coefficient of the substrate between the phases. Furthermore, the product might be in situ extracted from the aqueous phase according to its partition coefficient, so that product-inhibition of the microorganism is limited [29, 36]. As the upstream processing (USP) and the downstream processing (DSP) shall be investigated simultaneously, the consequences using different additives during fermentation need to be considered. To achieve an optimal solution for the whole process the USP and the DSP shall be harmonized and not optimized individually.

In the present case study, the first step was the decision of the additive during USP. Different vegetable oils and cyclodextrins are investigated to improve the substrate solubility. Considering both the fermentation performance and the tasks for the DSP, vegetables oils are preferred to cyclodextrins as they are cheaper, biodegradable and contain naturally small amounts of phytosterols [37, 38] and can be easily separated and recycled. Hence, for both the USP and the DSP, vegetable oils seem more promising, although the viscosity and flow behavior of the oil phase can be challenging. After screening several oils, soybean oil is evaluated as the best oil for a two-phase fermentation process to produce the target steroid precursor AD. The resulting fermentation can be differentiated in an oily (top) phase, an aqueous (middle) phase and a solid biomass-rich (bottom) phase. This multiphase system is the starting point for the following downstream process to recover the target steroid precursor AD. In general, the DSP development starts with an extensive literature research to find similar separation challenges, for instance. In case literature lacks of DSP-strategies for the target molecule, adequate downstream process alternatives are generated based on expert knowledge, namely heuristics collected from literature. However, before the heuristics can be used to generate alternative trees of separation techniques, the phase systems and molecules physicochemical properties need to be determined and analytics implemented to quantify target compounds and impurities. Then alternative trees for each separation step can be phrased and evaluated via additional heuristics and scouting experiments. Promising alternatives are investigated in more detail with screening experiments, where operating conditions for a certain separation technique are varied to determine the optimal parameter set. The methods for analysis and experimental procedures as well as the heuristics used for the DSP development are given in the next sections. The main focus is put on unit operations such as solid–liquid extraction, liquid–liquid extraction, adsorption–desorption, and precipitation to recover the target steroid precursor AD from the multiphase fermentation.

2 Materials

All chemicals used for the experiments are produced for analysis.

1. Centrifuge for 1.5 and 50 mL tubes (e.g., Eppendorf Centrifuge 5415R and 5804R).
2. Vortex.
3. Flexible tube pump (e.g., Watson Marlow 505 S).
4. Incubator (shaking and temperature control).
5. Drying oven.
6. Overhead shaker (e.g., Grant-Bio PTR-60).
7. HPLC equipment: Ultra High-Performance-Chromatography device with diode array detector (DAD) (e.g., Agilent 1290 Infinity, Agilent Technologies).
8. 1.5 and 50 mL tubes.
9. Solvents for HPLC analysis, extraction and desorption experiments should be of high purity (>99%). Fresh demineralized ultrapure water is used for precipitation.
10. AD fermentation broth.
11. Chromatographic column: Nucleosil© C18 endcapped stationary phase with a length of 250 mm, an inner diameter of 4 mm and a particle size of 5 μm with a pore size of 350 Å.
12. Mobile phase: water–acetonitrile. Gradient elution: elution ratio in v/v listed in Table 1.
13. Syringe filter (pore size around 0.22 μm) (*see* **Note 1**).
14. AD standard (Sigma) (*see* **Note 2**).
15. ADD standard (Sigma) (*see* **Note 2**).

Table 1
HPLC method

Parameter	Unit	Output
Mobile phase	–	Water–acetonitrile
Operation mode	–	Gradient
Eluent ratio	Vol.-%	0–10 min: 60/40 10–20 min: 60/40–0/100 20–30 min: 0/100 30–31 min: 0/100–60/40 31–35 min: 60/40
Volume flow	mL/min	1.1
Injection volume	μL	10
Oven temperature	°C	50
Detection	–	UV
Wavelength	nm	240

3 Methods

In the following sections the method to develop a downstream process to recover AD from a multiphasic fermentation broth is presented. However, the general strategy to isolate molecules from their mixtures or the experimental procedures to select appropriate additives are universally applicable to other separation tasks. For general hints for the preparation of the experiments (*see* **Note 3**).

3.1 HPLC Analysis for the Determination of the Steroid Compounds

A robust analyzing method for target and impurity molecules is fundamental for an efficient process synthesis [39]. In this case study, AD is the target, and ADD and (20S)-21-hydroxy-20-methylpregn-4-en-3-one, here referred to as 22-OH-AD, are the main impurities to be removed. A HPLC method is developed to analyze all the compounds necessary to calculate the yield and purity reached after a separation step. For the chromatographic method a gradient is applied to force elution of nonpolar components that strongly interact with the stationary phase and to flush the column (*see* Table 1).

3.2 Preliminary Considerations

Before starting with any experiment an extensive literature screening is mandatory to collect all kinds of data including physicochemical properties of the compounds. In addition, physicochemical properties can be determined by using the freeware ChemSketch [40], where parameters such as molecular weight and 1-octanol–water partition coefficient ($\log P_{ow}$) can be calculated. Furthermore, the polarity can be evaluated via the Hansen (δ_{Hansen}) parameter (hydrogen bonds (δ_h), dispersion forces (δ_d), intermolecular forces (δ_m), and the Hildebrand parameter ($\delta_{Hildebrand}$) which can be estimated from the chemical structure in combination with a group contribution method such as suggested by Krevelen [41]. Beside compound properties, literature shall be screened on already published purification strategies for the target molecule or related separation strategies similar to either the compounds or the phase system. These data are needed throughout the process development.

Next, identifying the distribution of target and impurity molecules in a multiphasic mixture is mandatory to define the separation task in early stages of process development. Physicochemical properties of the target and impurity molecules already collected from literature or calculated based on their structure and/or determined via experiment are useful to get first ideas of distribution of compounds. For instance, the solubility and the polarity of the steroid precursors AD, ADD, and 22-OH-AD determined give information about the distribution of compounds in the system. As the compounds under investigation are nonpolar substances (Polarity order with decreasing polarity: ADD–AD–22-OH-AD) they may distribute in the likewise nonpolar oily (top) phase or precipitate/crystallize as their solubility is exceeded in the aqueous (middle) phase and appear in the solid (bottom) phase after phase separation [1, 14, 24]. To get a deeper insight, the phases shall be investigated individually (*see* **Notes 4–6**). An accurate separation is necessary to analyze the oily (top) phase, aqueous (middle) phase, and solid biomass-rich (bottom) phase, which are obtained after sedimentation of the broth for approximately 24 h at room temperature (*see* **Note 7**). For a reliable analysis of the single phases (concentration of compounds and physicochemical properties of the phase) a more rigid separation is necessary. Several techniques can be applied. Based on the equipment available in standard laboratories (pump: e.g., flexible tube pump; filter: e.g., nutsche filter) the following procedure is developed:

1. Leave the fermentation broth for 24 h at room temperature to separate the phases.
2. Aspirate the oily (top) phase at low rotational speed as the oil phase is highly viscous (*see* **Note 8**).
3. To remove the residual water in the oil phase, centrifuge a sample of TP1 in falcon tubes at 2,300 × *g*. Freeze the centrifuged system at −20 °C and remove the water phase by sawing the falcon tube (repeat procedure two times). Mix the water-free oil phase to have a homogeneous phase.
4. Aspirate the aqueous (middle) phase at higher rotational speed after changing or cleaning the tube or flushing the pump, respectively (*see* **Note 9**). Do not separate the whole aqueous

phase as a certain amount shall be left to ensure a pumpable suspension/slurry of the solid (bottom) phase.

5. Aspirate the suspension with a tube pump and dispense on a filter paper on a nutsche filter to separate the solid and the aqueous phase (*see* **Note 10**). Here no cleaning or tube change is necessary as the aqueous and solid is separated on the filter. After filtration, the retentate is the solid phase and the collected permeate is considered to be the aqueous phase.
6. To remove the residual water in the solid phase, the solid is dried for 5 days in a drying oven at 60 °C. Afterward the dried solid is crushed to get a homogeneous mixture.
7. Measure the volume and mass of each phase to determine the phase-related mass of target and impurity compound(s) (g_i/g_{phase}) and the overall amount in the fermentation broth (g_i/g_{FB}). When evaluating different separation techniques the yield is calculated based on these values (*see* **Note 11**). The index: i = target(s), contaminant(s). Keep in mind that flushing and changing the tubes come along with mass loss.
8. In case the whole multiphase fermentation broth is used for the experiments, prepare the samples by mixing the broth intensively and decanting representative sample volumes in laboratory flasks (*see* **Note 12**).

The compounds distribution and the total amount in the fermentation broth are of main importance. To experimentally investigate the compound absolute masses and distribution leaching experiments are performed for each phase of the fermentation broth (phase-related mass (g_i/g_{phase})) and the whole fermentation broth. As liquid and solid phases need to be leached, extraction is recommended as a leaching method.

Execute leaching experiments to extract the complete technical available amount of the compounds from the oil and solid phases. Extraction in cross-flow mode, where the raffinate phase is extracted with fresh solvent on every extraction stage, is recommended as high yields are reached and a complete leaching of the material can be achieved. For evaluating the leaching experiments, it can be assumed that no solvent remains in the raffinate (complete phase separation) and the density of the extract is equal to the density of the pure solvent. The total mass of i = target(s) or impurity(ies) is defined as the sum of all masses obtained per extraction stage.

If solid materials are extracted, the extraction kinetic may be diffusion controlled. The rate of diffusion increases with decreasing particle diameter [42], so that the use of (equally) powdered solid material is recommended for the experiments [39]. As the diffusion depends on solvent viscosity and temperature [42] higher temperatures are useful. However, the target molecule(s) stability and the temperature limits of the equipment need to be taken into account. In general, a temperature of 20 °C (room temperature) can be used to give first hints and compare the solvents. The liquid/solid-ratio is also an important factor. On the one hand, low ratios result in high extract concentration but inefficient mixing and reduced extraction yields. On the other hand, high ratios result in good mixing conditions, higher yields but diluted extracts. To compare different solvents, which also might have different swelling behaviors, a liquid–solid ratio of 4–10 ($g_{solvent}/g_{phase}$) shall be tested. In case the material strongly tends to swell, ratios of up to 40 are needed. [39]. To avoid kinetic effects the extraction time shall be set to 24 h for each stage. The number of stages is hardly predictable, but at least five stages are recommended. Solvents used for extraction are selected in Subheading 3.3. Samples of the extracts and the raffinates are analyzed via the HPLC method mentioned above.

The experimental procedure for the leaching experiments for each phase is as follows:

1. For the aqueous phase leaching experiments are not necessary as this phase can immediately be measured via HPLC. However, a filtration step with a syringe filter (pore size 0.22 μm) before injection is mandatory.
2. For the oil and solid phase leaching experiments are carried out via cross-current extraction experiments.
3. Therefore, extract the water-free oil phase and solid phase separately with EtOH in 1.5 mL tubes at a phase ratio (φ) of 6 $g_{solvent}/g_{phase}$ for 15 min at 20 °C in an overhead shaker at 20 rpm.
4. Separate the phases via centrifugation at 16,100 × *g* for 15 min at 20 °C. Remove the extract phase with a syringe and filter the sample with a syringe filter (max. pore size 0.22 μm) and analyze it via HPLC.
5. The residual oil or solid phase is extracted again and **steps 3** and **4** are repeated four times.
6. Count all five extraction phases gained and calculate the total amount of target(s) and impurities produced in the bioconversion via the calibration curve and the absolute masses of the phases. For findings of the present case study *see* **Note 13**.

3.3 Generation of Downstream Process Alternatives

Physical and chemical properties such as polarity, boiling and melting temperature, size, and charge can be used to separate compounds from each other, which differ in at least one of these properties. Separation steps which exploit big property differences (*see* **Note 14**) are found at the beginning, while difficult separation tasks, in which the property differences are small, are executed late in DSP (*see* **Note 15**). It shall be named in this context, minimizing the number of purification steps increases the yield (*see* **Note 16**)

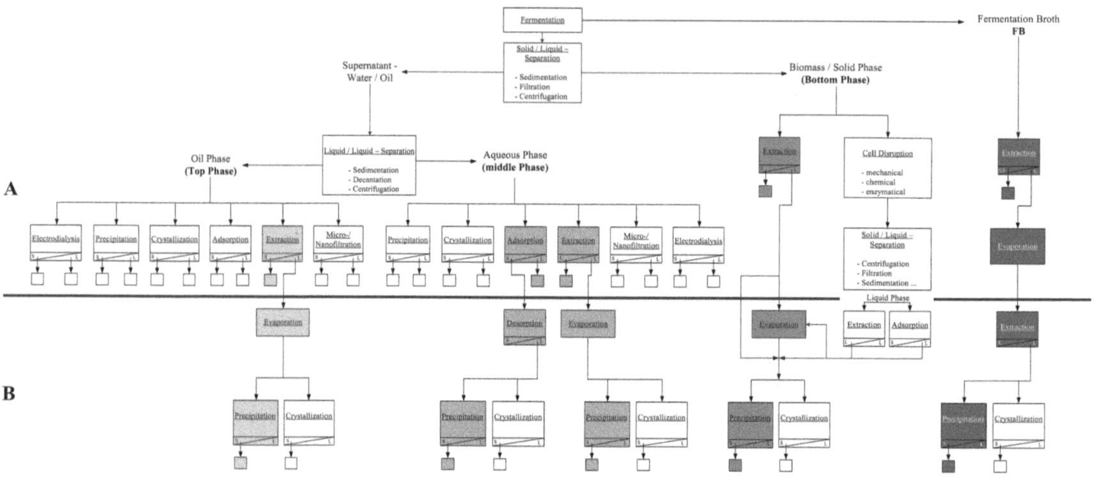

Fig. 1 Tree of downstream process alternatives

and the process productivity by means of reduced operating and investment costs (*see* **Note 17**). This can be achieved by the combination of purification steps, that exploit orthogonal properties for separation (*see* **Notes 18** and **19**). To find adequate separation techniques expert knowledge, which is collected from literature [43] and listed in Subheading 4, is used. With different heuristics and the data already gained process alternatives, as presented in Fig. 1, are generated. An overview how this alternative tree is built up is summarized in the following (Fig. 1, part A):

1. The phases are separated and the solid and oil phase are used as received after water removal (see procedure above).
2. The separated phases are stored in the freezer at −20 °C.
3. The phases of the fermentation broth are treated individually as they contain different amounts of target. Two heuristics justify this decision: (1) In case the system consists of several phases and the product is not located in each, the phases without target should be removed (*see* **Note 18**). (2) The quantitative biggest contamination should be removed first (*see* **Note 19**). In case one phase lacks of target product this phase is the main contamination. In addition, processing the more valuable phases, the overall volume of the material system gets reduced and with this the scale of apparatuses and machines.
4. For the recovery and purification of AD from the liquid phases (oil and aqueous phase) can be executed by adsorption or extraction processes. These unit operations use the difference in polarity of the target toward the impurities to separate them. Another alternative is the use of compounds solubility as driving force for unit operations like crystallization or precipitation. In case molecules have different sizes a filtration

(nanofiltration/microfiltration) might be applicable to separate either the target or the impurity molecules. Additionally, if components have different charge or charge can be induced, by changing the pH value of the system for instance, they can be separated using electrodialysis or comparable techniques.

5. For the generation of process alternatives for the isolation of AD from the solid biomass-rich (bottom) phase additional considerations and scouting experiments are necessary as the product can be differently distributed in the solid phase. It is possible that the molecule(s) under investigation are completely or partially secreted into the fermentation media or stay inside the microorganism. To generate alternatives the location of the product need to be known. If the product is extracellular, the biomass phase is not interesting for DSP and is separated from the supernatant (*see* **Notes 18** and **19**) by solid/liquid-separations like centrifugation or filtration. Is the product intracellular, the cells have to be disrupted to attain the molecules of interest (*see* **Note 20**). This can be done by use of enzymes, chemicals, extreme process parameters like high or low pH-values, or mechanical approaches like using a ball mill (*see* **Notes 21–24**). After cell disruption the solid phase and the liquid phase can be separated or processed together. Commonly extraction or adsorption–desorption steps follow, where the target compound is transferred to a new liquid phase for subsequent purification. In case the product is partly secreted into the medium the ratio of extracellular and intracellular product need to be determined. On basis of this data the effort needed to get the intracellular amount recovered has to be set against its benefit and is commonly expressed in costs. Make sure the increased size or number of fermentation batches, to reach the desired product quantity has to be taken into account for the calculation. Another possibility is that the target component is secreted completely into the medium but is attached to the biomass or crystallizes instantaneously and appears in solid aggregate state together with the biomass. For this scenario the whole solid phase (biomass and solid product) need to be processed and extraction as recovery step is favored. The solvent either extracts or solves the product and transfers it to a more preferable phase for DSP.

As one separation step is mostly not sufficient to purify the product additional steps are necessary to reach the purity requirement (Fig. 1, part B). Analog to the former alternative generation the property differences of the molecules present in the system after the first separation step and heuristics are used to generate the alternatives for the second purification step.

The generation of the alternatives for the second purification step is adapted to the present case study. Extraction and adsorption–desorption are selected as alternatives for separation step 1 (*see* **Note 25**) and appropriate techniques for the second separation step are discussed in detail below (Fig. 1, part B):

1. As for adsorption–desorption and extraction experiments the impurity profiles are similar (see following sections) a series connection of these steps seems not promising to reach the desired purity. However, for other separation tasks that may be checked if a combination cannot remove a certain impurity. As size and charge of the target AD are similar to the main impurities ADD and 22-OH-AD separation steps which involve this driving forces are not considered. Another molecule property used for separation is the solubility of the compounds and can be applied in processes like precipitation and crystallization, where the target compound is transferred into a solid aggregate state. This is preferable as biotechnological products may precipitate or crystallize in a highly pure form (*see* **Note 26**). Although yields of Υ = 100% in one stage are not expected these unit operations may be tested as alternatives to purify AD (*see* **Note 27**).
2. In case the product is highly diluted in the extract or desorbent phase, the solvent can be (partly) evaporated to increase the concentration. This step is efficient, especially, for processes such as precipitation and crystallization, where the yield is dependent on the starting concentration of the target molecule.
3. After precipitation and crystallization the solids need to be separated from its supernatant or mother liquor, respectively by a solid/liquid separation step. To increase the purity of precipitate or crystals washing steps, with solvents selectively solving the impurities from the solid material, can be applied. After that drying of the solids is the last process step.
4. Make sure at any time of DSP-synthesis the scalability of the purification methods selected is given (*see* **Notes 28** and **29**).
5. In case the product is already available on the market, a high product quality is mandatory to be competitive (*see* **Note 30**).

Another alternative for the purification of AD as a second or third purification step is chromatography, which is neither presented in Fig. 1 nor discussed as the costs of this step are too high compared to the product value of AD. However, to make this section more general, some notes about chromatography are given. This process step shall be applied late in DSP due to its low capacities and high operating costs (*see* **Note 31**). The main advantage of this technique is the ability to selectively separate compounds from complex mixtures [purities of >99% can be reached for each molecule (*see* **Note 32**)]. However, this performance is accompanied by a high dilution of the product fraction (*see* **Note 33**) making a concentration step necessary to formulate the product (e.g., via crystallization). In general, if the purity requirement cannot be reached by other process steps, chromatography shall be investigated (*see* **Notes 34** and **35**).

3.4 Evaluation of Downstream Process Alternatives

During the characterization of the single phases crystals are observed in the solid as well as in the oil phase. Scouting experiments showed that the crystals can be assigned to the steroid precursors. Hence, no cell disruption is needed to recover AD from the solid phase and the solid phase is processed as received after phase separation. In addition, as product crystals are present in the oil phase, techniques like adsorption, crystallization, precipitation or electrodialysis are not applicable as the product is not solved. Recovering the crystals via filtration or centrifugation the oil phase behavior (viscosity and density) limits the performance and a lot of energy has to be put into the system to separate the phases. However, for a certain separation task: If two or more phases are present, which contain different amounts of product(s), the first choice is to separate the phases as a first separation step (*see* **Notes 18** and **19**).

3.4.1 Extraction to Recover AD

Extraction plays a major role as every phase of the fermentation broth can be processed with this technique (*see* Fig. 1). In addition, the total amount of AD and ADD is gained via leaching extraction experiments (see above). The selection of adequate solvents is crucial for an efficient DSP and mass determination. Different approaches exist in literature to determine a good extraction solvent [44–46]. Especially in early stages of process development, where thermodynamic data and information about impurities are limited, a target-oriented method for solvent selection is necessary. Robust methodologies for the solubility estimation are given by using the 1-octanol–water partition coefficient or the Hildebrand and Hansen solubility parameter [41, 47–49]. However, the validity and performance of the prediction methods shall be investigated by experiments. According to [39] a standard procedure for solvent selection is summarized:

1. Collect physical and chemical properties of solvents from literature or the supplier.
2. Use the physical and chemical properties of the target and impurity molecules collected before.
3. It should be checked if the field of product application limits the selection of solvents, e.g., food or pharma industry. Another limitation can be the use of green solvents only.
4. Define the operating window by means of upper and lower temperature limits. Thus, solvents with melting points higher than the extraction temperature can be excluded [39].

Table 2
Selection of extraction solvents

Amyl acetate	Octanol
Toluene	Ethyl acetate
Cyclohexane	Methyl benzoate
n-Hexane	Chloroform
Petrol ether	1-Decanol
n-Butyl acetate	1-Hexanol
Diethyl ether	*Methanol*
n-Heptane	*Ethanol*

Depending on following process steps like evaporation, solvents with boiling points higher than the temperature limit to ensure a stable product are dismissed as the target molecule will get damaged (*see* **Note 36**).

5. Select solvents which differ slightly in Hansen and Hildebrandt parameter or the $\log P_{ow}$ as this indicates a good solubility of the target molecule in this solvent (*see* **Note 37**).
6. Determine if the feed phase is immiscible with the solvent phase. In case the liquid phases are completely soluble in each other, no two phase system is build and extraction is not possible.
7. Solvents of the final ranking shall be evaluated with extraction experiments which are executed following the standard protocol to compare the solvents appropriately (see leaching experiments).
8. If good solvents are determined vary the operating conditions (temperature, phase ratio, and extraction time) to optimize the process.

In the following part, the selection of appropriate solvents to recover AD from the different phases is described. For this case study, no real restrictions are present and all data are already collected.

Due to the nonpolar character of the target molecule AD, nonpolar extraction solvents are identified by the methods using the 1-octanol–water partition coefficient and the Hansen and Hildebrand solubility parameter (compare Table 2).

The listed solvents can be appropriate to extract AD from the aqueous phase, but for the oil phase the binary solubility between the nonpolar solvents and the oil need to be considered. As none of the solvents of Table 2 are suitable, findings from the extensive literature screening at the beginning of the downstream process synthesis are used. Some studies determined the good solubility of the target AD in lower alcohols such as methanol (MeOH) or ethanol (EtOH) [50–52]. As the alcohols are immiscible with the soybean oil used they are suitable extraction solvents for the oil phase. As they have a high solubility toward AD the alcohols are also tested to extract/dissolve AD from the solid phase.

A procedure on how the extraction solvents are screened and the approach to select the optimal process parameters are given below:

1. Extract the oil phase and the solid phase separately with MeOH and EtOH under standard conditions ($\varphi = 6\ g_{solvent}/g_{phase}$ for 15 min at 20 °C in an overhead shaker at 20 rpm). After centrifugation (e.g., at 15,700 × *g* for 5 min) to accelerate the phase separation, remove the extract phase with a syringe. Samples are filtered with a syringe filter (max. Pore size 0.22 μm) and analyzed via HPLC.
2. To investigate the time to reach quasi equilibrium a sample for each extraction time, defined $t = 2, 15, 20, 35, 65, 95$ min, is prepared in a separate vial. The time for phase separation is counted as part of the extraction time because contact between the two phases is still maintained. Experiments are executed for methanol and ethanol with a phase ratio of 6 ($g_{Solvent}/g_{Phase}$) at room temperature.
3. Extraction temperature is an important process parameter. On the one hand, the extraction at higher temperatures often results in higher solute concentration in the extract phase, and on the other hand, higher temperatures can lead to an enhanced extraction of contaminations, which reduce the purity of the extract [53]. According to the extraction kinetics, the temperature is varied ($T = 20$, 30, and 40 °C) while keeping the phase ratio [$\varphi = 6$ ($g_{Solvent}/g_{Phase}$)] and time ($t = 30$ min) constant. The experiments are carried out for both alcohols.
4. For optimal extraction processes the phase ratio is evaluated. The influence of the phase ratio is investigated in the sequence of 2, 4, 6, 8, 12 ($g_{Solvent}/g_{Phase}$) ratios. Based on the previous results, the experiments are executed at 30 °C for 20 min.
5. The extraction performance is best using ethanol at 30 °C, a phase ratio of 6 ($g_{Solvent}/g_{Phase}$), and 20 min for the oil and solid phase (*see* **Note 38** for the results of the oil phase extraction and **Note 39** for the solid phase extraction).

The solvents presented in Table 2 can be tested to recover AD from the aqueous phase. For this the miscibility of the aqueous phase and the tabled solvents are determined. Suitable solvents are investigated to recover AD from the aqueous solution. Although this phase lacks of target and therefore is not interesting for DSP, experiments are carried out to get information about the systems behavior. Optimization of the process conditions is in accordance to the oil and solid phase extraction.

1. Best extraction performance is determined for using butyl acetate at a phase ratio of 10 ($\mu L_{solvent}/\mu L_{phase}$), at 50 °C for 20 min (*see* **Note 40**). The phases are separated by centrifugation (e.g., at 9,300 × *g* for 10 min). Extract samples are diluted five times prior to HPLC injection.

3.4.2 Adsorption–Desorption Process to Recover AD

During an adsorption–desorption process in capture mode, the target molecules located in a fluid mixture selectively bind to a solid phase (adsorbent) and are selectively desorbed with a solvent. Suitable adsorbents are characterized by a large surface area and high capacity and selectivity toward the target compound. In addition, stability against chemical, physical, or thermal stress is advantageous. From economic point of view, the adsorber material shall be easy regenerable for multiple use and cheap. A variety of adsorber materials are available on the market, which can differentiate between carbonaceous-, oxidic-, and polymer-based adsorbents [54–56].

Before the adsorbents can be used a pretreatment is necessary:

1. Wash particles with fresh demineralized water twice. Therefore particles can be placed in a sealable vial and mixed intensively. Phase separation can be done by sedimentation or centrifugation.
2. Wash particles with pure methanol and separate the phases again.
3. Dry the wet adsorbents in a drying oven, e.g., at 40 °C for 24 h.

An adsorption–desorption process to purify AD is most advantageous for the aqueous phase as the main amount of AD in the oil and the solid phase are not dissolved. Different adsorber particles according to their polarity are chosen as presented in Table 3. This adsorber particle selection can be used for general purposes.

Adsorption and desorption experiments are executed with the same internal adsorbent surface. Thus, the amount of the adsorbents is varied while the volume of the feed liquid is kept constant [57]. The aqueous phase is used for the experiments to show the potential of an adsorption–desorption step to recover AD. The data can be used to select adsorption–desorption as a second purification step.

1. Vary the adsorption surface to 1, 5, and 10 m^2/mL (or higher) and add a volume of 5 mL aqueous phase. For screening experiments the temperature (25 °C) and the experimental time (7 h) are kept constant. Phase separation is done by centrifugation at high rotational speeds (e.g., at 15,700 × *g* for 20 min). The feed

Table 3
Selection of adsorbent materials

Adsorbent	Material	Internal surface (m^2/g)	Polarity (qualitative)
Amberlite XAD-16	Polystyrene–divinylbenzene	800	Nonpolar
Amberlite XAD-761	Polyformaldehyde polycondensate	200	Polar
Sepabeads SP850	Polystyrene–divinylbenzene	930	Very polar
Sepabeads SP207	Polystyrene–divinylbenzene	590	Nonpolar
Diaion HP2MG	Polymethacrylate	500	Polar
Amberlite XAD-7	Polymethacrylate	450	Polar
Norit NRS	Activated carbon	1125	Nonpolar
Norit Row 0.8	Activated carbon	1000	Nonpolar
YMC*Gel Cyano	Silica gel	330	Slightly polar

solution and the depleted supernatant are analyzed via HPLC to identify the adsorbed amount of the target molecule.

2. For the determination of the adsorption kinetic small surface areas are recommended to detect different adsorption states precisely, e.g., use an adsorbent surface of 1 m^2/mL. Samples for each sampling time, intervals of 1, 5, 10, 30, 60, 90, 120, 180, 300, and 420 min, are prepared in separate vials.
3. Desorption experiments are used to identify a suitable desorbent solvent. The selection of desorbents is done in accordance to the selection of extraction solvents. Hence, the solvents presented in Table 2 are investigated. For desorption experiments adsorption experiments with an adsorbent surface of 5 m^2/mL and 300 min are carried out first. The supernatant is separated and fresh desorbent is added to the adsorbent. A volume of 7 $mL_{Desorbent}/m^2_{Adsorbent}$ is used. Experiments are executed at 25 °C for 300 min.
4. Phase separation is done by centrifugation at high rotational speeds (e.g., at 15,700 × *g* for 20 min). The desorbent solvent is analyzed via HPLC to identify the desorbed amount of the target molecule (*see* **Note 41**).

3.4.3 Crystallization to Recover AD

Crystallization is induced by the supersaturation of the target molecule by cooling or evaporating the surrounding solution or adding an antisolvent. After crystallization the product is present in crystalline form of different habitus (needle and plates). To exactly design crystallization processes the solubility line and the metastable area of the solute in the target containing solution is determined via experiment or literature data (*see* **Note 42**). Crystallization processes are preferred, where no additives are used as no subsequent solvent recovery and recycle are necessary (*see* **Note 43**). Although crystallization is a powerful separation technique, it is not experimentally investigated during the present process development.

3.4.4 Precipitation to Recover AD

Precipitation can be induced by addition of organic solvents, acids, bases, salts, or antisolvents [58–60]. Considering the nature of the compounds under investigation namely neutral, hydrophobic and low molecular weight, the addition of an antisolvent seems most appropriate to initiate a precipitation process. Requirements to an antisolvent are a low solubility of either the target or the impurity molecules depending on which molecule(s) shall be separated (*see* **Note 44**) and the complete solubility in the molecules containing solution (here extract or desorption solvent). The products of a precipitation are mostly amorphous or semicrystalline (*see* **Note 45**), and the purity is probably lower compared to crystallization processes (*see* **Note 46**). However, the precipitation is faster and often easier to perform.

An antisolvent precipitation, as the second purification step, is investigated for extracts of the oil and solid phase. As the target component AD is hardly soluble in water (*see* **Note 11**) and since water and lower alcohols are completely miscible, water is used as antisolvent.

1. As the precipitation is dependent on the starting concentration of the target molecule in solution a concentration step is useful to increase the yield. Different starting concentrations shall be screened to identify the critical concentration beyond precipitation is achievable and to determine the yield dependency. In the case of this study a critical concentration of 7 $g_{AD}/L_{Extract}$ is determined.
2. Experiments with varying water–extract volume ratios (ξ) in a range of 0.5, 1, 2, 4, 5, 6 ($g_{water}/g_{Extract}$) at room temperature are carried out to find the best ratio in terms of purity and yield. After the addition of water, the samples are centrifuged, e.g., at 15,700 × *g* for 15 min. A ratio of 5 ($g_{water}/g_{Extract}$) resulted in best precipitation results.
3. The weight of each sample is determined to calculate the precipitated mass. The supernatant is analyzed by HPLC and the precipitate is redissolved in the same volume of the previous solvent used. With this, equal conditions of the sample before and after precipitation are generated (*see* **Note 47**).

3.5 Downstream Process for Separated Phases

Considering the alternative tree presented in Fig. 1 and the experiments done, the fermentation shall be separated and the aqueous phase discarded. The oil and the solid phase shall be extracted using ethanol. Therefore, it is recommended to pool the extract phases of the oil and the solid phase and concentrate the mixed extract to precipitate AD and finish the downstream process. The first two extraction stages of the solid phase extraction and the first extraction stage of the oil phase extraction are pooled and concentrated to 35 $g_{AD}/L_{Mixed\ extract}$. An overall process yield of $Y_{Process,AD} = 0.82$ and a purity of $X_{Process,AD} = 0.89$ are achieved.

3.6 Downstream Process for the Whole Fermentation Broth

Although the heuristics lead to separate the phases to recover AD, the processing of the whole fermentation broth can be more economical. Based on the experimental investigations with the separated phases a process route to recover AD from the whole fermentation broth is developed (*see* Fig. 1). Starting with an extraction using butyl acetate as extractant AD is extracted from the fermentation broth (findings from the extractant screening for the aqueous phase). As butyl acetate is partly miscible with the soybean oil, oil is coextracted. In a second step, the solvent is changed, e.g., by evaporation, and coextracted oil phase including high concentrations of AD remain. The subsequent purification task is comparable to the purification of the oil phase, so that an extraction with ethanol is applicable to further enrich the target component.

The last purification is done by adding water to the concentrated extract phase and AD is precipitated. An overview of the process parameters is given in the following:

1. To have sufficient material to execute the whole process route, the scale of the first extraction is increased. A phase ratio of 2 ($g_{solvent}/g_{phase}$) is used and the complete fermentation broth is extracted with butyl acetate at 23 °C for 30 min and one stage.
2. After centrifugation (e.g., at 15,700 × g for 5 min) to accelerate the phase separation, a sample for HPLC analysis is taken. Then the complete extract phase is removed by solvent evaporation in a rotary evaporator at 20 mbar and 60 °C. Select gentle conditions for solvent evaporation, otherwise small amounts of the product are dragged along with the evaporating solvent and the product gets lost.
3. After the evaporation step an oily residue remains. A sample of both, the evaporated butyl acetate and the oily residue, is taken for HPLC analysis.
4. The second extraction step is executed with ethanol, at a phase ratio of 6 ($g_{solvent}/g_{phase}$), at 30 °C, and for 20 min. After centrifugation (e.g., at 15,700 × g for 5 min) to accelerate the phase separation, a sample is taken for HPLC analysis.
5. The extract is concentrated via a second evaporation step in a rotary evaporator at 100 mbar and 40 °C to achieve concentrations above 7 $g_{AD}/L_{Extract}$. To reach high yields, concentrations around 40 $g_{AD}/L_{Extract}$ are recommended.

6. The concentrated extract is precipitated using water as antisolvent at a phase ratio (ξ) of 5 ($g_{water}/g_{Extract}$) and temperature of 25 °C. After the addition of water, the samples are centrifuged, e.g., at 15,700 × *g* for 15 min to separate precipitate and supernatant.

An overall process yield of $\Upsilon_{Process,AD} = 0.68$ and a purity of $X_{Process,AD} = 0.87$ is achieved.

As stated in (*see* **Note 14**) the more process steps are involved, the lower the yield (compare performance of the DSP processing the separated phases). To rank the best process the costs for each process route need to be calculated and compared. Moreover, the feasibility and scalability have to be taken into account (*see* **Note 29**), especially for the three phase separation (DSP route with separated phases).

4 Notes

1. To avoid clogging of the HPLC device, it is recommended to filter the samples before injection. In addition, samples taken from experiments might be diluted or concentrated before analysis. For this study, the samples were diluted 1/50 with acetonitrile.
2. A calibration with external standards for the molecules of interest (e.g., AD) is necessary to calculate yield and purity accurately. If no standards are available the purity and yield can be calculated via the ratio of the chromatographic peak area. In this study, AD and ADD in concentration ranges of 0.02–1.5 g/L are calibrated. For 22-OH-AD no standard is available. Hence, all purity values given in the text are referred to the ratio of the AD peak area to the sum of all peak areas in the chromatogram (excluding solvent peaks).
3. To make sure that experiments are reliable, consider the following advices:

   (a) Sufficient mixing of the phases need to be guaranteed.

   (b) The temperature has to be precisely adjusted and maintained for accurate experiments.

   (c) The scale of experiments depends on the amount of samples available ranging from small scale experiment, e.g., in 1.5 mL Eppendorf, to larger scale experiments, e.g., 15 mL/50 mL Falcon tubes. Depending on the scale, different equipment is used. If comparing results from different scales, be sure that the mixing behavior, the temperature and the performance of the phase separation are comparable.

   (d) Make sure the scale for the experiment is not too small. For instance, running an adsorption to recover AD from the aqueous phase, the ratio of feed phase to adsorber particles shall be as high to allow sufficient wetting of the particles and good mixing of the phases.

(e) Maintain samples at conditions to preserve their stability. Here, samples of the fermentation broth and separated phases are stored at −20 °C and thawed at room temperature before use.

(f) Experiments shall be executed at least in duplicate to assure certain accuracy.

4. (Heuristic ID.1-10 in [43]) "For the development, the optimization, and the evaluation of process alternatives the upstream processing (USP) and downstream processing (DSP) have to be treated as one coherent system."
5. (Heuristic ID.3-17 in [43]) "If possible do not use complex media in USP as it increases complexity and effort in DSP."
6. (Heuristic ID.3-19 in [43]) "The fermentation medium should be defined precisely in its chemical composition."
7. First separation of the phases appears after a few minutes. However, the separation is more rigid if the resting time is chosen higher.
8. Using high rotational speeds the aqueous phase may be aspirated and an incomplete separation of the phases is the consequence. If the separation using the tube pump is not efficient enough to separate oil and aqueous phase, the samples shall be centrifuged and either the water is removed directly. This procedure can be repeated two to three times.
9. The oil phase is very viscous and sticks to every surface. Hence, parts of the oil may remain in the tube and contaminating the aqueous and solid phase if the same or uncleaned tube is used during the phase separation. As the oil sticks in the tube and the one is exchanged take care to calculate the mass loss of this phase.
10. Particle retention of the filter paper should be at least 1 μm, as the microbial cells are in the scale of 1–10 μm [61] and other solid structures of the bottom phase appear significantly larger in investigations with a microscope.
11. The yield is defined as the mass of a component before and after a purification step. To calculate the yield of the first purification step, the initial mass of AD has to be determined.
12. It is important that decantation is executed directly after mixing the fermentation broth to avoid the de-mixing of the phases.
13. In the bioconversion using from 70 $g_{Phytosterol}$ a total amount of 34.49 g_{AD} and 1.17 g_{ADD} are determined. AD and ADD are located to 47.1 and 72.6% in the oil phase (top phase), while

approximately 52.2% (AD) and 19.7% (ADD) are located in the solid phase (bottom phase). Amounts around 0.64% of AD and 7.70% of ADD are identified in the aqueous phase (middle phase), which is comparably low. Thus, the oil (top phase) and solid phase (bottom phase) are of special interest for downstream process synthesis as they contain around 99.3% of the total mass of AD.

14. (Heuristic ID.2-46 in [43]) "Choose purification steps which exploit big differences of physico-chemical properties between the product and impurities."
15. (Heuristic ID.2-47 in [43]) "Choose purification steps which exploit small differences of physico-chemical properties between the product and impurities late in DSP."
16. (Heuristic ID.0-5 in [43]) "The more process steps are needed, the lower is the maximum yield, especially for sensitive products."
17. (Heuristic ID.0-4 in [43]) "In order to maximize the productivity and minimize investment and operation costs, the minimization of process steps is mandatory."
18. (Heuristic ID.0-3 in [43]) "The sequence of process steps (the "ideal" linkage of unit operations) has a great influence on the result of process synthesis."
19. (Heuristic ID.2-48 in [43]) "Link purification steps which exploit different physico-chemical properties between the product and impurities."
20. (Heuristic ID.2-50 in [43]) "In case different phase already exist from USP and the product is located just in one of them, this phase should be separated from the other phases first."
21. (Heuristic ID.2-49 in [43]) "Separate the quantitative biggest impurity first."
22. (Heuristic ID.3-57 in [43]) "In case the product is located intracellular and cannot be extracted from the cells, they have to be disrupted to make the product accessible for DSP."
23. (Heuristic ID.3-61 in [43]) "Mechanical methods should be used for cell disruption in large scale processes."
24. (Heuristic ID.3-76 in [43]) "Cell disruption by applying ultrasound is not applicable for large volumes."
25. (Heuristic ID.3-77 in [43]) "Cell disruption by freezing and thawing is not applicable for technical applications."
26. (Heuristic ID.3-78 in [43]) "Cell disruption based on physical methods, such as freezing/thawing or osmotic shock, are time-consuming and thus uneconomical for large scale processes."

27. (Heuristic ID.3-134 in [43]) "To purify non-polar components such as primary- or secondary metabolites, solvent extraction might be successful."
28. (Heuristic ID.3-191 in [43]) "Lots of fine chemicals and biotechnological products reach high purity in solid aggregate state, so that precipitation/crystallization is mandatory in the process concept"
29. (Heuristic ID.3-192 in [43]) Crystallization or precipitation processes never reach a yield of 100%, a small amount of the solute will stay in the supernatant.
30. (Heuristic ID.0-8 in [43]) "Process development is based on engineering expertise to avoid the development of purification methods, which are unable to scale up or which do not reach the performance desired in large scale."
31. (Heuristic ID.3-3 in [43]) "The scalability of purification steps should be considered even in early phases of process development."
32. (Heuristic ID.1-3 in [43]) "In case the product is not unique and somehow already available on the market, the product quality is a central aspect to be competitive on the market and even more important than shorter development times."
33. (Heuristic ID.3-157 in [43]) "Due to low capacities of adsorption processes, they should only be applied if purification based on other methods is insufficient."
34. (Heuristic ID.3-159 in [43]) "Purification by chromatography should be applied if high purities >99% have to be reached."
35. (Heuristic ID.3-160 in [43]) "The product is highly diluted after chromatographic separation."
36. Solvents with a higher melting point than room temperature appear as solid and cannot be used for solid/liquid extraction. Solvents with a higher boiling point than the component stability are excluded as product degradation appears if the solvent is evaporated.
37. It is recommended to estimate the solubility for each parameter separately and generate three solvent rankings, starting with the 1-octanol–water partition coefficient ($\log P_{ow}$), and afterward the Hansen (δ_{Hansen}) and the Hildebrand parameter ($\delta_{Hildebrand}$). All values can be weighted, e.g., one-third of each parameter to apply a balanced weighing, and condensed to a final average value for each solvent or a final ranking of all solvents, respectively.
38. Extraction experiments of the oil phase with methanol and ethanol. Yield for AD is calculated for a single extraction stage under the experimental conditions described before. Purity is

calculated via the percentage peak area ratio of the AD peak to the sum of all other peaks in the chromatogram except the solvent peak. Extraction using MeOH: $\Upsilon_{MeOH,AD} = 0.77 \pm 0.001$ and $X_{Extr,AD} = 0.75 \pm 0.001$. Using EtOH: $\Upsilon_{EtOH,AD} = 0.86 \pm 0.025$ and $X_{Extr,AD} = 0.75 \pm 0.005$

39. Extraction experiments of the solid phase with methanol and ethanol. Yield for AD is calculated for a single extraction stage under the experimental conditions described before. Purity is calculated via the percentage peak area ratio of the AD peak to the sum of all other peaks in the chromatogram except the solvent peak. Extraction using MeOH: $\Upsilon_{MeOH,AD} = 0.77 \pm 0.001$ and $X_{Extr,AD} = 0.89 \pm 0.002$. Using EtOH: $\Upsilon_{EtOH,AD} = 0.86 \pm 0.003$ and $X_{Extr,AD} = 0.88 \pm 0.0004$. It is particularly mentionable that the bottom phase consists of highly concentrated AD, so that during extraction experiments with a phase ratio ≤ 4 $(g_{solvent}/g_{phase})$ the solvent is saturated with AD which results in a decreased extraction performance.

40. Extraction experiments of the aqueous phase with butyl acetate. Yield for AD is calculated for a single extraction stage under the experimental conditions described before. Purity is calculated via the percentage peak area ratio of the AD peak to the sum of all other peaks in the chromatogram except the solvent peak. A yield of $\Upsilon_{BuAc,AD} = 0.87 \pm 0.018$ and a purity of AD of $X_{Extr,AD} = 0.29 \pm 0.006$ are achieved.

41. The adsorbents “Amberlite XAD-7” and “Diaion HP2MG” show the best adsorption performance. As desorbent solvents chloroform and decanol reached good results. Final purity of $X_{XAD7\text{-}Chloroform,AD} = 0.46 \pm 0.002$ ($X_{XAD7\text{-}Chloroform,ADD} = 0.34 \pm 9E-05$) and $X_{Diaion\text{-}Decanol,AD} = 0.42 \pm 0.006$ ($X_{Diaion\text{-}Decanol,ADD} = 0.34 \pm 0.002$) could be reached in the adsorption–desorption step. The yield is around $\Upsilon_{XAD7\text{-}Chloroform,AD} = 0.94 \pm 0.047$ ($\Upsilon_{XAD7\text{-}Chloroform,ADD} = 0.71 \pm 0.039$) and $\Upsilon_{Diaion\text{-}Decanol,AD} = 0.80 \pm 0.011$ ($\Upsilon_{Diaion\text{-}Decanol,ADD} = 0.66 \pm 0.022$).

42. (Heuristic ID.3-196 in [43]) “To design a crystallization process, the solubility line as well as the metastable area should be investigated.”

43. (Heuristic ID.3-4 in [43]) “Separation steps which do not need additives/auxiliary substances for purification should be favoured towards those which need additives/auxiliary substances such as extraction, adsorption, etc.”

44. (Heuristic ID.3-194 in [43]) “In case precipitation of the target component is an interesting process alternative, one may also consider to precipitate the impurities instead.”

45. (Heuristic ID.2-44 in [43]) “Products of precipitation steps are normally amorphous or semi-crystalline.”

46. (Heuristic ID.3-8 in [43]) "In case several products exist, crystallization and adsorption/desorption processes might be successful."
47. Precipitation experiments of the solid phase extract (ethanol). Prior to precipitation the extract is concentrated to 43.55 $g_{AD}/L_{Extract}$. Yield for AD is calculated for a single precipitation step under the experimental conditions described before. Purity is calculated via the percentage peak area ratio of the AD peak to the sum of all other peaks in the chromatogram except the solvent peaks. A yield of $\Upsilon_{waterc,AD} = 0.92 \pm 0.092$ and a purity of AD of $X_{Precipitate,AD} = 0.90 \pm 0.035$ are achieved.

Acknowledgment

The development of a downstream process to recover AD from a fermentation broth is part of the ERA-IB 3 project "MySterI—Novel industrial bioprocesses for production of key valuable steroid precursors from phytosterol" (European Research Area ERA-NET Scheme of the 7th EU Framework Program in an Industrial Biotechnology) and received funding from the Federal Ministry of Education and Research (ERA-IB, FKZ 031A221).

References

1. Malaviya A, Gomes J (2008) Androstenedione production by biotransformation of phytosterols. Bioresour Technol 99:6725–6737
2. Westfechtel A, Behler A (2006) Ethercarboxylic acid ester of sterol or stanol. US Patent 0,183,723, 20060183723 A1
3. Martin C (1977) Microbial cleavage of sterol side chains. Adv Appl Microbiol 22:29–58
4. Perez C, Falero A, Hung BR, Tirado S, Balcinde Y (2005) Bioconversion of phytosterols to androstanes by mycobacteria growing on sugar cane mud. J Ind Microbiol Biotechnol 32:83–86
5. Schmid A, Dordick JS, Hauer B, Kiener A, Wubbolts M, Witholt B (2001) Industrial biocatalysis today and tomorrow. Nature 409:258–268
6. Ostlund RE Jr, Racette SB, Okeke A, Stenson WF (2002) Phytosterols that are naturally present in commercial corn oil significantly reduce cholesterol absorption in humans. Am J Clin Nutr 75:1000–1004
7. Perez C, Perez I, Herve E (1995) Isolation and partial characterization of a new mutant for sterol biotransformation in *Mycobacterium* sp. Biotechnol Lett 17:1241–1246
8. Mitra RB, Kapadia VH (1988) Isolation of sterols from sugarcane wax. IN Patent 1,63,626
9. Nuissier G, Bourgeois P, Grignon-Dubois M, Pardon P, Lescure MH (2002) Composition of sugarcane waxes in rum factory wastes. Phytochemistry 61:721–726
10. Szykula J, Hebda C, Orpiszewski J, Saganska K (1991) Microbial transformation of neutral fraction and upgraded neutral fraction of Polish tall oil. Biotechnol Lett 13:917–922
11. Kutney JP, Milanova RK, Vassilev CD, Stefanov SS, Nedelcheva NV (2000) Process for the microbial conversion of phytosterols to androstenedione and androstadienedione. US Patent 6,071,714, US Patent and Trademark Office, Washington, DC
12. Welzel P, Robert K, Ponty A, Milkova T (1983) 4-Androsten-3,17-dione from cholesterol, beta-sitosterol, and campesterol by chemical degradation. Tetrahedron Lett 24:3199–3202
13. Xu Y-G, Guan Y-X, Wang H-Q, Yao S-J (2014) Microbial side-chain cleavage of phytosterols by mycobacteria in vegetable oil/aqueous two-phase system. Appl Biochem Biotechnol 174:522–533

14. H. Sahm, G. Antranikian, K.-P. Stahmann, R. Takors (Eds.), (2013) Steroide und Aromastoffe Industrielle Mikrobiologie, Klein, A, Hilmer J-M., Springer Berlin.
15. Arima K, Nagasawa M, Bae M, Tamura G (1969) Microbial transformation of sterols: Part II. Cleavage of sterol side chains by microorganisms. Agric Biol Chem 33:1644–1669
16. Arima K, Nagasawa M, Bae M, Tamura G (1969) Microbial transformation of sterols: Part I. Decomposition of cholesterol by microorganisms. Agric Biol Chem 33:1636–1643
17. Gyuere I, Lenkey B, Szentirmai A (1993) Propionyl-CoA elimination may be a rate-determining step of selective cleavage of sterol side chain. Biotechnol Lett 15:925–930
18. Fujimoto Y, Chen CS, Szeleczky Z, Ditullio D, Sih CJ (1982) Microbial degradation of the phytosterol side chain. I. Enzymic conversion of 3-oxo-24-ethylcholest-4-en-26-oic acid into 3-oxochol-4-en-24-oic acid and androst-4-ene-3,17-dione. J Am Chem Soc 104:4718–4720
19. Fernandes P, Cruz A, Angelova B, Pinheiro H, Cabral J (2003) Microbial conversion of steroid compounds: recent developments. Enzym Microb Technol 32:688–705
20. Mahato SB, Mukherjee A (1984) Steroid transformations by microorganisms. Phytochemistry 23:2131–2154
21. Marsheck WJ, Kraychy S, Muir RD (1972) Microbial degradation of sterols. Appl Microbiol 23:72–77
22. Liu WH, Lee CY (1990) Microbial transformation of cholesterol to androsta-1,4-diene-3,17-dione by *Mycobacterium sp.* NRRL B-3683 in a cholesterol-synthetic medium. J Chin Agric Chem Soc 28:166–174
23. Lo C-K, Pan C-P, Liu W-H (2002) Production of testosterone from phytosterol using a single-step microbial transformation by a mutant of Mycobacterium sp. J Ind Microbiol Biotechnol 28:280–283
24. Goetschel R, Bar R (1992) Formation of mixed crystals in microbial conversion of sterols and steroids. Enzym Microb Technol 14:462–469
25. Borrego S, Niubó E, Ancheta O, Espinosa ME (2000) Study of the microbial aggregation in mycobacterium using image analysis and electron microscopy. Tissue Cell 32:494–500
26. Dias AC, Cabral J, Pinheiro HM (1994) Sterol side-chain cleavage with immobilized Mycobacterium cells in water-immiscible organic solvents. Enzym Microb Technol 16:708–714
27. Sedlaczek L, Lisowska K, Korycka M, Rumijowska A, Ziolkowski A, Dlugonski J (1999) The effect of cell wall components on glycine-enhanced sterol side chain degradation to androstene derivatives by mycobacteria. Appl Microbiol Biotechnol 52:563–571
28. Korycka-Machala M, Ziolkowski A, Rumijowska-Galewicz A, Lisowska K, Sedlaczek L (2001) Polycations increase the permeability of *Mycobacterium vaccae* cell envelopes to hydrophobic compounds. Microbiology 147:2769–2781
29. Smith M, Zahnley J, Pfeifer D, Goff D (1993) Growth and cholesterol oxidation by *Mycobacterium* sp. in tween 80 medium. Appl Environ Microbiol 59:1425–1429
30. Shity H, Bar R (1992) New approach for selective separation of dilute products from simulated clostridial fermentation broths using cyclodextrins. Biotechnol Bioeng 39:462–466
31. Singh M, Sharma R, Banerjee U (2002) Biotechnological applications of cyclodextrins. Biotechnol Adv 20:341–359
32. Hesselink PG, van Vliet S, de Vries H, Witholt B (1989) Optimization of steroid side chain cleavage by *Mycobacterium* sp. in the presence of cyclodextrins. Enzym Microb Technol 11:398–404
33. Khomutov S, Donova MV (2011) Nanodimer cyclodextrin ligands with high affinity to steroids. J Incl Phenom Macrocycl Chem 70:353–357
34. Donova MV, Nikolayeva VM, Dovbnya DV, Gulevskaya SA, Suzina NE (2007) Methyl-beta-cyclodextrin alters growth, activity and cell envelope features of sterol-transforming mycobacteria. Microbiology 153:1981–1992
35. Stefanov S, Yankov D, Beschkov V (2006) Biotransformation of phytosterols to androstenedione in two phase water-oil systems. Chem Biochem Eng 20:421–427
36. Donova MV (2007) Transformation of steroids by actinobacteria: a review. Appl Biochem Microbiol 43:1–14
37. Hammond EG, Johnson LA, Su C, Wang T, White PJ (2005) Soybean oil: Bailey's industrial oil and fat products, 6th edn. Wiley, New York
38. Phillips KM, Ruggio DM, Toivo JI, Swank MA, Simpkins AH (2002) Free and esterified sterol composition of edible oils and fats. J Food Compos Anal 15:123–142
39. Bergs D, Merz J, Delp A, Joehnck M, Martin G, Schembecker G (2013) A standard procedure for the selection of solvents for natural plant extraction in the early stages of process development. Chem Eng Technol 36:1739–1748
40. ACDLabs (2010) ChemSketch freeware. ACDLabs, Toronto
41. van Krevelen DW (1990) Properties of polymers: their correlation with chemical structure, their numerical estimation and prediction from additive group contributions, 3rd edn. Elsevier, Amsterdam

42. Baehr HD, Stephan K (2004) Wärme- und Stoffübertragung, 4th edn. Springer, Berlin
43. Bauer K, (2012) Synthese von downstream Prozessen. PhD-thesis, TU Dortmund University, Dr. Hut, München
44. Sarker SD, Latif Z, Gray AI (2005) Natural products isolation, 2nd edn. Humana, Totowa
45. Pfennig A, Bol JB (2012) Wissensbasierte Designmethode zur Auslegung von maßgeschneiderten Feststoffextraktoren auf der Basis von Laborversuchen. AIF-Schlussbericht, Technische Universität Graz 16146N
46. Kassing M, Jenelten U, Schenk J, Hänsch R, Strube J (2012) Combination of rigorous and statistical modeling for process development of plant-based extractions based on mass balances and biological aspects. Chem Eng Technol 35:109–132
47. Meylan WM, Howard PH (1995) Atom/fragment contribution method for estimating octanol–water partition coefficients. J Pharm Sci 84:83–92
48. Hansen CM (ed) (1999) Hansen solubility parameters. CRC Press, Boca Raton
49. Hansen CM (2007) Hansen solubility parameters: a user's handbook, 2nd edn. CRC Press, Boca Raton
50. Casas A, Rodríguez JF, del Peso GL, Rodríguez R, Vicente G, Carrero A (2014) Liquid–liquid phase equilibria for soybean oil methanolysis: experimental, modeling, and data prediction. Ind Eng Chem Res 53:3731–3736
51. Follegatti-Romero LA, Lanza M, da Silva CAS, Batista EAC, Meirelles AJA (2010) Mutual solubility of pseudobinary systems containing vegetable oils and anhydrous ethanol from (298.15 to 333.15) K. J Chem Eng Data 55:2750–2756
52. Tang W, Xie C, Wang Z, Wu S, Feng Y, Wang X, Wang J, Gong J (2014) Solubility of androstenedione in lower alcohols. Fluid Phase Equilib 363:86–96
53. Prabhudesai RK (1988) Leaching. In: Schweitzer PA (eds) Handbook of separation techniques for chemical engineers 2nd edn. McGraw-Hill, New York
54. Bathen D, Breitbach M (2001) Adsorptionstechnik. Springer, Berlin
55. GoedeckeR(ed)(2011)Fluidverfahrenstechnik: Grundlagen, Methodik, Technik, Praxis. Wiley-VCH Verlag GmbH & Co, KGaA, Weinheim
56. Stach H (ed) (1990) Adsorption aus waessrigen Loesungen. Deutscher Verlag für Grundstoffindustrie, Leipzig
57. Winkelnkemper T, (2012) Key performance indicators for economic downstream process development. PhD Thesis, TU Dortmund University, Dr. Hut, München
58. Ladisch MR (2001) Bioseparations engineering: principles, practice, and economics. Wiley, New York
59. Mullin JW (2000) Crystallization and precipitation. In: Ullmann's T (ed) Ullmann's encyclopedia of industrial chemistry. Wiley-VCH Verlag GmbH & Co. KGaA, Weinheim, pp 582–630
60. Mersmann A, Kind M, Stichlmair J (2011) Thermal separation technology: principles, methods, process Design. Springer, Berlin
61. Velayati AA, Farnia P (2012) *Mycobacterium tuberculosis* RD-1 secreted antigens as protective and risk factors for tuberculosis. In: Ganguly N, Sharma P (eds) Morphological characterization of *Mycobacterium tuberculosis*. INTECH Open Access Publisher, Rijeka, Croatia

Chapter 24

Analysis of Intermediates of Steroid Transformations in Resting Cells by Thin-Layer Chromatography (TLC)

Govinda Guevara, Julián Perera, and Juana-María Navarro-Llorens

Abstract

Thin-layer chromatography (TLC) is a useful and convenient method for the analysis of steroids due to: its simple sample preparation, low time-consuming process, high sensitivity, low equipment investment and capacity to work on many samples simultaneously. Here we describe a TLC easy protocol very useful to analyze steroid molecules derived from a biotransformation carried out in wild-type and mutant resting cells of *Rhodococcus ruber* strain Chol-4. Following this protocol, we were able to detect the presence or the absence of some well-known intermediates of cholesterol catabolism in *Rhodococcus*, namely AD, ADD, and 9OHAD.

Key words Resting cells, Detection, Thin-layer chromatography, TLC, Steroids, AD, ADD, 9OHAD

1 Introduction

1.1 The TLC Technique

Thin-layer chromatography (TLC) is a technique that accomplishes the separation of a mixture of nonvolatile compounds on the basis of their differences in solubility in a two-phase system. The stationary phase is usually a thin, uniform layer of silica gel (SiO_2) or alumina powder (Al_2O_3) (about 0.25 mm thick) coated onto a piece of glass, metal or rigid plastic. On the other hand, the mobile phase is a liquid single or mixed solvent that slowly rises up the TLC plate by capillary action. The stationary phase may contain a fluorescent powder that allows the visualization of the spots under short-wave UV light ($\lambda_{254\ nm}$) [1]. Samples are applied onto the TLC plate near its bottom side; then the plate is placed standing on a container filled with the mobile phase which starts rising up the plate. When the solvent reaches the sample spot, different interactions are established among the components of the sample mixture, the solid phase and the solvent. These interactions are the result of a wide array of forces (hydrogen bonding, electron pair donor–electron pair acceptor (charge transfer), ion–ion, ion–dipole,

José-Luis Barredo, Ignacio Herráiz (eds.), *Microbial Steroids: Methods and Protocols*, Methods in Molecular Biology vol. 1645, DOI 10.1007/978-1-4939-7183-1_24, © Springer Science+Business Media LLC 2017

and van der Waals interactions). Depending on the particular ratio of affinities for both the solvent phase (solubility) and the stationary phase (adsorption), every compound in the sample (analyte) is carried to a definite height in the plate. Therefore, the choice of a proper eluent respect to the mixed compounds to be separated is a key decision in order to get a successful outcome of the chromatographic process [2].

The separation of the compounds in the sample for a particular combination of solvent and stationary phase is considered to be completed when the solvent almost reaches the top of the plate. Then, individual sample compounds will appear as vertically separated spots. Each compound has a retention factor (*Rf*) which is equal to the distance migrated over the total distance covered by the solvent. *Rf* values must be between 0 (for the spot that remains on the starting place) and 1.0 (for the spot that migrates with the front of the mobile phase). Compounds identification in TLC is based on the comparison of their *Rf* value to those of reference standards run in the same conditions, much better if both, sample and standards are separated in the same TLC plate [3]. Several factors can affect the *Rf* values: dimensions and type of chamber, nature and size of the layer, direction of the mobile phase flow, volume and composition of the mobile phase, equilibration conditions, humidity, and sample preparation methods preceding TLC.

TLC is a very useful analytical tool (e.g., for monitoring the progress of a reaction) but it also allows the purification of a small amount of a certain compound. TLC advantages are simplicity, simultaneous use of many samples, relatively low cost, high sensitivity (that can reach around the nanogram range, or even the picogram range if fluorescence is used), and speed of separation (it takes about 5–10 min. a run). Moreover, multiple separation techniques and detection procedures can be applied after this process [1, 2]. For instance, mass spectrometry detection is usually coupled to TLC steroid analysis [4–6]. For all these reasons, this technique is a valuable tool in the laboratory, and it has deserved a big deal of literature on their basic principles and applications since the second half of the twentieth century [1, 2, 7–13].

1.2 Steroid Metabolism and Resting Cells

In the last few years, the bacterial catabolism of cholesterol has been extensively studied searching to improve its biotechnological applications [14–18]. In the cholesterol upper catabolic pathway, the carbon side chain is degraded following a beta-oxidation process yielding 4-androstene-3,17-dione (AD). The steroid ring structure is broken through the medium pathway, beginning with the C1–C2 dehydrogenation in the A ring and the hydroxylation of C9 carbon in the B ring. The products of the A and B rings cleavage are finally degraded through the lower catabolic pathway [19]. This complex process involves more than 80 genes and is highly regulated [20]. Some valuable steroid compounds, such as

AD and some of its derivatives, are intermediaries in the cholesterol catabolic pathway. AD, the first intermediate of the medium pathway, is the substrate of the 3-ketosteroid-Δ1-deydrogenase (KstD) and the 3-ketosteroid 9α-hydroxylase (Ksh) activities that transform it into 1,4-androstadiene-3,17-dione (ADD) and 9α-hydroxy-4-androsten-3,17-dione (9OHAD). These steroid molecules are of great interest for the pharmaceutical industry as starting materials for the production of commercial steroids.

Intermediate metabolites of steroid catabolism are generally present at rather low concentration levels in active bacterial cells because they are substrates for other activities and, sometimes they can even be toxic for the cell at higher concentration [21, 22]. This fact makes difficult the analysis of some steroid intermediates in growing bacterial cultures. The use of nongrowing but metabolically active resting cells is widely used to evaluate the presence and the concentration of steroid metabolites during resting cell biotransformation [23, 24]. This approach also allows the independent optimization of both the cell growth and the biotransformation process. Moreover, it reduces the risk of undesirable bacterial contamination during the biotransformation process and improves the extraction of the reaction products [25, 26].

Resting cells have been already proved to be useful in biosurfactant and ADD production [27, 28] or in processes such as the enantioselective hydrolysis of pharmaceutically interesting 2-arylpropionitriles [29]. Finally, resting cells may be recycled and reused [30].

1.3 Analysis of Steroid Compounds by TLC

Due to the biological and pharmaceutical importance of steroids and their metabolites, it is often necessary to analyze them from biological or chemical samples. Owing to its many advantages, the TLC technique has been widely used for qualitative analysis of steroids [31–34].

In this chapter, we describe a method that allows the detection of several compounds that are intermediaries in the complex pathway of the bacterial catabolism of steroids. We successfully applied this procedure to carry out a semiquantitative analysis of AD, ADD, and 9OHAD, the three major intermediaries of the cholesterol catabolism in *Rhodococcus ruber* strain Chol-4. This protocol includes four steps: (1) preparation of the cellular system to search for the target compounds; (2) sample preparation and plate spotting; (3) TLC progress/development; and (4) spot visualization.

When cellular samples contain potential interfering substances (e.g., proteins, DNA, and lipids), as it is the case of the bacteria we are using in this protocol, a step of purification of steroids is necessary. Moreover, this purification step can simultaneously help to lead to the concentration of the searched steroid into the detection window of the assay. Several methods have been reported for the steroid purification from the samples [35]: solvent extraction, solid-phase extraction (using microparticulate silica coated with octadecasilane packed into syringes or cartridges), hydrolysis of steroid conjugates, immunoaffinity extraction, or selective extraction using molecularly imprinted polymers (e.g., cholesterol extraction using molecularly imprinted cyclodextrin microspheres [36]). In this protocol we propose to extract cellular samples with chloroform at freezing temperature in order to get the maximum recovery of the steroids.

Depending on the steroid we are working with, different mobile phases should be used in the TLC. Optimization of the composition of this phase for a certain steroid is not an easy task [31, 37, 38], and sometimes the best alternative is to follow some "try and error" approach. In this particular protocol, a mix of hexane and ethyl acetate (10:4, v/v) will be used for a good separation of the cholesterol, AD, ADD, and 9OHAD molecules. The TLC is generally run until the mobile phase nearly reaches the top of the plate as mentioned before.

Some steroids contain a polyconjugated structure visible by UV light (e.g., cholestenone, AD, and ADD) and then no special visualization method is required. Using plates coated with a fluorescence indicator, these compounds can be easily detected by plate inspection under 254 nm UV light. In other cases (e.g., cholesterol), there are a wide variety of destructive methods to reveal the spots on TLC plates [37]. After the application of the reagent by spraying or dipping, the optimum color development is frequently obtained by heating [35, 39]. The chemicals most often used for steroids detection on TLC plates are sulfuric acid, phosphoric acid, hydrochloric acid, antimony (III) chloride and iodine vapor [40]. In the protocol shown here, both UV and sulfuric acid visualization of the TLC spots yield good results for the molecules that we are searching for.

Using the protocol that we present here, we have proved that a *kshB Rhodococcus ruber* strain Chol-4 mutant accumulates ADD from AD due to the KstD action and the lack of an active Ksh.

2 Materials

All solutions should be prepared with ultrapure water, analytical grade reagents and stored at room temperature (unless indicated otherwise). All work with volatile solvents should be done under a fume hood. The cell culture and preparation must be done under sterile conditions using a laminar flow hood (*see* **Note 1**).

2.1 Strains, Cell Culture and Preparation

1. Bacterial strains: *Rhodococcus ruber* wild type [41] and *Rhodococcus ruber kshB*, a mutant that lacks Ksh activity (*laboratory collection*).

2. Luria–Bertani (LB) medium: Dissolve 10 g tryptone, 5 g yeast extract, and 10 g NaCl in H_2O; adjust pH to 7.5 with NaOH and volume to 1 L; autoclave at 120 °C for 15 min.
3. HEPES buffer: 25 mM HEPES (Sigma), pH 7.0; autoclave at 120 °C for 15 min.
4. AD (TCI, >99.0%): 20 mg of powder. Whenever possible, autoclave them at 110 °C for 15 min.
5. Nalidixic acid stock solution (1000×): Prepare a 15 mg/mL solution in water, sterilize by filtration, aliquot and store at −20 °C.
6. Chloramphenicol stock solution (1000×): Prepare a 34 mg/mL solution in ethanol, aliquot and store at −20 °C.
7. Cold sterile distilled water.
8. Sterile flasks of 100 and 250 mL.
9. 15 mL sterile disposable tubes.
10. 50 mL sterile centrifuge tubes.
11. Thermostatic orbital shaker with tubes, racks, and bottle adapters.
12. Bench centrifuge for Falcon tubes.
13. Spectrophotometer.
14. Freezer.
15. Laminar flow cabinet.
16. Ice or a 4 °C chamber.

2.2 Steroid Extraction

1. Chloroform.
2. 50 mL Falcon tubes.
3. Eppendorf tubes of 2 and 1.5 mL.
4. Vortex mixer.
5. Dry block heater.
6. Fume hood.
7. Ice.

2.3 Thin-Layer Chromatography (TLC)

1. Stationary phase: Aluminum TLC Silica gel 60 F254 sheets (Merck) (*see* **Note 2**).
2. Mobile phase (to separate AD, ADD, and 9OHAD): Hexane–ethyl acetate (10/4, v/v) (*see* **Note 3**).
3. Standard control samples (AD, 9-OH-AD, and ADD): Dissolved in chloroform at 1 μg/μL.
4. Visualization reagent: Methanol–sulfuric acid (90/10, v/v).
5. A 100 mL or a 250 mL beaker covered with a glass lid (watch glass).
6. Tweezers.

7. UV lamp ($\lambda_{254\ \mathrm{nm}}$).
8. Hotplate stirrer.
9. Hair dryer.

3 Methods

3.1 Resting Cells Preparation and Intermediates Production

1. In a 15 mL sterile disposable tube add 3 mL of liquid LB medium supplemented with nalidixic acid to 15 μg/mL (most *Rhodococcus* strains are resistant to this antibiotic).
2. Inoculate *Rhodococcus ruber* strain Chol-4 strain and incubate for 48 h at 30 °C and 250 rpm.
3. Add 0.5 mL of the liquid culture to 50 mL of LB supplemented with 15 μg/mL nalidixic acid in a 250 mL flask.
4. Incubate at 30 °C in a shaker at 250 rpm for 48 h. Depending on the *Rhodococcus* strain a bigger volume of culture could be necessary (*see* **Note 4**).
5. Add chloramphenicol stock solution to the cell culture to a final concentration of 34 μg/mL and incubate at 30 °C in a shaker at 250 rpm for 1 h (*see* **Note 5**).
6. Measure the OD of the culture and calculate the volume needed in order to get, after cell pelleting (*see* **step** 7) and washing (*see* **steps 8** and **9**), a 20 mL cell resuspension with an OD_{600} between 18 and 20 units (about 3×10^{19} cells) (*see* **step 10**). In the case of *Rhodococcus ruber* strain Chol-4 the whole 50 mL volume of the culture in **step 5** is a quite right amount.
7. Transfer the culture to 50 mL sterile centrifuge tubes and centrifuge for 10 min at 4000 × *g* and 4 °C to pellet the cells.
8. Discard the supernatant and add 3 mL of cold sterile water; it is also possible to use sterile physiological serum. Gently resuspend the pellet with a 1 mL pipette. It should be clean of debris. Work on ice or in a 4 °C chamber. Fill up the tube to 25 mL of sterile water and centrifuge for 10 min at 4000 × *g* and 4 °C to pellet the cells.
9. Repeat **step 8**.
10. Resuspend the cells in 20 mL of cold 25 mM, pH 7.0 HEPES buffer, take to 34 μg/mL chloramphenicol and transfer the cell suspension to a 100 mL flask (*see* **Note 6**).
11. Label a set of 1.5 mL tubes to collect the culture aliquots.
12. Add the 20 mg of AD to the cell suspension, mix and take 500 μL into the first labeled 1.5 mL tube as the starting point of the conversion (*see* **Note** 7).
13. Incubate the cell suspension in a shaker at 150 rpm and 30 °C and collect 500 μL from the flask at different times (e.g., 0, 6, 24, 48, and 120 h) (*see* **Note 8**).

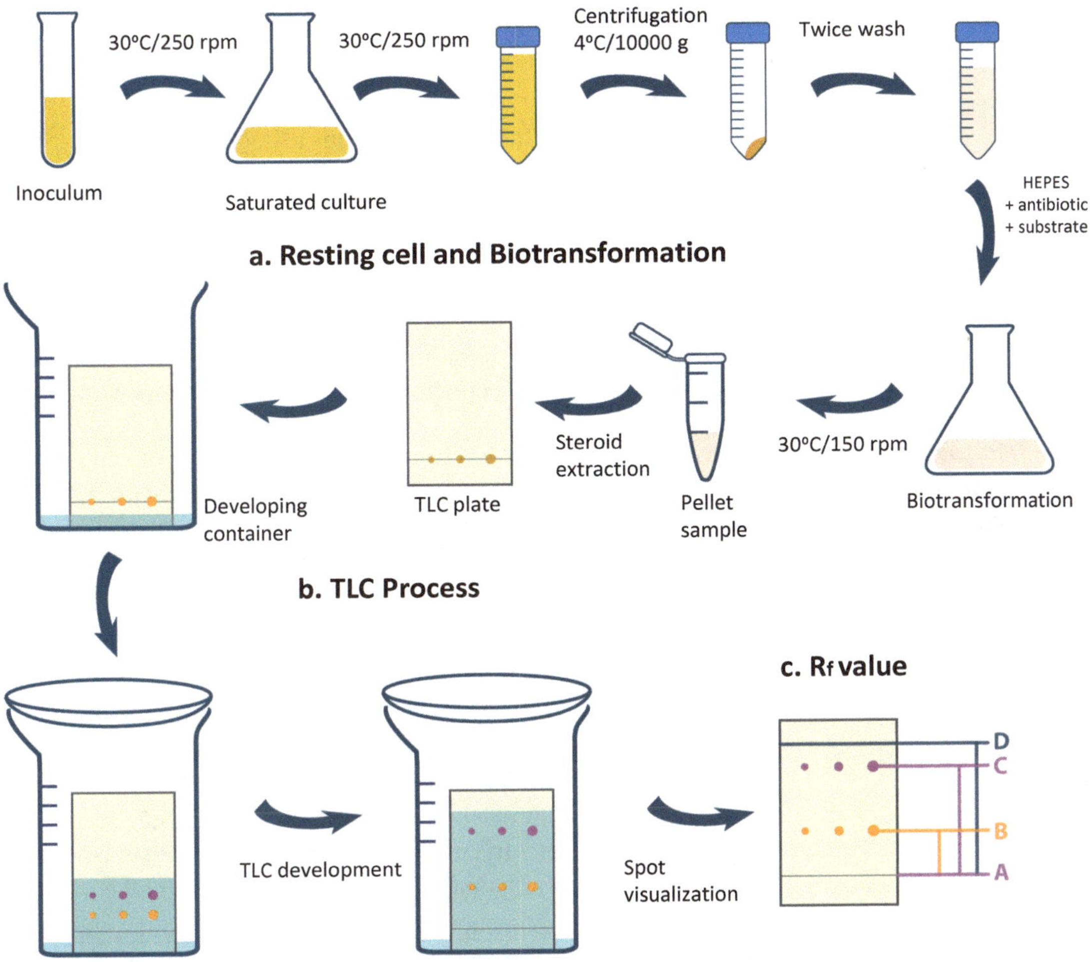

Fig. 1 Basic scheme of resting cells biotransformation and steroid metabolites analysis on an aluminum silica gel TLC plate. (**a**) Resting cells and biotransformation. *R. ruber* strain Chol-4 cells, both wild-type and mutant, are grown in LB at 30 °C and 250 rpm. After 2 days of growth cells are centrifuged and washed twice with sterile water. The pellet is suspended in HEPES buffer 50 mM, pH 7 containing a suitable antibiotic, in such a final volume to reach an OD_{600nm} of 18–20 (around 20 mL). The mix is kept at 30 °C and 150 rpm. The biotransformation starts at the moment the substrate is added; samples can be taken at different times. (**b**) TLC Process. The samples are applied into the TLC plate. Once the applications are dried, the plate is placed vertically in the developing container. (**c**) *Rf* value determination. After visualization of the spots, *Rf* values are equal to the ratio between the distance from the center of the spot to the start line (AB or AC) and the distance from the solvent front to the start line (AD). The *Rf* obtained for a compound can be compared to those of known reference materials

14. Freeze aliquots at −20 °C. A scheme of the whole process is depicted in Fig. 1.

3.2 Steroid Extraction

1. Let the frozen samples melt on ice.
2. Transfer the samples to a 50 mL Falcon tube and add two volumes of chloroform (1 mL) (*see* **Note 9**). From now on, work under a fume hood.

3. Close tightly the tubes and vortex vigorously for 30 s.
4. Freeze samples at −80 °C for 15 min and label 2 mL tubes (*see* **Note 10**).
5. Carefully disrupt the frozen aqueous phase with a 1 mL pipette tip and transfer the chloroform phase to a labeled 2 mL tube. Avoid any melting of the aqueous phase (*see* **Note 11**).
6. Set the uncovered 2 mL tubes into a 65 °C dry block heater and let the chloroform to evaporate completely so that samples can be later prepared as much concentrated as desired.
7. Transfer the tubes containing the dry samples to a rack and close them tightly. Keep the samples at room temperature.

3.3 Thin-Layer Chromatography (TLC) Development

1. Prepare the developing container pouring 5 mL of the mobile phase into a 100 mL or a 250 mL beaker (*see* **Note 12**) and cover it with a watch glass or a similar lid, swirl it gently and let it still. The depth of mobile phase must be less than 8 mm.
2. Prepare the TLC plate cutting the TLC Silica gel 60 F254 sheet into pieces of 5 cm high by the proper width (*see* **Note 13**).
3. Using a pencil, carefully draw a horizontal line across the plate 1 cm above its bottom. Mark a point every 8 mm in the pencil line to be used as a reference to put the samples (*see* **Note 14**). Lightly mark under the line the name of each sample to be laid.
4. Resuspend the samples obtained in Subheading 3.2, **step** 7, in 100 μL of chloroform or a similar volatile solvent such as hexane, ethyl acetate, or methylene chloride (*see* **Note 15**).
5. Take 5 μL of the sample solution (*see* **Note 16**) and gently and slowly put them onto the proper marked location on the TLC plate. Put 1 μL and let it dry before applying a new 1 μL drop. This way of loading allows the wet areas on the plate to remain small. Don't let the spot become too large (*see* **Note 17**).
6. Similarly, apply 1 μL of every standard control (1 μg/μL) on the same location on the plate, letting the spot dry between consecutive applications.
7. With a pair of tweezers, carefully place the TLC plate standing into the developing container with the marked line down, cover the beaker with the watch glass, and leave it undisturbed (*see* **Note 18**). The solvent (mobile phase) will rise up the TLC plate by capillarity.
8. Allow the TLC to develop until the solvent is about half a centimeter below the top of the plate; then remove it from the container. Immediately mark the solvent front with a pencil (Fig. 1).

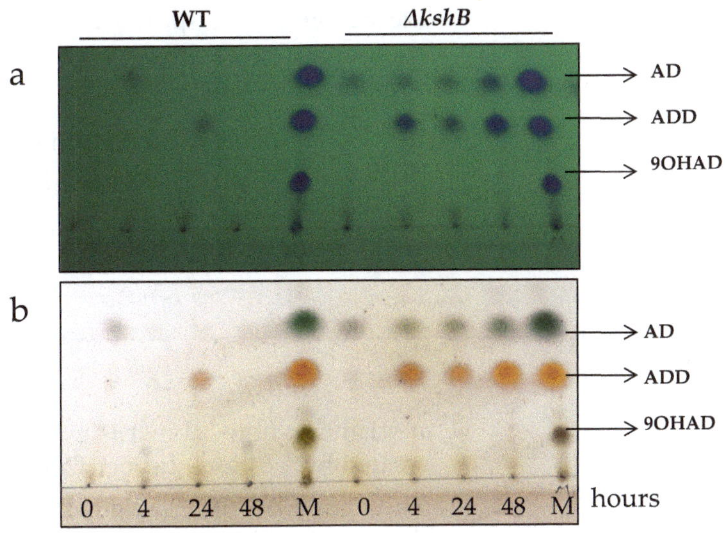

Fig. 2 Analysis of a mix of steroid metabolites in an aluminum silica gel TLC plate. Different aliquots were taken at different times of resting cells AD biotransformations using wild-type *R. ruber* strain Chol-4 (WT) and a derived *kshB* mutant (Δ*kshB*) strain. M shows the lane of standards steroids. (**a**) The TLC plate was put under a $\lambda_{254\ nm}$ UV light after the TLC development to visualize the spots; (**b**) the TLC plate was stained with a visualization reagent (methanol–sulfuric acid; 90:10, v/v) and heated. When heated, AD, ADD, and 9OHAD acquired different colors. The Δ*kshB* mutant strain showed an accumulation of ADD due to the presence of KstD activities and the lack of Ksh activities. Moreover, the TLC plate showed a slower consumption of AD in the mutant sample in comparison to the WT

3.4 Visualization of the Spots

1. Let the plate dry completely in the fume hood.
2. Put the dried plate under a $\lambda_{254\ nm}$ UV light: spots of UV absorbing compounds will be visible; mark them with a pencil (Fig. 2).
3. To see other compounds, pour in a beaker an amount of visualization reagent enough to cover the TLC plate. With the help of a pair of tweezers briefly dip the plate into the visualization reagent and let it drain. Deeply dry the plate by hot air (*see* **Note 19**).
4. Heat up the dried plate by putting it on a 230 °C hotplate-stirrer; in a few seconds spots will start to appear. Keep heating as much as needed in order to clearly see all the spots, but be careful not to burn the plate (Fig. 2).
5. Once every spot has been seen and marked, calculate its *Rf* value. Take the distance from the line where the spot was applied (starting point) to the highest point the solvent reached. Then, measure the distance from the starting point to

the center of the spot. Divide the distance traveled by the compound by the distance traveled by the solvent to get the compound *Rf* value.

$$Rf = \frac{\text{Spot distance}}{\text{Solvent Distance}}$$

6. Compare the *Rf* of every compound with those of the standards (*see* **Note 20**) (Fig. 1).

4 Notes

1. If a laminar flow cabinet is not available, a sterile field can be created by a Bunsen burner flame.
2. The silica TLC plate (original size 20 × 20 cm) has a 200 μm thick layer with a mean particle size of 10–12 μm. It can easily be cut to the desired size with scissors. It contains inorganic fluorescent indicators for UV detection of colorless compounds at an excitation wavelength of $\lambda_{254\ \text{nm}}$. Any other stationary phase with similar traits could be used; in the case it does not allow the UV detection, a visualization reagent could be used.
3. The solvents used in the mobile phase are chosen on the basis of the kind of steroid to be separated. Depending on the polarity of the compounds you should try a more or less polar mixture of solvents until a good separation is observed (an *Rf* value in the range of 0.3–0.7 is advisable). To start with, the most common solvents to mix are hexane (or petroleum ethers/ligroin) as the nonpolar component and ethyl acetate or alcohols (methanol, ethanol) as the polar component. Acetic acid can be used, usually in a very small quantity, when an increase in polarity is needed. Acetone can be used as a solvent of intermediate polarity.
4. LB seems to be able by itself to induce the expression of proteins involved in the steroids metabolic pathways, so there could be no need of using any additional inductor. If induction is not wanted, change the LB to Minimal Medium (e.g., 457 DSMZ minimal medium: 11.1 mM KH_2PO_4, 17.1 mM Na_2HPO_4, 3.78 mM $(NH_4)_2SO_4$, 0.8 mM $MgSO_4$, 0.45 mM $CaCl_2$, 13 μM EDTA, 0.34 μM $ZnSO_4$, 7.2 μM $FeSO_4$, 0.15 μM $MnCl_2$, 0.06 μM $CuCl_2$, 0.84 μM $CoCl_2$, 0.08 μM $NiCl_2$, 0.12 μM Na_2MoO_4, and 4.8 μM H_3BO_3) with an appropriate carbon source to get maximum growth (e.g., 2 g/L sodium acetate for *Rhodococcus*). On the contrary, if induction of gene expression is needed, add the inducer when the culture is on the mid-exponential growth phase ($OD_{600\ \text{nm}}$ = 0.6). The time needed to reach this growth level

depends on the bacterial strain (e.g., *R. erythropolis* needs 6–8 h to reach an $OD_{600\ nm}$ of 0.4–0.6).

5. Chloramphenicol is used to stop protein synthesis and avoid the subsequent catabolism of the biotransformation products [42]: this broad-spectrum antibiotic irreversibly binds to a receptor site on the 50S subunit of the bacterial ribosome, inhibiting the peptidyltransferase activity and therefore provoking the inhibition of the protein synthesis. A concentration of 50 μg/mL of streptomycin could be used instead of chloramphenicol.
6. In the presence of a carbon source (glucose, isopropanol, glycerol, and acetate) the biotransformation could be improved due to the redox potential generated by the catabolism of the carbon source.
7. Sometimes, complexing compounds (e.g., methyl-β-cyclodextrin) are used to enhance steroid solubility. In these cases, it should be taken into account that these complexing compounds are not separated from the steroid in the extraction phase and TLC does not run properly due to these compounds interfere with the process. An efficient washing of cells before steroid extraction should avoid this problem.
8. Depending on the speed of the reaction, times to take aliquots may vary. The times suggested here are a first approach.
9. A tube with a wide diameter is necessary to make easier the rupture of the ice layer with the pipette tip.
10. The freeze-drying method (**steps 4–6** in Subheading 3.2) is a useful protocol to remove the water and the hydrophilic compounds in this phase (e.g., sugars and proteins).
11. When working with several resting samples simultaneously, be careful to handle a limited number of them to avoid melting (e.g., three tubes at a time). In case the samples are going to be analyzed later by HPLC, be sure to add an internal standard before extraction and repeat the extraction step at least twice. The HPLC signal of the internal standard should not interfere with signals of the steroids to be analyzed.
12. The developing container can be a specially designed chamber, a jar with a lid or any glass container with a lid. The diameter size depends on the width of the TLC plate and the bigger the chamber the bigger the quantity of mobile phase to use. A folded piece of filter paper as tall as the beaker can be put standing into the beaker and let it get soaked during the process in order the air inside the beaker be saturated with the solvent vapors. This trick will improve the sample resolution and the *Rf* value determination.

13. The more the samples to run, the wider the plate must be. Separation between samples should be at least 8 mm. Handle the plates carefully, avoid disturbing the coating of adsorbent by any vibration or getting it dirty. Use always gloves and very sharp scissors. A longer size of the plate will not give a better resolution of the compounds but it will increase the time required to run TLC. Note that only the powdered side of the silica gel plate should be spotted.
14. Use a blunt pencil and take care not to press hard with it, otherwise the adsorbent will be disturbed. You will need this line as a guide to load all the samples at the same height and to calculate *Rf* values.
15. Resuspend and apply samples on the plate one by one, otherwise the solvent will slightly but continuously evaporate and the steroid concentration will vary from sample to sample.
16. Otherwise, the use of 10 μL microcaps could make it easier to deposit the sample by capillarity.
17. The sample solvent quickly evaporates and leaves a small spot of the solute on the plate. Usually 1% of the total sample is enough for TLC analysis. If the sample solution is too concentrated, compounds leave a smear or streak behind them, while if its concentration is too low, drop detection will be difficult. A trial and error procedure could be needed in order to obtain well-sized, easy-to-read spots.
18. The plate must stand upright into the mobile phase. If the positioning of the plate into the mobile phase is not even, the TLC developing will be distorted. Also be sure that the solvent does not cover the line where the sample spots were applied.
19. It is advisable to use at least two different visualization techniques to be certain about the nature of the compounds in the spots. Take into account that mixed components with similar polarities will migrate together as a single spot. To dry the plate it is possible to use a hair dryer. If the plate is not very well dried, it will tear apart when it is heated.
20. After sulfuric acid developing, spots of different colors appear. Spots with the same *Rf* as the standards should display also the same color to be sure that they are the same compound.

Acknowledgment

This work was supported by project PROBIODES (RTC-2014-2249-1) from the Ministerio de Economía y Competitividad (Spain). We are grateful to Prof. Jose Luis García and his group for scientific support. G.G. is a recipient of a predoctoral contract from the Ministerio de Economía y Competitividad. Thanks to Elena Navarrete Francés for technical support.

References

1. Jork H, Funk W, Fischer W, Wimmer H (1994) Thin layer chromatography: reagents and detection methods, vol Ib. VCH, Weinheim, Germany
2. Sherma J, Fried B (2003) Handbook of thin-layer chromatography. In: Cazes J (ed) Chromatographic science series, vol 89, 3rd edn. Marcel Dekker, New York
3. Touchstone JC, Dobbins MF (1983) Practice of thin layer chromatography. John Willey & Sons, New York
4. Kollerov V, Fokina V, Sukhodolskaya G, Shutov A, Donova M (2015) 11β-Hydroxylation of 6α-fluoro-16α-methyl-deoxycorticosterone 21-acetate by filamentous fungi. Appl Biochem Microbiol 51(2):169–179
5. Naumoska K, Vovk I (2015) Analysis of triterpenoids and phytosterols in vegetables by thin-layer chromatography coupled to tandem mass spectrometry. J Chromatogr A 1381:229–238
6. Yao L, Gerde JA, Lee S-L, Wang T, Harrata KA (2015) Microalgae lipid characterization. J Agric Food Chem 63(6):1773–1787
7. Dyer WG, Gould JP, Maistrellis NA, Peng T-C, Ofner P (1963) Thin-layer chromatography of some $C_{(19)}$-steroids: use of a simple standardized procedure. Steroids 1(3):271–285
8. Hahn-Deinstrop E (2007) Applied thin-layer chromatography: best practice and avoidance of mistakes, 2nd edn. John Wiley & Sons, Germany. doi:10.1002/9783527610259
9. Kumar S, Jyotirmayee K, Sarangi M (2013) Thin layer chromatography: a tool of biotechnology for isolation of bioactive compounds from medicinal plants. Int J Pharm Sci Rev Res 18(1):126–132
10. Poole C (2014) Instrumental thin-layer chromatography. Elsevier, USA
11. Stahl E (1969) Thin-layer chromatography. A laboratory handbook (trans: Ashworth MRF), 2nd edn. Springer-Verlag, Berlin, Heidelberg, Germany. doi:10.1007/978-3-642-88488-7
12. Touchstone JC (1992) Practice of thin layer chromatography, 3rd edn. John Wiley & Sons, USA
13. Wall PE (2005) Thin-layer chromatography: a modern practical approach. In: Smith RM (ed) RSC chromatography monographs. Royal Society of Chemistry, Cambridge
14. Donova MV, Egorova OV (2012) Microbial steroid transformations: current state and prospects. Appl Microbiol Biotechnol 94(6): 1423–1447
15. Doukyu N (2009) Characteristics and biotechnological applications of microbial cholesterol oxidases. Appl Microbiol Biotechnol 83(5):825–837
16. García JL, Uhia I, Galán B (2012) Catabolism and biotechnological applications of cholesterol degrading bacteria. Microb Biotechnol 5(6):679–699
17. McLean K, Hans M, Munro AW (2012) Cholesterol, an essential molecule: diverse roles involving cytochrome P450 enzymes. Biochem Soc Trans 40(3):587–593
18. Yam KC, Okamoto S, Roberts JN, Eltis LD (2011) Adventures in *Rhodococcus* – from steroids to explosives. Can J Microbiol 57(3):155–168
19. Bergstrand LH, Cardenas E, Holert J, Van Hamme JD, Mohn WW (2016) Delineation of steroid-degrading microorganisms through comparative genomic analysis. mBio 7 (2). doi: 10.1128/mBio.00166-16
20. Van der Geize R, Yam K, Heuser T, Wilbrink MH, Hara H, Anderton MC, Sim E, Dijkhuizen L, Davies JE, Mohn WW, Eltis LD (2007) A gene cluster encoding cholesterol catabolism in a soil actinomycete provides insight into *Mycobacterium tuberculosis* survival in macrophages. Proc Natl Acad Sci U S A 104(6):1947–1952
21. Malaviya A, Gomes J (2008) Androstenedione production by biotransformation of phytosterols. Bioresour Technol 99(15):6725–6737
22. Philipp B, Erdbrink H, Suter MJ, Schink B (2006) Degradation of and sensitivity to cholate in *Pseudomonas* sp. strain Chol1. Arch Microbiol 185(3):192–201
23. Epifanova OI, Terskikh VV, Polunovskiĭ VA (1982) Active metabolic processes in resting cells. Tsitologiia 24(11):1259–1273
24. Sudo SZ, Dworkin M (1973) Comparative biology of prokaryotic resting cells. Adv Microb Physiol 9:153–224
25. Angelova B, Mutafov S, Avramova T, Dimova I, Boyadjieva L (1996) 9 alpha-hydroxylation of 4-androstene-3,17-dione by resting *Rhodococcus* sp. cells. Process Biochem 31(2):179–184
26. Mutafov S, Angelova B, Avramova T, Boyadjieva L, Dimova I (1997) The inducibility of 9α-steroid hydroxylating activity in resting *Rhodococcus* sp. cells. Process Biochem 32(7):585–589
27. Kosaric N, Cairns WL, Gray NC (1987) Biotechnology and the surfactant industry. In: Kosaric N, Cairns WL, Gray NC (eds) Surfactant science series, vol 25. Marcel Dekker, New York, USA, pp 1–19

28. Wang Z, Zhao F, Chen D, Li D (2006) Biotransformation of phytosterol to produce androsta-diene-Dione by resting cells of *Mycobacterium* in cloud point system. Process Biochem 41(3):557–561
29. Effenberger F, Graef BW (1998) Chemo- and enantioselective hydrolysis of nitriles and acid amides, respectively, with resting cells of *Rhodococcus* sp. C3II and *Rhodococcus erythropolis* MP50. J Biotechnol 60(3):165–174
30. Julsing MK, Kuhn D, Schmid A, Buhler B (2012) Resting cells of recombinant *E. coli* show high epoxidation yields on energy source and high sensitivity to product inhibition. Biotechnol Bioeng 109(5):1109–1119
31. Cimpoiu C, Hosu A, Hodisan S (2006) Analysis of some steroids by thin-layer chromatography using optimum mobile phases. J Pharm Biomed Anal 41(2):633–637
32. Musharraf SG, Gulzar U (2012) Effective separation and simultaneous analysis of anabolic androgenic steroids (AAS) in their pharmaceutical formulations by a validated TLC-densitometry method. Chem Cent J. doi:10.1186/1752-153X-6-54
33. Musharraf SG, Arfeen QU, Mazhar W, Kanwal N (2013) A validated stability-indicating TLC-densitometric method for the determination of stanozolol in pharmaceutical formulations. Chem Cent J. doi:10.1186/1752-153X-7-142
34. Wang L, Wang X, Yuan X, Zhao B (2011) Simultaneous analysis of diosgenin and sarsasapogenin in *Asparagus officinalis* byproduct by thin-layer chromatography. Phytochem Anal 22(1):14–17
35. Makin HL, Honour JW, Shackleton CH, Griffiths WJ (2010) General methods for the extraction, purification, and measurement of steroids by chromatography and mass spectrometry. In: Gower DB, Makin HL (eds) Steroid analysis. Springer, Netherlands, pp 163–282. doi:10.1023/b135931_3
36. Egawa Y, Shimura Y, Nowatari Y, Aiba D, Juni K (2005) Preparation of molecularly imprinted cyclodextrin microspheres. Int J Pharm 293(1):165–170
37. Bhawani SA, Sulaiman O, Hashim R, Ibrahim MNM (2010) Thin-layer chromatographic analysis of steroids: a review. Trop J Pharm Res 9(3):301–313
38. Stoenoiu CE, Bolboacă SD, Jäntschi L (2006) Mobile phase optimization method for steroids separation. Appl Med Inform 18(1):17–24
39. Edwards RWH (1969) Steroids. Methods for the detection of biochemical compounds on paper and thin layer chromatograms. In: Dawson RMC, Elliott DC, Elliot WF, Johns KM (eds) Data for biochemical research, 2nd edn. Clarendon, London, pp 567–578
40. Görög S (1983) Quantitative analysis of steroids. Elsevier, Amsterdam
41. Fernández de Las Heras L, García Fernández E, Navarro Llorens JM, Perera J, Drzyzga O (2009) Morphological, physiological, and molecular characterization of a newly isolated steroid-degrading actinomycete, identified as *Rhodococcus ruber* strain Chol-4. Curr Microbiol 59(5):548–553
42. Traut RR, Monroe RE (1964) The puromycin reaction and its relation to protein synthesis. J Mol Biol 10(1):63–72

Chapter 25

A Ligand-Binding Assay to Measure the Affinity and Specificity of Sterol-Binding Proteins In Vitro

Rabih Darwiche and Roger Schneiter

Abstract

Sterols are major constituents of the plasma membrane of eukaryotic cells and serve as a precursor for several classes of signaling molecules, including steroids and hydroxy sterols. They maintain the functionality and permeability barrier of the plasma membrane through lipid–lipid and lipid–protein interactions. The *S. cerevisiae* *p*athogen-*r*elated *y*east proteins 1, 2, and 3 (Pry) belong to a large protein superfamily known as CAP/SCP/TAPS. Members of this superfamily have been implicated in a wide variety of processes, including immune defense in mammals and plants, pathogen virulence, sperm maturation and fertilization, venom toxicity, and prostate and brain cancer. Pry proteins bind and export sterols in vivo and the purified Pry1 protein binds sterols and related small hydrophobic compounds in vitro. Here we describe a method to determine lipid binding of a purified protein in vitro.

Key words Cholesterol, Lipids, Sterol binding proteins, Pathogen-related yeast 1 (Pry1), CAP/SCP/TAPS superfamily, *Saccharomyces cerevisiae*

1 Introduction

Proteins belonging to the CAP/SCP/TAPS superfamily (*c*ysteine-rich secretory proteins, *a*ntigen 5 (Ag5), *p*athogenesis related 1 proteins (PR-1)/*s*perm *c*oating *p*roteins/*T*px-1/*A*g5/*P*R-1/*S*c7; Pfam accession number PF00188) are present in all kingdoms of life and they have been implicated in many different physiological processes. CAP proteins are mostly secreted and they are stable in the extracellular space over a wide range of conditions. All members of this superfamily share a common CAP domain of approximately 150 amino acids, which adopts a unique α–β–α sandwich fold. The conserved structure of this domain suggests that CAP proteins exert a fundamentally similar function. The molecular mode of action of the CAP proteins, however, has remained enigmatic [1–3].

Saccharomyces cerevisiae has three members belonging to this family termed Pry proteins (*p*athogen *r*elated in *y*east). Pry1 and

José-Luis Barredo, Ignacio Herráiz (eds.), *Microbial Steroids: Methods and Protocols*, Methods in Molecular Biology vol. 1645, DOI 10.1007/978-1-4939-7183-1_25, © Springer Science+Business Media LLC 2017

Pry2 are secreted glycoproteins whereas Pry3 is predicted to be a glycosylphosphatidyl inositol (GPI)-anchored surface protein associated with the fungal cell wall. We recently showed that the yeast Pry proteins are required for the secretion of cholesteryl acetate in vivo and that Pry proteins bind cholesterol and cholesteryl acetate in vitro [4]. Cells lacking both *PRY1* and *PRY2* have a complete block in secretion of the acetylated lipid and Pry1 and Pry2 proteins bind both free cholesterol and cholesteryl acetate in vitro. The functions of Pry1 and Pry2 in lipid binding and export are partially redundant, and the conserved CAP domain of Pry1 is necessary and sufficient for lipid export and sterol binding. Expression of a human CAP superfamily member, the *c*ysteine-*r*ich *s*ecretory *p*rotein 2 (CRISP2), or the *Schistosoma mansoni* *v*enom *a*llergen *l*ike protein 4 (SmVAL4) rescues the phenotype of yeast mutants lacking Pry function and purified CRISP2 or SmVAL4 bind cholesterol in vitro, indicating that lipid binding is a conserved function of the CAP superfamily proteins [4–6].

Biochemical, optical, biophysical, or structural methods can be used to study lipid–protein interactions [7, 8]. One elegant way to measure sterol binding to proteins in real time is to make use of a fluorescent sterol, such as dehydroergosterol or NBD- or BODIPY-labeled cholesterol and to monitor fluorescence quenching upon binding of the ligand to the protein [9, 10]. Similarly, quenching of tryptophan fluorescence or fluorescence energy transfer (FRET) between the fluorescent sterol and a tryptophan acceptor can be used to monitor ligand binding [11–13]. An alternative method is to measure ligand–protein interactions using surface plasmon resonance (SPR) spectroscopy, which enables real-time quantification of lipid-binding affinities without the need for lipid labeling [14, 15]. A relatively new technique to study lipid–protein interaction relies on native protein mass spectrometry (MS) [8, 16, 17]. Native MS gives information about the composition, topology, dynamics, and structural properties of protein complexes. It offers the advantage of low sample consumption, high sensitivity and selectivity.

This paper discusses a traditional in vitro ligand-binding assay using isotope labeled cholesterol and ion exchange chromatography to separate the unbound ligand from the one that has bound to the protein [4, 18, 19]. Binding of related, nonlabeled compounds can be assessed in competition binding studies in which the unlabeled ligand competes for protein binding with the radiolabeled ligand.

2 Materials

2.1 Equipment

1. Ultrasonic bath.
2. Vortex.
3. Bench centrifuge.

4. Shaker.
5. Incubator.
6. Liquid scintillation counter.

2.2 Reagents and Supplies

1. Deactivated Clear Glass Vials (12 × 32 mm) (Waters, USA).
2. Q-Sepharose beads (GE Healthcare, USA).
3. Scintillation vials.
4. Scintillation cocktail, OptiPhase Hisafe (PerkinElmer, Waltham, MA).
5. 15 mL Falcon tubes.
6. (^{3}H)-cholesterol, 1 mCi/mL (American radiolabeled chemicals, Inc. St. Louis, MO) (*see* **Note 1**).
7. Cholesterol 99.9% pure (Sigma-Aldrich, St. Louis, MO).
8. Silanized glass vials (*see* **Note 2**).
9. Purified Pry1 protein.

2.3 Buffers

1. Binding buffer: 20 mM Tris pH 7.5, 30 mM NaCl, and 0.05% Triton X-100 (*see* **Note 3**).
2. Washing buffer: 20 mM Tris pH 7.5.
3. Elution buffer: 20 mM Tris pH 7.5 and 1 M NaCl.

3 Methods

3.1 In Vitro Lipid-Binding Assay

Each binding reaction is carried out in a final volume of 100 μL of binding buffer, containing the indicated amounts of purified Pry1 protein and (^{3}H)-cholesterol (50 pmol) delivered in ethanol (*see* **Note 4**). To determine nonspecific binding, an assay is performed in the absence of protein in the binding reaction.

1. Add 50 pmol (^{3}H)-cholesterol in ethanol to silanized glass vials.
2. Add the appropriate volume of binding buffer to the labeled cholesterol (the sum of volume added in **steps 1**, **2**, and **4** should be 100 μL).
3. Vortex, sonicate in the ultrasonic bath for 3 min, and vortex again.
4. Add 100 pmol of purified Pry1 protein to the binding buffer. Use vial without added protein as negative control.
5. Incubate vials for 1 h at 30 °C.
6. Pre-equilibrate 20 μL of Q-Sepharose beads with washing buffer. Wash the beads two times with 1 mL of washing buffer, centrifuge the beads at 3000 × *g* for 1 min in between washes. Pour off the supernatant.

7. Add 20 μL of the washed beads to the binding reaction and incubate for 10 min.
8. Place vials into a 15 mL Falcon tube and centrifuge at 3000 × *g* for 1 min.
9. Wash beads with 1 mL of washing buffer three times. Remove supernatant.
10. Add 500 μL of elution buffer to elute the protein from the Q-Sepharose beads and incubate for 5 min (*see* **Note 5**).
11. Place vials into a 15 mL Falcon tube and centrifuge at 3000 × *g* for 5 min.
12. Take 200 μL of the supernatant and mix well with 2 mL of scintillation cocktail.
13. Vortex for 1 min.
14. Incubate samples for 1 h at room temperature with shaking.
15. Quantify the protein-bound radioligand by liquid scintillation counting.

3.2 Competitive Binding Assay

1. Add 50 pmol (^{3}H)-cholesterol to silanized vials.
2. Add an equal amount of unlabeled sterol or steroid (*see* **Note 6**).
3. Each reaction is performed in a final volume of 100 μL of binding buffer. Add 100 pmol of the purified protein to the binding reaction and incubate for 1 h.
4. Separate the protein from the unbound ligand by adsorption to washed Q-Sepharose beads by following **steps 6–12** described above (*see* Subheading 3.1).
5. Quantify the protein-bound radioligand by scintillation counting.

4 Notes

1. Great care is needed when handling radioactive materials. Precautions include: wearing protective clothing (gloves and lab coat), keeping radioactive materials in protective and properly labeled containers, and arranging for proper waste disposal.
2. Sterols have the ability to stick to plastic tubes. It is therefore recommended to use silanized glass vials for the binding reactions.
3. Binding of Pry1 to cholesterol is pH-dependent. Since both ligand–protein interaction and adsorption of the ligand–protein complex to the anion-exchange beads are pH-dependent, it is important to first establish the optimal pH for the binding assay and the anion-exchange step. Using binding-, washing-, and elution buffers of different pH, the optimal condition for Pry1 to bind cholesterol is Tris buffered to pH 7.5 (*see* Fig. 1).

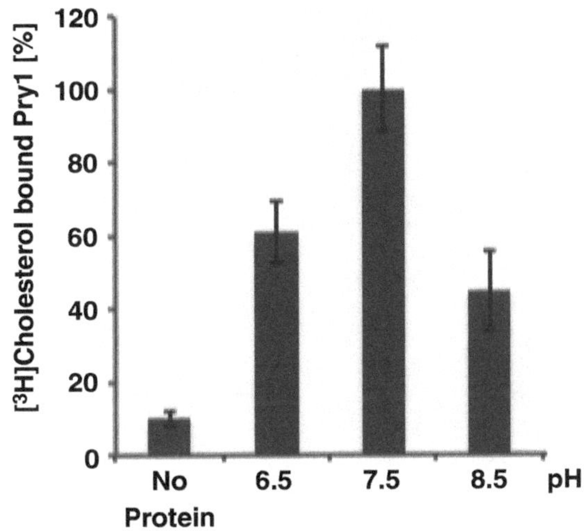

Fig. 1 Cholesterol binding by Pry1 is pH-dependent. Binding reactions were carried out using binding, wash, and elution buffers of the indicated pH. Values are means ± standard deviations of three independent experiments

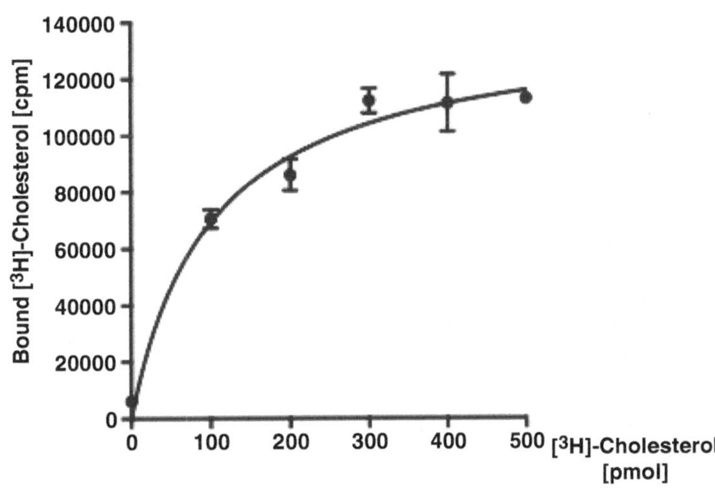

Fig. 2 Dose-dependent binding of cholesterol by Pry1. A constant amount of Pry1 (100 pmol) was incubated with increasing concentrations of tritiated cholesterol and lipid binding was quantified by scintillation counting. Values are means ± standard deviations of three independent experiments

4. Ligand-binding affinity of Pry1. To determine the affinity for the ligand, binding reactions with increasing concentrations of the ligand (0–500 pmol) and constant concentrations of the protein (100 pmol) were performed (*see* Fig. 2). These experiments revealed a concentration-dependent saturable binding of the ligand with an apparent K_d of 1.25 μM.

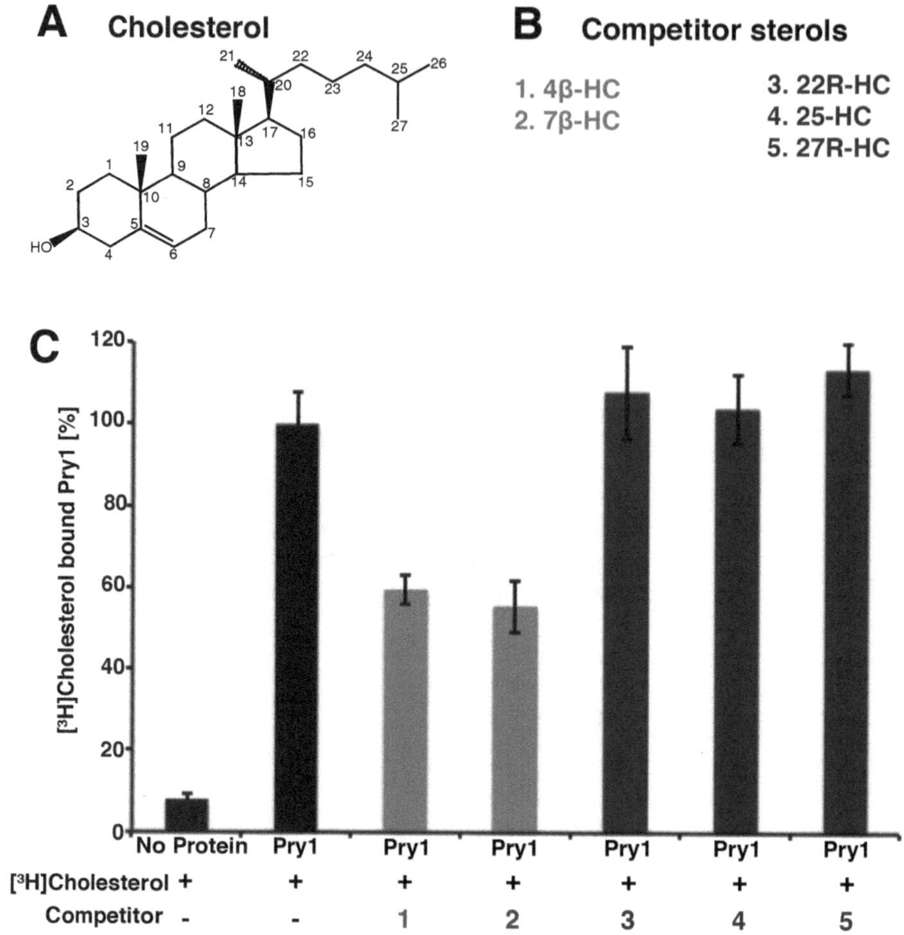

Fig. 3 Specificity of sterol binding by Pry1. (**a**) Structure of cholesterol. (**b**) unlabeled hydroxyl sterols (HC) tested for competition with (^{3}H)-cholesterol for binding to Pry1. *Light grey* hydroxyl sterols denote those that compete, whereas those that do not compete for binding are indicated in *black*. (**c**) competitive binding of the indicated hydroxyl sterols, 4β-hydroxycholesterol (4β-HC), 7β-hydroxycholesterol (7β-HC), 22(R)-hydroxycholesterol (22R-HC), 25-hydroxycholesterol (25-HC), and 27(R)-hydroxycholesterol (27R-HC). Each reaction contained 50 pmol of the indicated hydroxyl sterol and an equal amount of radiolabeled (^{3}H)-cholesterol. Values are means ± standard deviations of three independent experiments

5. To confirm the presence of the protein in the Q-Sepharose eluate, the eluted material is precipitated with trichloroacetic acid (TCA 10%) and the presence of the protein confirmed by Western blot.
6. Ligand-binding specificity of Pry1. To study the binding specificity of Pry1, we tested a series of unlabeled hydroxy sterols for competitive binding with 50 pmol labeled cholesterol. Sterols with modifications either to the tetracyclic sterol nucleus or the isooctyl side chain were selected (*see* Fig. 3).

Sterols with hydroxyl groups at position 4 or 7 of the ring nucleus reduced binding of the radiolabeled cholesterol to around 50%, indicating that they compete with cholesterol for binding to Pry1. On the other hand, sterols with hydroxyl groups at position 22, 25, or 27 of the side chain failed to compete with (^{3}H)-cholesterol for binding to Pry1. These results thus indicate that the isooctyl side chain is important for cholesterol binding by Pry1, whereas at least part of the tetracyclic ring structure tolerates hydrophilic substituents.

Acknowledgment

We thank Stéphanie Cottier for critically reading the manuscript. This work was supported by the Swiss National Science Foundation.

References

1. Gibbs GM, Roelants K, O'Bryan MK (2008) The CAP superfamily: cysteine-rich secretory proteins, antigen 5, and pathogenesis-related 1 proteins—roles in reproduction, cancer, and immune defense. Endocr Rev 29:865–897
2. Cantacessi C, Campbell BE, Visser A, Geldhof P, Nolan MJ, Nisbet AJ, Matthews JB, Loukas A, Hofmann A, Otranto D, Sternberg PW, Gasser RB (2009) A portrait of the "SCP/TAPS" proteins of eukaryotes—developing a framework for fundamental research and biotechnological outcomes. Biotechnol Adv 27:376–388
3. Schneiter R, Di Pietro A (2013) The CAP protein superfamily: function in sterol export and fungal virulence. Biomol Concepts 4:519–525
4. Choudhary V, Schneiter R (2012) Pathogen-Related Yeast (PRY) proteins and members of the CAP superfamily are secreted sterol-binding proteins. Proc Natl Acad Sci U S A 109:16882–16887
5. Choudhary V, Darwiche R, Gfeller D, Zoete V, Michielin O, Schneiter R (2014) The caveolin-binding motif of the pathogen-related yeast protein Pry1, a member of the CAP protein superfamily, is required for *in vivo* export of cholesteryl acetate. J Lipid Res 55:883–894
6. Kelleher A, Darwiche R, Rezende WC, Farias LP, Leite LC, Schneiter R, Asojo OA (2014) *Schistosoma mansoni* venom allergen-like protein 4 (SmVAL4) is a novel lipid-binding SCP/TAPS protein that lacks the prototypical CAP motifs. Acta Crystallogr D Biol Crystallogr 70:2186–2196
7. Scott JL, Musselman CA, Adu-Gyamfi E, Kutateladze TG, Stahelin RV (2012) Emerging methodologies to investigate lipid-protein interactions. Integr Biol (Camb) 4:247–258
8. Saliba AE, Vonkova I, Gavin AC (2015) The systematic analysis of protein-lipid interactions comes of age. Nat Rev Mol Cell Biol 16:753–761
9. Mikes V, Milat ML, Ponchet M, Ricci P, Blein JP (1997) The fungal elicitor cryptogein is a sterol carrier protein. FEBS Lett 416:190–192
10. Liu R, Lu P, Chu JW, Sharom FJ (2009) Characterization of fluorescent sterol binding to purified human NPC1. J Biol Chem 284:1840–1852
11. Schroeder F, Butko P, Nemecz G, Scallen TJ (1990) Interaction of fluorescent delta 5,7,9(11),22-ergostatetraen-3 beta-ol with sterol carrier protein-2. J Biol Chem 265:151–157
12. Holt A, de Almeida RF, Nyholm TK, Loura LM, Daily AE, Staffhorst RW, Rijkers DT, Koeppe RE, Prieto M, Killian JA (2008) Is there a preferential interaction between cholesterol and tryptophan residues in membrane proteins. Biochemistry 47:2638–2649
13. de Saint-Jean M, Delfosse V, Douguet D, Chicanne G, Payrastre B, Bourguet W, Antonny B, Drin G (2011) Osh4p exchanges sterols for phosphatidylinositol 4-phosphate between lipid bilayers. J Cell Biol 195:965–978
14. Stahelin RV, Cho W (2001) Differential roles of ionic, aliphatic, and aromatic residues in membrane-protein interactions: a surface plasmon resonance study on phospholipases A2. Biochemistry 40:4672–4678
15. Sheng R, Chen Y, Yung Gee H, Stec E, Melowic HR, Blatner NR, Tun MP, Kim Y,

Källberg M, Fujiwara TK, Hye Hong J, Pyo Kim K, Lu H, Kusumi A, Goo Lee M, Cho W (2012) Cholesterol modulates cell signaling and protein networking by specifically interacting with PDZ domain-containing scaffold proteins. Nat Commun 3:1249

16. Zhou M, Morgner N, Barrera NP, Politis A, Isaacson SC, Matak-Vinković D, Murata T, Bernal RA, Stock D, Robinson CV (2011) Mass spectrometry of intact V-type ATPases reveals bound lipids and the effects of nucleotide binding. Science 334:380–385
17. Contreras FX, Ernst AM, Haberkant P, Björkholm P, Lindahl E, Gönen B, Tischer C, Elofsson A, von Heijne G, Thiele C, Pepperkok R, Wieland F, Brügger B (2012) Molecular recognition of a single sphingolipid species by a protein's transmembrane domain. Nature 481:525–529
18. Im YJ, Raychaudhuri S, Prinz WA, Hurley JH (2005) Structural mechanism for sterol sensing and transport by OSBP-related proteins. Nature 437:154–158
19. Infante RE, Radhakrishnan A, Abi-Mosleh L, Kinch LN, Wang ML, Grishin NV, Goldstein JL, Brown MS (2008) Purified NPC1 protein: II. Localization of sterol binding to a 240-amino acid soluble luminal loop. J Biol Chem 283:1064–1075

Index

José-Luis Barredo, Ignacio Herráiz (eds.), *Microbial Steroids: Methods and Protocols*, Methods in Molecular Biology vol. 1645, DOI 10.1007/978-1-4939-7183-1, © Springer Science+Business Media LLC 2017

O

P

Q

R

S

T

U

V

W

Z

FSC
www.fsc.org
MIX
Papier | Fördert
gute Waldnutzung
FSC® C083411

Kolibri 360 GmbH
Ferdinand-Jühlke-Straße 7
99095 Erfurt, Deutschland
produktsicherheit@kolibri360.de